Read   p 140-154?

p 197. Dom. man. co.

SYSTEM DYNAMICS MODELLING

A PRACTICAL APPROACH

SYSTEM DYNAMICS MODELLING

A PRACTICAL APPROACH

R. G. COYLE
Cranfield University, UK

CHAPMAN & HALL/CRC

Boca Raton   London   New York   Washington, D.C.

Library of Congress Cataloging-in-Publication Data

Catalog record is available from the Library of Congress

Visit the CRC Press Web site at www.crcpress.com

© 1996 by R.G. Coyle
First edition 1996
First CRC Press reprint 2001

No claim to original U.S. Government works
International Standard Book Number 0-412-61710-2
Library of Congress Card Number 95-71233
Printed in the United States of America 1 2 3 4 5 6 7 8 9 0
Printed on acid-free paper

For Julie,
who makes the sun shine,
and for
Rhian and Patricia,
who do the same for my sons.

Errata

Users of COSMIC should replace the exponent **2 in the objective functions on pages 243, 254, 255 (three occurrences), 280 and 393 with INT(2) and make the corresponding changes in the models on the disk (FIG8-5M.COS, FIG9-6.COS and FIG9-12.COS – there is no model on the disk for Problem 15). Users of other packages should consult their manual. Depending on the package, an execution error may occur with X**N if X is negative and N is treated as a real number. COSMIC avoids the problem because its INT function converts real numbers to integers. An alternative approach is to use the absolute value of X: ABS(X)**N. If the package does not support exponents, just use ABS(X). If in doubt, try your package and see what happens, or contact its supplier.

Contents

Foreword

More and more people are realizing that social and business systems are too complex to be understood by intuition, compromise, and superficial debate. But, the traditional social and managerial sciences are providing little help in designing policies for better behaviour of large dynamic systems. The rapidly growing field of system dynamics is increasingly seen as the best hope for dealing with multiple-feedback-loop, nonlinear systems that extend across many different intellectual disciplines.

Conferences on the application of system dynamics to business are becoming larger and more frequent. Many system dynamics models deal with interactions between people and the environment. Use of system dynamics is under way in economics, government and the management of universities.

The field needs an expanding literature in each of the many areas where people are concerned with how things change through time. In this book, Professor Geoff Coyle has written a welcome addition to system dynamics. He brings to this book an extensive background as a leader in both the academic and operational sides of system dynamics. At the University of Bradford, Coyle founded one of the early academic programs in system dynamics. The extensive practical aspects of his career are reflected here in experiences with systems in business and government. His professional background yields insights regarding both systems and the political and psychological aspects of working with clients.

Beyond modelling of specific systems, Coyle also is explicit about his modelling philosophy and raises issues that deserve increased attention and debate. Some of his modelling practices are of great importance but are often overlooked, some are controversial and need more discussion, and some represent unsettled territory. His stress on consistency of units in equations is a welcome emphasis on a fundamental step in good modelling that is too often overlooked. Coyle presents an approach to computer-optimization of models, which is an area of intriguing promise but where the strengths and limitations are not yet clear.

Geoff Coyle has given us a well-written and useful contribution to the practical side of system dynamics. He sets an example and opens questions that can shape future work in the field.

Jay W. Forrester

Preface

I have spent more than 25 years using system dynamics for modelling complex problems, mainly in management and defence, and this book is the result. System dynamics (SD) has proved to be particularly powerful for analysing why social, economic, ecological and other managed systems do not always behave as we wish them to. It sheds light on why management actions can have negligible or destructive effects, rather than the beneficial ones which were hoped for, and why systems sometimes behave in ways which are contrary to intuition.

The system dynamics method has also proved to be particularly good at supporting a *strategic* point of view, in the sense of matching very closely the concerns of top-level decision-makers. By supporting the modelling of the forces underlying a system's evolution into the future, it allows the analyst to offer advice on how evolution can be directed so that the system can be made *robust*, in that it is enabled to defend itself against setbacks and to exploit opportunities.

Much of the strength of SD comes from its ability to be used in two related, but different, ways. On the one hand, it can be used *qualitatively* to portray the workings of a system as an aid to thinking and understanding. On the other, the diagram can be turned into a simulation model for *quantitative* simulation and optimization to support policy design.

The purpose of the book is to introduce and explain these powerful ideas, to illustrate them with some cases and, above all, to teach how these models are built and used. It is, in short, a model-builder's textbook and the aim is to bring the reader to the point at which he or she can build, analyse and optimize system dynamics models of serious problems. As such, it should be useful on graduate and undergraduate courses in the management sciences, economics and other disciplines in which the student learns to build and analyse models of problems involving human intervention in social and economic systems. It should also be useful to academics in other disciplines and to practitioners in industry, business and government who wish to acquire a new skill.

I have long since lost count of how many SD models I have worked with, but it must be hundreds. They range from simple teaching models through the work of numerous research students at three universities and into my own practical work as a consultant to numerous business firms and governments and in NATO. The experience has convinced me that mistakes are very easy to make when one is modelling a complex problem and that it is easy to overlook errors. The book therefore places much emphasis on techniques for avoiding errors. The treatment is developed through simple models to illustrate ideas and more complicated case studies to develop understanding and skill. Throughout the book important points are printed in **bold** type when they are first introduced and *italic* type when it is necessary to re-emphasize them.

My ideas about, and understanding of, system dynamics have developed considerably during the 15 years since I wrote *Management System Dynamics*. This is, therefore, a completely new work. As with all books, it has benefited from the comments of colleagues and students too numerous to mention. I am, however, particularly grateful to Professor Jay Forrester, who introduced me to SD many years ago. Professor Eric Wolstenholme was a stimulating colleague with whom I worked happily for a long time. Many of my colleagues and students at Shrivenham have made helpful comments on earlier drafts of the book. Mr Ken McNaught and Mr Don Valler have been meticulous in reading drafts and finding typing errors and infelicities of explanation. Those that remain are my responsibility.

My thanks are also due to Arlene Curtis and Stephanie Muir who put up with my occasional distractedness, guarded my door and telephone when I was trying to work and did endless photocopying in the midst of running a very busy departmental office.

The people who, above all, have supported the years of work which led to this book are my wife and sons. When my children were young they, and my wife, endured many hours of 'Daddy is working'. They, and above all my wife Julie, have never failed to encourage me when the work wasn't coming out. My elder son, Jonathan, makes his living from SD and has proved to be extremely good at it.

RGC
Shrivenham

CHAPTER 1

Introduction to system dynamics

SYSTEM DYNAMICS AND MANAGEMENT SCIENCE

This chapter introduces system dynamics and gives the reader a preview of what is to come. We start, however, with a history lesson, as it is usually helpful to be able to put topics into a wider context. In this case, we want to relate the system dynamics modelling methodology to other approaches to the study and analysis of managed systems so that we can see how system dynamics complements and competes with other methods.

During the 1940s, formal analysis, often involving mathematical and statistical techniques, had been applied to the problems of fighting a war and, subsequently, to the running of industries and business firms. These methods were eventually formalized into the disciplines of operational research and management science, on which there are any number of good textbooks. For example, the mathematics of linear programming (LP) was evolved and applied to such problems as how to plan production in an oil refinery so that given quantities of a range of end products could be manu-factured at minimum cost by a very complicated process from crude oil inputs of varying chemical composition and price. This is a problem of considerable importance to oil companies, and LP, and its later variants, are very useful for solving it. Similarly, techniques of decision theory were developed for tackling problems of choice when there are a range of uncertain outcomes and possibly several conflicting criteria. There are several more management science methodologies for different categories of problem.

These methods have proved to be excellent and powerful for dealing with certain classes of problems: in essence those in which the ability of a system to adjust to changing circumstances as time passes is not a significant aspect of the management problem. However, the behaviour of a system as time passes and new decisions have to be made is a significant type of management problem which requires the analyst to tackle the issues of how a system (a term we shall define in a moment) reacts to **dynamic** forces and how those reactions shape its behaviour as it moves into the future.

A good example of a dynamic system is the ordinary domestic central heating system. The home owner sets a desired temperature on the thermostat and the system does the rest, controlling itself as time passes and circumstances change. If the outside temperature falls, the room eventually cools down to the point where the heating is turned on until the house is comfortable again. If the weather is warm, the system turns itself off until the house cools down and heat is required again. A more sophisticated system might also include air conditioning, which will be applied, automatically, to cool the house to the owner's desired setting. A system which is vastly more complex in operation, but comparable in purpose, is the automatic pilot on an airliner, which is capable of accurately controlling the aircraft to follow a pre-set flight plan over great distances despite the occurrence of unpredictable forces such as adverse headwinds. In some cases, the aircraft can be controlled during the final approach to land without human intervention. There are many other cases of the same type.

These **control systems** have common features which we shall emphasize in bold type. They recognize, or **sense**, the effect the external environment has produced on the actual **state** of the system, such as the temperature of the room in response to the weather outside. They **compare** the actual state with the **desired state** and they employ 'rules' or **policies**, specifying what to do in given circumstances, such as, perhaps, that the room can be 2 degrees cooler than the required temperature before the heating is turned on and 3 degrees warmer before it is turned off. They nearly always involve **delays** before the action has an effect; it takes time to warm a house. They depend on **information feedback**; when the house is sensed as having warmed up, the heating goes off.

It is intuitively obvious that poorly designed rules would give poor performance from a central heating system. If the point at which the heating is turned on is set too low, the house will not be comfortable. If the off point is too high, heat will be wasted. If the two points are too close together the heating system will probably not work very well because of the time lags in the system. However, this does not happen in practice because the skill of designing control systems has been well learned.

In the early 1960s it was recognized that this control systems viewpoint could be very powerful if applied to business firms. To take a very simple example, if demand for its products starts to rise, how quickly should a firm recruit more workers to increase production? There will almost certainly be a delay before the new staff can be recruited and trained, so there is a danger that the work force will overshoot the level required to meet the apparent increase in demand, not least because the delay in increasing production may mean that the demand cannot be met and will go elsewhere. Even worse, the new staff will have to be paid even before they are productive and the resulting drain on cash might mean cutting the marketing budget, which may be a very good way of killing off the demand. The firm might, of course, use sub-contractors, though usually at less profit and with less control over quality.

Under what circumstances, then, should the firm use different **policies** to control its behaviour as time passes and circumstances change? How can its policies be **designed** so that the firm becomes **robust** against change and is able to create and exploit opportunities and avoid, or defend itself against, setbacks? How can the organization design its **information feedback structure** to ensure that effective policies become possible? These are the issues that system dynamics addresses.

In practice, most problems, whether they are in business, economics, the environment or whatever, are *much* more complicated than the workforce expansion problem discussed above. The reader might care to pause for a moment and jot down a few more of the influences which might arise in that problem, just to get a feel for how complex it can become. Because problems can be so complicated, it is necessary to have a methodology for dealing with them, and that is what system dynamics provides and what this book will explain. System dynamics can, therefore, be interpreted as the branch of management science which deals with the **dynamics and controllability of managed systems**.

The recognition that this type of problem exists in the business world, and that the concepts of control theory could be applied, is due to Professor Jay Forrester, of the Massachusetts Institute of Technology. He further realized that the mathematical techniques of control theory do not apply to managed systems because they tend to be much more complicated than engineering problems. The final stage in Forrester's intellectual *tour de force* was, therefore, to define the structure of specialized computer simulation languages to enable the calculations in system dynamics to be performed quickly and easily.

Since Forrester's pioneering steps, some 35 years ago, system dynamics has been applied to an enormous range of problems, some of which will be reviewed in Chapter 10. For the present, we must get on with learning about what system dynamics does, how it works, and the techniques required.

THE PARADIGM OF SYSTEM DYNAMICS

The paradigm, or basic viewpoint and associated methodologies, of system dynamics requires that we first define a 'system'. There are many definitions of this elusive idea, most of which imply that a 'system' is a collection of parts which interact in such a way that the whole has properties which are not evident from the parts themselves. Thus, the behaviour of an angry human being cannot be explained from the chemical composition of the body, or, at any rate, not yet. Such definitions sometimes tend towards an almost mystical stance, and we shall adopt a more pragmatic point of view and define a system as:

a collection of parts organized for a purpose

This is simple and lays emphasis on the ideas that there are 'parts', such as labour recruitment and marketing in the earlier example, which are 'organized', in the sense that they relate to each other in some way, and they attempt to achieve or contribute to a purpose, such as the survival and prosperity of the firm. This is all very obvious and the real subtlety lies in the idea that **the system may fail to achieve its purpose**. It may fail because its parts are poorly designed, because they are poorly connected, because it is knocked off course by an external shock, because the purpose is inherently incapable of being achieved or because its attempts to adjust to change, its **policies**, are badly designed, even to the extent of making matters worse.

In principle, any of these failures could arise in a central heating system, but they rarely do because such systems are well designed. The fundamental purpose of system dynamics is to achieve comparable quality of design, and hence performance, in managed systems.

System dynamics achieves this purpose through adopting the viewpoint shown in Fig. 1.1 and supporting the viewpoint with a systematic procedure, which is covered later in this chapter, and a range of techniques which form the subject matter of the rest of the book.

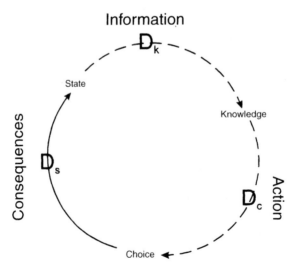

Fig. 1.1   The information/action/consequences paradigm of system dynamics.

Figure 1.1 can be understood by first following the loop. This shows that the 'state' of the system, such as the size of the workforce, derives from 'choices' made by its managers, such as recruiting more people, making them redundant or not replacing those who leave. The large D on the link from Choice to State reminds us that there is usually a substantial delay between a choice being made and the effect being felt in the state. The subscript on the D simply makes it clear that the delays in the different stages may be different. The next step is that the 'state' becomes known to the controllers of the system, perhaps in the form that people are leaving at an average rate of 20 per week from natural wastage. There will, however, be a knowledge delay before the fact that people actually are leaving at the rate of 20 per week becomes realized and accepted. This knowledge then leads, after a delay for choice, to new actions, such as to attempt to recruit 30 per week, because it is desirable to increase the workforce smoothly, but not by too much.

The words round the outer edge emphasize the ***information/action/ consequences*** paradigm. Information produces actions which have consequences, generating further information and actions, and so on. This sequence of the three components as time passes is **dynamic behaviour.** The state will change continually as time goes by and the pattern of behaviour will depend on how well the information and actions are tuned to each other and to the way in which consequences arise in the particular system. The tuning must take account of the delays. The techniques of system dynamics are the tools by which the tuning is achieved to make the pattern of behaviour as acceptable as it can be.

The second point about Fig. 1.1 is that the central loop *is* a loop; that is that there is a closed chain of cause and effect in which information about the result of actions is fed back to generate further action. These **feedback loops** are a central element both of engineering control theory and of system dynamics. This connection now allows us to offer a preliminary definition of system dynamics as:

the application of the attitude of mind of a control engineer to the improvement of dynamic behaviour in managed systems.

This involves analysing a managed system so as to:

- model the ways in which its information, action and consequences components interact to generate dynamic behaviour
- diagnose the causes of faulty behaviour
- tune its feedback loops to get better behaviour.

A control engineer uses sophisticated mathematical tools, supported by computer simulation, to go about those tasks for an engineering system. System dynamics makes use of control theory insights, simulation and optimization to deal with managed systems.

The tuning of a managed system means adapting its policies *and* the structure of its loops to achieve better behaviour. The dotted lines for information and action in Fig. 1.1 imply that these parts of a system are capable of being changed. As we shall see when we analyse models, loops can be abolished or created, and it is this ability to redesign parts of the system which makes system dynamics a viable methodology for managed systems.

It is tempting to assume that a managed system is always a business firm, and there have indeed been numerous examples of the study of such systems. System dynamics has, however, also been applied to the dynamics of decline in cities, to oscillation in ecological systems, the growth and collapse of civilizations, the behaviour of national social and economic systems, the future of the world, the criminal justice system, and to a range of military problems as well as to many other topics. All of these have two things in common.

The *first* is a wish to make matters better by suggesting how **people** can act upon the system, which is why we term them 'managed systems'. What, for example, would be a good policy for controlling fishing so that food is obtained but stocks are not depleted? How can the criminal justice system be made more 'effective'? What type of labour recruitment policy should a firm adopt? These questions all imply that there must be a set of criteria by which we shall assess the quality of performance. Evidently, criteria are not a simple matter, as some aspects may be more important than others and it may be necessary to trade off one against another. We shall have much more to say about this in the chapters on optimization.

The idea of trading off between criteria means that we must know what is supposed to be achieved, and that may depend on the circumstances. A firm which is experiencing growth in demand may wish to expand as fast as possible, though without over-taxing itself and getting too big for the market. If, on the other hand, the same firm faced a declining market it might well seek to contract in size in an orderly fashion. The required behaviour in these two cases would be different, but it would be desirable for the behaviour to be the best that could be achieved in the circumstances. This is expressed in the important **principle of robustness**, which states that:

the system should always perform as well as the circumstances allow, regardless of what the circumstances are.

In particular, a managed system should be able to defend itself against, or recover from, shocks, and it should be able to create and exploit opportunities.

The *second* thing that managed systems have in common is the ubiquitous presence of feedback loops in all these systems. There are, however, two types of feedback loop: goal-seeking, or **negative** loops and growth-producing or **positive** loops.

POSITIVE AND NEGATIVE FEEDBACK

The differences between the two types of loop are quite crucial in understanding dynamic behaviour.

Central heating systems, aircraft autopilots and any other system, including managed systems, which seeks to achieve a target *must* embody at least one **negative or goal-seeking** feedback loop. The term 'negative feedback' is the standard one and is used throughout this book, but goal-seeking is what the loops try to achieve[1]. The standard form of a negative loop is shown in Fig. 1.2(a).

The essential idea of negative feedback is that, when there is a difference between the desired and actual states of the system, actions are generated,

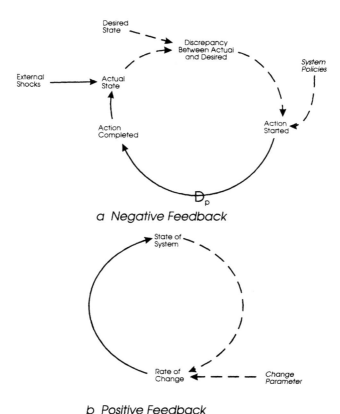

a **Negative Feedback**

b **Positive Feedback**

Fig. 1.2   Negative and positive feedback loop structures.

---

[1] Some authors refer to negative loops as 'balancing' loops. This is acceptable, though it implies that the loop will succeed in its balancing act. A badly-designed negative loop can, however, **unbalance** a system, so the term goal-seeking is probably better. The traditional term of 'negative feedback' will be used throughout this book.

under the influence of the system's policies, so as to attempt to *eliminate* the difference; the loop seeks the goal of having no difference between desired and actual conditions and the quality of the performance of the system as a whole is assessed by how well that is achieved as time passes.

In a simple case, such as a central heating system, the desired and actual states are the temperature set on the room thermostat and how warm one feels. If the two match quite closely as time passes, despite rises and falls in the outside temperature, the system is performing well. In another example, a football team has a desired state of aiming to win matches and an actual state of the matches it does win. Its policies would affect the choosing of the team for successive games or the purchase of new players. If the policies are successful, the team will perform well by being close to achieving its targets throughout the season. An ill-judged policy of changing the team before it has had a chance to settle down might easily make matters worse.

In Fig. 1.2(a), the policies are shown in *italic*, suggesting that they are fixed 'rules' specifying what is to be done in given circumstances. Later, we shall meet policies which vary, in the sense that they may be chosen from a menu of options. However, there is nearly always a delay before the action which was started (turning on the central heating boiler or starting to recruit workers) is completed, and it is the completed action which affects the actual state of the system. That state is, however, affected, perhaps continually and before actions can take effect, by external shocks such as changes in the weather or another firm offering better wages. The diagram shows the influences at work in negative feedback and is, in fact, a simple version of the **influence diagrams** which we shall study in the next chapter, at which point we shall introduce + and − signs on the links to give a much more precise meaning to terms such as 'actual state' and 'discrepancy'. Notice, though, that solid lines represent the consequences parts of the paradigm and dashed lines represent feedback of information and the generation of actions.

Ideally, negative feedback should produce a stable reaction to external shocks. If a discrepancy arises, it should be smoothly eliminated within a reasonable period of time. In fact, unless the policies are well designed in relation to the 'physics' of the consequences in the system, the loop may not succeed, and a discrepancy may oscillate between the workforce being too large and too small and the oscillations may go on forever. In severe cases, the oscillations may get progressively larger!

Positive feedback, on the other hand, acts as a growth-generating mechanism. The state of the system, such as one's bank balance, grows continually larger as interest payments act as the rate of change. The change parameter is the percentage interest paid by the bank. This is sometimes called a virtuous circle, as opposed to the vicious circle which arises when the balance becomes negative and one gets deeper and deeper into debt as interest is added to the debt. Positive feedback is quite common in managed systems and may be valuable as an engine of growth. In an engineering system, however, positive feedback is undesirable and is designed out, which is one reason why the mathematical techniques of control engineering are of little help in designing managed systems.

DEFINITIONS OF SYSTEM DYNAMICS

Having discussed some basic ideas, we are now in a position to put forward some definitions of system dynamics. The earlier remark about the application of control engineering to managed systems makes it clear that the aim is to achieve in managed systems the same standards of performance that control engineers can attain in very complex systems such as the autopilot for a Boeing 747. It does not, however, tell us much about how system dynamics goes about its task.

Forrester, the founder of system dynamics, defined it as (Forrester, 1961):

... the investigation of the information-feedback characteristics of [managed] systems and the use of models for the design of improved organizational form and guiding policy.

In a previous book, this author's version was (Coyle, 1979):

A method of analysing problems in which time is an important factor, and which involve the study of how the system can be defended against, or made to benefit from, the shocks which fall upon it from the outside world.

or, alternatively

System dynamics is that branch of control theory which deals with socio-economic systems, and that branch of Management Science which deals with problems of controllability.

Wolstenholme (1990) offered:

A rigorous method for qualitative description, exploration and analysis of complex systems in terms of their processes, information, organizational boundaries and strategies; which facilitates quantitative simulation modelling and analysis for the design of system structure and behaviour.

None of these is completely satisfactory. Forrester does not say what type of models are involved. Neither Forrester's nor Wolstenholme's definitions refer to time, and mine does not mention information feedback. Wolstenholme rightly mentions qualitative analysis but does not include the powerful optimization methods that we shall study later in this book.

Perhaps something on the following lines will suffice to start us on our study:

System dynamics deals with the time-dependent behaviour of managed systems with the aim of describing the system and understanding, through qualitative and quantitative models, how information feedback governs its behaviour, and designing robust information feedback structures and control policies through simulation and optimization.

That is a shade on the long side, but readers should accept it for the moment. At the end of the book we shall challenge you to define system dynamics for yourself.

A STRUCTURED APPROACH TO SYSTEM DYNAMICS ANALYSIS

Having discussed some of the underlying ideas of what system dynamics is about, we turn to the underlying five-stage approach for its application, which is shown in Fig. 1.3.

The **first stage** is to recognize the problem and to find out which people care about it, and why. It is rare for the right answers to be found at this stage, and one of the attractive features of system dynamics as a management science methodology is that one is often led to re-examine the problem that one is attempting to solve.

Secondly, and the first stage in system dynamics as such, comes the description of the system by means of an **influence diagram**, sometimes referred to as a 'causal loop diagram'. This is a diagram of the forces at work in the system which appear to be connected to the phenomena underlying people's concerns about it. Influence diagrams are constructed following well-established techniques which will be explored in the next chapter. The arrow that leads from Stage 2 to Stage 4 will be explained in a moment.

Having developed an initial diagram, attention moves to **Stage 3**, 'qualitative analysis'. The term simply means looking closely at the influence diagram in the hope of understanding the problem better. This is, in practical system dynamics, a most important stage which often leads to significant results. Indeed, the problem is sometimes solved at this stage and there is no need to go on to the others, which is the meaning of the dotted line between Stages 3 and 4.

In **Stage 3**, the analyst draws on so-called bright ideas and pet theories. The former arise from experience with other problems. One may have seen something like the set of feedback loops for this problem in some other case, and what was learned then may be applicable here. That, of course, does not help the inexperienced analyst who may have to rely on looking for obvious inadequacies in the system. If, for instance, it contains negative feedback loops which have no clear desired states then it is very unlikely to be successful in eliminating discrepancies and may well be showing evidence of being out of control; managers talking about never having the right

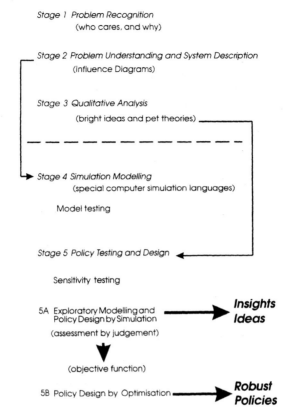

Stage 1 Problem Recognition
         (who cares, and why)

Stage 2 Problem Understanding and System Description
         (Influence Diagrams)

Stage 3 Qualitative Analysis
         (bright ideas and pet theories)

Stage 4 Simulation Modelling
         (special computer simulation languages)

Model testing

Stage 5 Policy Testing and Design

Sensitivity testing

5A Exploratory Modelling and
    Policy Design by Simulation   **Insights**
                                                    Ideas
    (assessment by judgement)

    (objective function)

5B Policy Design by Optimisation   **Robust**
                                                    Policies

Fig. 1.3   The structure of the system dynamics approach.

number of people available could be a symptom of that. Similarly, a system which should be growing ought to have positive loops. If they cannot be found in the influence diagram then a diagnosis of the problem may be emerging.

Pet theories are frequently even more useful. They are the views of experienced people in the system as to what is wrong with it. The views themselves may be found to be wrong on deeper analysis, and the reasons why they are wrong are usually of great interest, but they are almost always a useful source of knowledge about the problem and should be searched for by the analyst.

If qualitative analysis does not produce enough insight to solve the problem, work proceeds to **Stage 4**, the construction of a simulation model. At this stage, we exploit the important property that, as we shall see in Chapters 2 and 3, the influence diagram can be drawn at different levels of aggregation. It is usually not even necessary to show every single detail, because, if the influence diagram has been properly drawn, **the simulation**

model can be written from it without a separate stage of flow charting. This derives from the use of special simulation languages for system dynamics and, in effect, the influence diagram and the simulation model are simply two versions of the same model; one written in arrows and words, the other in equations and computer code. This property is of fundamental importance in system dynamics as it gives rise to some powerful practical consequences.

The first is speed. The equations can be written in any order the modeller chooses. In practice, an order is chosen which is easy to understand and explain, but the modeller does not have to worry about the order in which calculations must take place; the specialist languages do that. This means that the model can usually be written 10 or even 100 times faster than the equivalent model in BASIC or any other high-level programming language. The time saving means that initial results from a model can often start to emerge within days or even hours of starting work, which is vitally important from the point of view of communicating with, and retaining the interest of, a client or sponsor.

The second is ease of revision and expansion of a model. It is a truism of practical modelling that one never understands the problem correctly at first and the initial attempt at a model will usually be inadequate in some way. Since equations can be written quickly, a model can easily be revised as better understanding develops. Similarly, it is easy to extend a model when the need arises without the existing code ceasing to work, something which almost invariably happens when a program written in a high-level language is extended or modified. Indeed, system dynamics modellers usually exploit this property by building small models to start with and allowing them to expand in a controlled fashion, rather than seeking to write a large model from the outset.

Thirdly, the fact that the model exists in the two equivalent forms of a diagram and a set of equations is a powerful aid to thinking about and understanding a problem. One uses whichever form is most conducive to effective thought and communication.

Naturally, this stage also includes the testing and debugging of the model; most important steps, which are often referred to as 'validation' of the model. This is a most inappropriate terminology as it implies that the model somehow becomes absolutely correct and true. It is far better to regard 'valid' as meaning: *well suited to a purpose and soundly constructed*.

To test the soundness of construction, system dynamics modellers make use of techniques such as dimensional analysis, mass balance checks, extreme value tests, and other approaches which we shall encounter when we build some models in later chapters. For the moment, the reader need simply be aware that testing a model to ensure that it is well constructed is something which is taken very seriously in system dynamics and that there are several approaches to carrying out the tests.

Stage 5 is where results based on quantitative analysis start to emerge. Initially, use is made of the insights from the bright ideas and pet theories from qualitative analysis, which is the meaning of the line joining the two stages.

In the first stage, the emphasis is on policy design and system testing by simulating potential changes to the system to see what effect they have and on trying to understand those effects by thinking about the system's loops. The assessment of whether behaviour has improved, and why, is done by qualitative judgement and looking at the output. This is, as it were, Stage 5A, in which the emphasis is heavily on exploring the behaviour of the system rather than predicting precise details. This stage represents exploratory modelling of the system's characteristic patterns of behaviour with the aim of enhancing understanding.

The large arrow going down to the words 'objective function' suggest that, as one's insights into, and ideas about, the problem develop during this simulation stage, it should be possible to develop an extra set of equations which capture the essence of what the system is trying to achieve, or what one would like it to achieve, and measure how well it succeeds. Objective functions exist in other branches of management science, such as linear programming, and system dynamics uses them to unleash the power of optimization software to design robust policies which address the issues identified in Stage 1. This phase of system dynamics is applicable whether the system is closely managed, as in a business firm or a defence problem, or whether the purpose is research into how a past civilization might have avoided collapse.

THE PROCESS OF SYSTEM DYNAMICS

Figure 1.3 showed the technical stages in the system dynamics methodology, but it is also useful to look at the same ideas in another way to emphasize the relationships between stages of work and the results to which they give rise. This is shown in Fig. 1.4.

The left-hand side shows 4 of the stages of system dynamics and the right-hand the results produced. Notice that Stage 3, qualitative analysis, leads from system description to understanding and ideas and that the latter may lead back to the problem definition stage. Similarly, Stage 5A corresponds to the use of the model to verify ideas generated earlier and to stimulate new ones. Finally, Stage 5B, Optimization, uses the model as its basis, but relies more heavily on the emerging ideas and insights which have developed down the right-hand side.

The double-headed arrow between optimization and robust policies is intended to emphasize that optimization does *not* automatically produce *the* single answer to the problem. Very often the nature of the policies suggested by the optimizer leads to further understanding of the problem

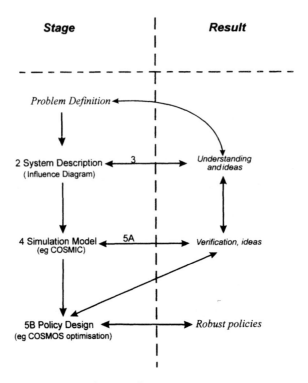

Fig. 1.4   The process of system dynamics.

and the workings of the system and hence to a richer vein of thinking leading to more profound policies being discovered.

Technically, as we shall see in a later chapter, the optimizer works by running the simulation model many times. This can be called **optimization by repeated simulation**. However, the double headed relationship by which optimization can lead to better optimization is similar to a simulation experiment making one think of a better experiment, and should make one think in terms of **simulation by repeated optimization**.

Figure 1.4 refers to two of the specialist system dynamics software support tools, COSMIC and COSMOS: the former for model building and testing and the latter for optimization. There are several commercial packages available and it must be emphasized that the most significant aspect of successful system dynamics is intelligent and careful use of the package to which one has access, not the use of any particular package. The models in this book were developed with COSMIC and COSMOS, but users of other packages should have no difficulty in using the models, with minor amendments. The software packages available for system dynamics, some of which do not optimize, are reviewed in Appendix A.

SYSTEM DYNAMICS AND TOP MANAGEMENT

We have seen several times that system dynamics concentrates on the policies and dynamic behaviour of the system in question, which are the strategic concerns of the top managers, whether they be chief executive officers of businesses, city managers, generals, politicians or whatever. Figure 1.5 brings together the characteristics of system dynamics, which we shall study as the book progresses, and these top management concerns.

The diagram can be read from left to right or vice versa. Thus, the system viewpoint of system dynamics and the 'whole company' concern of a senior manager equate to one another. In particular, the transparency of the influence diagram and the simple simulation technique provide for the top manager to understand what is going on. This is precisely the point made earlier about the duality of the influence diagram and the simulation model as two representations of the same problem.

The last point in the diagram is worth expanding upon. Quite large system dynamics models typically take only a minute or two to compile and run on a standard PC. Making the 20 to 30 runs usually required for an optimization experiment normally takes only another minute or so. Such fast model running capabilities are ideal for study periods in which the

SD Characteristic	Top management concerns
System viewpoint	Whole company
Feedback analysis	Consequences of actions
Dynamic modelling	Concern with future Testing of ideas
Optimisation	Robustness against uncertainty
Transparency of Influence Diagram Simple simulation technique	Understanding, input and control
Fast simulation	Study periods

Fig. 1.5 System dynamics and top management.

client team, with some assistance from an analyst, can work with the model themselves, testing their own ideas and developing their own insights, over a period of a day or so. Practical experience of these sessions suggests how enormously valuable they can be, both in drawing out ideas and insights from the problem proprietors and in giving them confidence that they understand and support what has been done and that 'their' problem has not been taken over by the analyst.

SUMMARY

This chapter has been an introductory survey of what system dynamics is for, how it relates to the rest of management science, its main processes and how it supports top level decision issues. It should have given the reader an idea of what to expect in the rest of the book. A number of claims have been made about system dynamics, which, in essence, are as follows:

- System dynamics is relevant to dynamic problems and therefore complements other management science approaches, which are more closely geared to static problems. The implication is *not* that static analysis is inferior or inappropriate; it *is* that dynamic problems are also important.
- System dynamics deals with the broad behaviour of the system and how it influences its own evolution into the future. These can be seen as the strategic issues which concern top management in the organization.
- System dynamics, like all good methodologies, has a systematic procedure; the five-stage method.
- System dynamics models come in two equivalent forms: the influence diagram and, where appropriate, a simulation model. These two forms make it possible to communicate easily and effectively with the client or sponsor of the work.
- The simulation technique of system dynamics makes it particularly easy and quick to build models.
- Once the model has been built and tested, the analyst can use it to test alternative policies and to redesign the system so that its policies become more effective. Powerful techniques of optimization support this process but do not replace thinking about what is going on in the model.
- System dynamics can be applied to a very wide range of problems.
- Last, but by no means least, as an intellectual pursuit and as a way of making a living, system dynamics is very challenging and enormous fun.

So far, all of these are no more than assertions which will be developed in detail in the ensuing chapters. Readers should, however, be constantly reminding themselves of them and continually evaluating them as they proceed.

REFERENCES

Coyle, R.G. (1979) *Management System Dynamics*, John Wiley & Sons, Chichester.
Forrester, J.W. (1961) *Industrial Dynamics*, MIT Press, Cambridge MA.
Wolstenholme, E.F. (1990) *System Enquiry*, John Wiley & Sons, Chichester.

CHAPTER 2

Influence diagrams

INTRODUCTION

In Chapter 1, we stated that influence diagrams are an essential tool in system dynamics; they are not only the foundation on which quantitative models are built but are also a valuable device in their own right for describing and understanding systems. This chapter will cover the techniques for constructing diagrams, and we shall first study a simple example so as to show the detailed conventions used in influence diagrams. Later, we shall show how diagrams can sometimes be simplified and study the important point that it is not only permissible, but also often desirable, to draw several diagrams at different levels of detail *for the same problem.*

Influence diagrams are sometimes called 'causal loop diagrams'. There is little or no difference, but causal loop diagrams are best thought of as influence diagrams drawn at a very broad level, and not showing the fine detail which can be included in an influence diagram. We shall discuss this in more detail later in the chapter when we consider the 'cone' of influence diagrams.

AN INFLUENCE DIAGRAM FOR A CENTRAL HEATING SYSTEM

We start by studying the completed diagram for the central heating system discussed in Chapter 1, the diagram being Fig. 2.1. Initially, we shall not worry about the solid and dotted lines, the + and − signs or any of the details, but concentrate on what the diagram shows, which are the **influences at work in the system, the interplay of which is the cause of its dynamic behaviour**.

For this system, the influences are fairly simple, or, at any rate, readers will think so when they have studied some of the more complex cases in this book. The householder has set a desired temperature on the thermostat, but the actual temperature may not be at that value. If it is not, the boiler is automatically turned on and heat starts to flow into the water pipes and radiators or into the hot air ducts. After a delay due to the length of the pipes, heat starts to reach the room, adding to the quantity of heat already

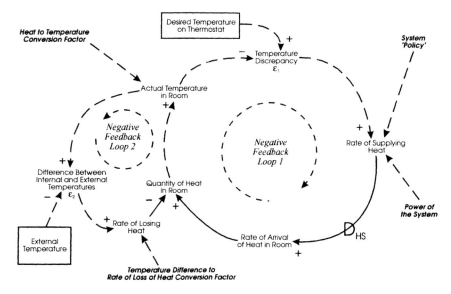

Fig. 2.1 Influence diagram for central heating system.

there. Depending on the size of the room, the heat *quantity* is sensed as *temperature* by people in the room and by the thermostat and, when the temperature is close to the desired value, the supply of heat stops. We shall suppose that this is also an air-conditioning system so, if the room is too warm, the system starts to subtract heat until the room cools down. Subtracting heat should be thought of as supplying *negative* heat, so the words on the diagram do not need to be changed.

This simple system shows the essential trick in developing a good diagram – '*think physics*'. In the information/action/consequences paradigm the consequences are always physical in the sense that something **flows** in the system – people are recruited, heat is supplied, goods are manufactured, casualties are incurred, money is transferred and so on. Even in a psychological case, unhappiness builds up or decreases. In all cases, therefore, *something flows*, and recognizing what flows in the system is the key to a good model, a good trick being to visualize the problem as a set of pipes and tanks through which water flows. In fact, flow analysis is the basis of one of the techniques of diagram construction which we shall examine later. We shall leave the question of what a 'good' diagram is until later in the chapter, when we have built some influence diagrams.

At the risk of labouring the point, in this case the heating system cannot supply *temperature*; it must supply *heat*, and the amount of heat which has accumulated in a room of a certain size is what determines the temperature, and it is temperature which the managers of this system wish to achieve.

Unfortunately, the higher the room temperature relative to the outside temperature, the greater will be the rate of loss of heat from the room. Again, if the external temperature is higher than that in the room, heat will flow in. That is a negative loss rate so, again, the words on the diagram are appropriate.

THE CONVENTIONS FOR INFLUENCE DIAGRAMS

It is now time to turn to the diagrammatic conventions. First, we shall state them, and then explain them in relation to Fig. 2.1.

- Solid lines show physical flows, and these are the *consequences* in the paradigm.
- Broken lines represent influences which are not physical flows and indicate the *information* and *action* parts of the paradigm. The actions may be deliberate choices by the managers of the system or they may be the ungovernable reactions of nature to the state of the system.
- The large D on the physical flow represents a significant time delay. A subscript may be attached to the D to distinguish one delay from another.
- A box denotes an external driving force over which the system has no control and to which it must respond. In this case, the External Temperature is such a driving force, changing from time to time for reasons of its own. The Desired Temperature on the Thermostat is also a driving force. It is set by human action but, having been set, the central heating system simply has to respond to whatever it is.
- A + sign means that when the variable at the tail of the arrow changes, the variable at the head always changes in the *same* direction. Thus, the greater the quantity of heat in the room the higher the temperature will be. On the other hand, if the quantity falls, so does the temperature.
- A − sign has the opposite effect: if the tail variable changes then the head variable changes in the *opposite* direction. Thus, the greater the rate of losing heat the less heat will remain in the room and, the smaller the rate of losing heat, the more heat will be left.
- Items in **bold italics** are constant factors, usually called **parameters**. (Underlining can also be used, especially when drawing by hand. The choice is immaterial, as long as a parameter is identified as such.) The dotted arrows leading from them emphasize that they are not physical flows, they are simply influences on the system. One might use a different pattern of broken line, but that would complicate the drawing of diagrams.
- It is not always obvious whether a parameter has a positive or negative effect on the variable it influences and, for that reason, signs on links from parameters are optional. The author sometimes does not include

them in diagrams, though that is partly a matter of personal style, and readers should develop their own styles as skill improves with practice.

- Feedback loops have attention drawn to them by a broken circle and a name for the loop. A loop exists when, starting from a given variable and following arrows *in the direction they lead*, it is possible to get back to the start, without going through any variable more than once.

- The sign, or **polarity**, of a loop is, theoretically, found by multiplying its signs together. Multiplying the signs on Loop 1, starting from Quantity of Heat in Room gives:

$$+ \times - \times + \times + \times +$$

which, by ordinary mathematics, is negative. A simpler method is to count the negative signs. If there is an *odd* number of $-$ signs, the loop is *negative*, or goal-seeking; otherwise it is a *positive*, growth-generating, loop.

These conventions are summarized in Fig. 2.2 with the addition of the averaging, or **smoothing** of information. In some systems, a variable, such as

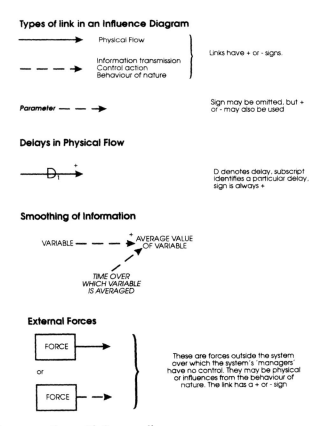

Fig. 2.2 The conventions of influence diagrams.

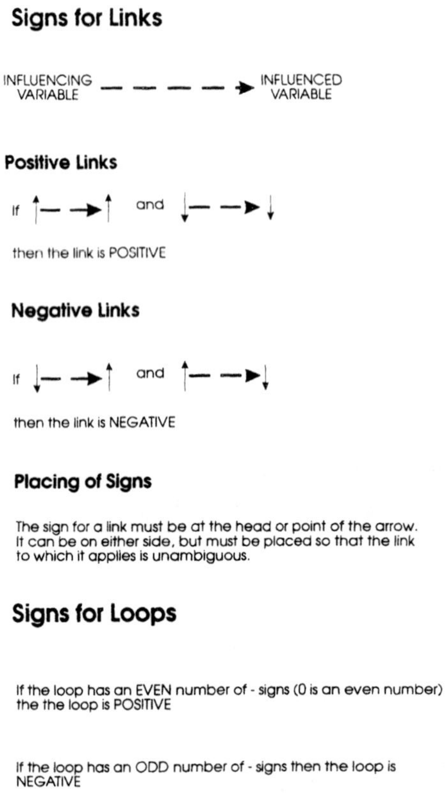

Signs for Links

INFLUENCING — — — — ▶ INFLUENCED
VARIABLE   VARIABLE

Positive Links

If ↑— ▶↑   and   ↓— —▶↓

then the link is POSITIVE

Negative Links

If ↓— ▶↑   and   ↑— —▶↓

then the link is NEGATIVE

Placing of Signs

The sign for a link must be at the head or point of the arrow.
It can be on either side, but must be placed so that the link
to which it applies is unambiguous.

Signs for Loops

If the loop has an EVEN number of - signs (0 is an even number)
the the loop is POSITIVE

If the loop has an ODD number of - signs then the loop is
NEGATIVE

Fig. 2.3 Signs for links and loops.

the rate of sales, might fluctuate quite sharply from day to day and one might want to smooth out that variation by taking the average sales rate over a period of time, such as a month. Thus, the variable influences its own average, but requires a parameter of the period over which the average is taken. Figure 2.3 summarizes the sign rules for links and loops.

THE CENTRAL HEATING SYSTEM REVISITED

Having explained the conventions, we can now explain some of the factors in the central heating problem a little more closely.

The **discrepancy**, or difference, between desired temperature and actual temperature is denoted by the Greek letter epsilon, ε, with a subscript 1 to distinguish it from other epsilons. The difference is calculated in the sense of:

Desired Temperature − Actual Temperature

so that an increase in desired temperature will increase ε_1 and an increase in actual temperature will reduce it. If actual temperature increased so much that it exceeded the desired temperature, then ε_1 would become negative, that is it would have continued to decrease, and the $-$ sign would still be correct.

It would not be *wrong* to define ε_1 in the sense of

Actual Temperature $-$ Desired Temperature

with the corresponding changes of sign, but it would be rather confusing, as it would imply that ε_1 had to be small before the heat was turned on, rather than the common-sense approach which says that ε_1 has to be large to turn on the heat. However, even if one did define ε_1 in this way, the net effect of the signs on Loop 1 would not change and it would still be a negative loop.

The rate of supplying heat is influenced by the discrepancy, by the 'policy' which states how large ε_1 can be before the heat is turned on and, obviously, by the heat-generating power of the system. Heat arrives in the room after a delay due to the size of the house. The actual temperature depends, as we have seen, on the quantity of heat and also on a parameter, the heat to temperature conversion factor, which depends on the size of the room.

The external temperature is shown in a box to signify that it is a driving force over which the managers of the system have no control. Such external forces are called **exogenous** factors, to distinguish them from the **endogenous** factors within the system itself. The difference between the room and external temperatures, ε_2, determines the rate at which heat will be lost and the heat loss represents an ungovernable force of nature, about which the home owner can do nothing. To represent this requires another parameter, the Temperature Difference to Rate of Loss of Heat Conversion Factor.

The parameter for the power of the system is obvious common sense, but why do we need to introduce these two conversion factors? Again, common sense supplies the answer. The amount of heat in the room must be measured in some sort of Heat Units and the rate at which heat is supplied and lost must be measured in Heat Units/Hour. Temperatures and ε_2, on the other hand, are measured in degrees so, to get the rate of loss of heat, one must have a conversion factor which will be measured in (Heat Units/Hour)/Degree. The ( ) are used here to emphasize the rate at which heat flows under the influence of the temperature difference. The reader should work out the units, or **dimensions** as they are called, for the Heat to Temperature Conversion Factor. It should now be clear that, had these two parameters been omitted, the model would have been incomplete, so the ability to recognize when a parameter is needed is crucial to getting a good influence diagram. Thinking in terms of the units involved is the key: a process called **dimensional analysis**, about which we shall learn more in Chapters 4 and 5.

Before we leave this diagram it should be noted that it contains **two** negative loops. Loop 1 is drawn to be larger than Loop 2, because it is the obvious one which attempts to control the system. As we pointed out in connection with Fig. 1.2, any goal-seeking system must contain at least one negative loop, and Loop 1 has all the necessary characteristics of the negative feedback loop; that is, it has both a desired and an actual state, and it has a discrepancy between the two which leads to action intended to eliminate the discrepancy.

Figure 2.1 has a second loop which is clearly negative from its signs and which acts as drain on the system, counteracting the efforts of the home owner to be comfortable. Loop 2 is nature's response to people's attempts and it is the source of the profits of utility companies. Its 'desired' state is the external temperature, so this loop is trying to control the room to the ambient temperature while Loop 1 is trying to control it to the owner's desires. Such conflict between loops is often the source of interesting dynamic behaviour.

Having studied an influence diagram, it is now time for the reader to try constructing one.

A PRODUCTION MANAGEMENT PROBLEM

In practical system dynamics an analysis often starts from a narrative description of a problem, furnished by a client, drawn from the analyst's notes of meetings or suggested by the literature in academic research. Figure 2.4 is such a description of a problem in production management[1]. We shall use this problem, with many additional factors, throughout this book. We shall build our diagram in small steps, to develop skill in diagramming.

Figure 2.4 should be carefully studied before reading further.

In two places, the narrative refers to eliminating discrepancies *over a period of 4 weeks*, and we must digress a little to understand that idea. In Fig. 1.2 we saw that a negative loop must generate control action to eliminate errors between actual and desired states. In Fig. 2.1, Loop 1 in the central heating system does exactly that by turning on the boiler, but the rate at which the room warms up is controlled by the fixed power of the system. People controlling managed systems have, however, freedom to choose how swiftly or sluggishly to respond to discrepancies, and this can be seen by again using the powerful concepts of **dimensional analysis**.

For this problem the actual and desired backlogs of washing machines, and the discrepancy between them, must all be measured in washing ma-

---

[1] This problem is a very simplified version of a consultancy assignment the author undertook for a manufacturing company, though the product has been changed. The real company did use the policies and parameter values given here. We shall revisit this problem several times throughout this book.

The Domestic Manufacturing Company, DMC, produces washing machines for sale to major retail companies. Customers tend to order large batches of machines, with delivery required about 6 weeks after the order so as to match the customer's own marketing plans. Making a machine is very simple, mainly involving the assembly of standard parts, and the production process takes a short time. Building machines uses up raw materials, fresh supplies being ordered from parts makers who deliver in about 6 weeks.

DMC have never been able to forecast the inflow of new orders and just have to try to cope with a very unpredictable order pattern. New orders accumulate into a backlog which the company tries to keep down to a target level.

The Production Manager considers two factors in setting the production rate. First, he aims to eliminate any discrepancies between actual and target backlog over a period of 4 weeks. New orders are averaged over a 4 week period and the backlog target is 6 weeks of this level of orders. Second, he tries to keep up with the current order level, so to the production rate which would eliminate the discrepancy within the planned time is added the average order rate.

The Raw Materials Manager tries to keep raw material stock up to a target level, ordering raw materials to eliminate any discrepancies within a period of 4 weeks. The target level is based on smoothing production variations over 4 weeks and aiming to have sufficient stocks to cover 8 weeks of average production. He, too, tries to keep up with current usage, so to the order rate which would eliminate the discrepancy is added the average usage rate of raw materials.

Fig. 2.4 Narrative of the production and raw material problem.

chines, or WMS (it is very bad modelling practice to measure objects in vague 'units' and it is far better to call the units what they are). The production rate must, however, be measured in washing machines/week, or WMS/week. If we wrote:

Production Rate = Discrepancy

we should be in error because the units on the left do not match those on the right of the =. It is evident that we must have:

Production Rate = Discrepancy/Time to eliminate the discrepancy

for the units to balance across the = sign. The 'time to eliminate the

discrepancy' is usually called the **time constant**. This simple analysis introduces two very important ideas.

The first is that this form of control is called **proportional control**:
the strength of the control action is proportional to the size of the discrepancy to be controlled. It is a very common mistake in system dynamics
models to omit the time constant. Dimensional analysis shows that the
time constant is essential to the equation, so not explicitly including
the time constant in the preceding equation is tantamount to including
it with the numerical value of 1. There is no earthly reason why it *has* to
be 1, nor is there any reason to suppose that a value of 1 will produce the
best behaviour in the system. Omitting the time constant is, therefore,
not only a dimensional mistake, it also denies the analyst the use of
what might be an important opportunity to design better behaviour into the
system.

The second idea is that the time constant is an aspect of the system's
policies, generating *action* in response to *information* about the *consequences* of earlier actions. It can and should be chosen to suit the system's
need for **robust** behaviour. In this case the managers have chosen to use
time constants of 4 weeks, which may, or may not, be good values. The
system dynamics analyst's task is to find out whether *x* weeks is a value
which makes the system work well and to propose another if it is not. As we
shall see, the modeller has a very large degree of freedom to redesign
systems to improve their behaviour, and it is that freedom which makes
system dynamics a practical possibility.

Reverting to the problem, it usually requires a little care before attempting to draw a diagram from a narrative. A good method is to highlight in
the narrative the names of the variables which will be needed, to pin-
point parameters and, above all, to identify the flows as an essential
precursor to the need to *think physics*. Figure 2.5 reproduces Fig. 2.4, but
with items noted in italics. This process is called 'parsing', by analogy with
grammar.

In practice, parsing is usually more difficult than Fig. 2.5 implies. Figure
2.4 was a completed narrative for a textbook problem, but practical
narratives are rarely complete. The attempt to parse reveals the incom-
pletenesses in the narrative and thus takes the analyst back to the problem
which has not been understood if the narrative is imperfect. This is a good
example of what was meant in Chapter 1 about the methodology of system
dynamics helping one to understand the problem more clearly.

Clearly, what flows is washing machines, but there seem to be two sub-
flows: orders and raw materials. One of the tricks of system dynamics is not
to get bogged down in too much detail in the early stages of an analysis. In
this case, the narrative gives no reason to worry about the difference between, say, motors and paint as raw materials, and we can assume that the
raw material manager knows his job and orders raw materials in 'washing

The Domestic Manufacturing Company, DMC, produces *washing machines* for sale to major retail companies. Customers tend to order large batches of machines, with delivery required about 6 weeks after the order so as to match the customer's own marketing plans. Making a machine is very simple, mainly involving the assembly of standard parts, and *the production process takes a short time*. Building machines uses up raw materials, fresh supplies being ordered from parts makers who *deliver in about 6 weeks*.

DMC have never been able to forecast the inflow of new orders and just have to try to cope with *a very unpredictable order pattern*. New orders accumulate into a *backlog* which the company tries to keep down to a *target level*.

The Production Manager considers two factors in setting the production rate. First, he aims to eliminate any *discrepancies between actual and target backlog* over a *period of 4 weeks*. New orders are averaged over *a 4 week period* and the *backlog target is 6 weeks of this level of orders*. Second, he tries to keep up with the current order level, so to the production rate which would eliminate the discrepancy within the planned time is *added the average order rate*.

The Raw Materials Manager tries to keep *raw material stock* up to a *target level*, ordering raw materials to eliminate any *discrepancies* within *a period of 4 weeks*. The target level is based on *smoothing production variations* over *4 weeks* and aiming to have *sufficient stocks to cover 8 weeks of average production*. He, too, tries to keep up with current usage, so to the order rate which would eliminate the discrepancy *is added the average usage rate* of raw materials.

Fig. 2.5 Parsed narrative of the production and raw material problem.

machine equivalents'. Our aim is to analyse the robustness of the overall policies for running the business, not to do detailed order planning.

Secondly, we note a significant delay in obtaining new materials, but not in the production process. This tells us that we need to distinguish between the ordering and the arrival of raw materials, but not between the rate of starting production and the rate of shipping from the factory to the customer. Finally, the policies for controlling production and raw material ordering are described and the parameters identified.

The first step in drawing the diagram is to identify the physical flows, and the reader should pause and attempt to draw them before going further.

The physics is shown in Fig. 2.6, the rather strange layout of which is to give consistency with the final diagram in Fig. 2.7. In Fig. 2.6 New Order Inflow Rate is drawn in a box to show that it is an unpredictable driving force, as was external temperature in the central heating diagram. Orders accumulate into a backlog, which is depleted by production. The production rate depletes raw material stock which is fed by raw material arrivals, in turn a delayed version of raw material orders.

Figure 2.6 is the skeleton of the system's *consequences* on to which we must impose the nerves and muscles of the *information* and *action* components. To emphasize that, Production Rate and Raw Material Order Rate have been enclosed in ellipses. These are not part of the influence diagram, but are temporary additions to draw attention to the fact that these two variables are the controlling flows for this system to which the information and action parts of the influence diagram should lead. Recalling the analogy of water flowing in a set of pipes and tanks, these are the valves which regulate flow. Again, the reader should try drawing the regulating influences before looking at Fig. 2.7.

Figure 2.7 shows the complete diagram. It is unlikely that the reader used these variable names, though they are not important in themselves. It is also unlikely that the reader's solution was as clear as this and involved no lines crossing. Laying out a diagram so that it is easy on the eye is important to give the impression of a smooth flow of control. The reader now has to go back to Fig. 2.5 and verify from the narrative that the factors in the diagram are those mentioned there and that the signs on the diagram are those required by the narrative.

For example, the narrative comment that the average usage rate (of raw materials) is added to the raw material orders required to eliminate the discrepancy requires that there be a + sign on the **direct** link from Average Production Rate to Raw Material Order Rate.

The pattern of signs on the links from Desired Backlog and Backlog to Backlog Discrepancy should be compared with those on the links from Desired Raw Material Stocks and Raw Material Stocks to Raw Material Stock Discrepancy. The Backlog Discrepancy takes the form:

Backlog − Desired Backlog

while the Raw Material Discrepancy takes the form:

Desired Raw Material Stock − Raw Material Stock

Again, the narrative makes this clear. The production task is to keep Backlog *down* to a level which will satisfy the customers, while the Raw Material Manager tries to keep Raw Materials Stock *up* to a level which will ensure that the factory can keep running. Great care is needed in ensuring that discrepancies are properly defined, but the influence diagram provides a vital clue. The influence of New Order Inflow Rate on Backlog is positive, that is it will tend to drive Backlog up and the manager has to try to bring

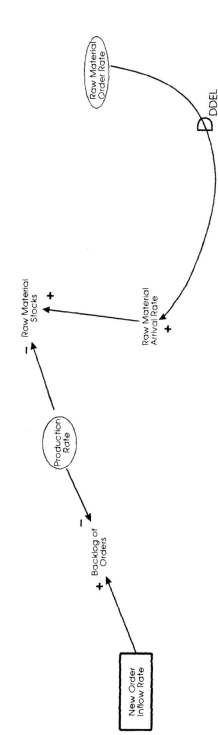

Fig. 2.6 The physics of the production and raw material problem.

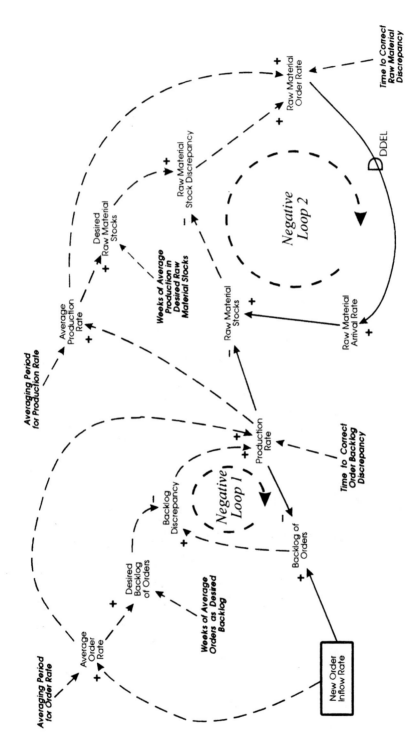

Fig. 2.7 The complete production and raw material problem.

it back down. On the other hand, the influence of Production Rate on Raw Material Stock is to drive it down, and the manager has to try to keep it up.

Finally, this diagram has only two loops. The reader should prove that starting, for example, at Production Rate and moving to Average Production Rate does not lead back to the starting point.

Both of the loops are negative, goal-seeking, mechanisms, as is shown by the pattern of signs. That is not a sufficient guarantee that they *should* be negative and we must apply the qualitative test of reasoning about what the loops are trying to do. Loop 1, for example, represents the Production Manager's efforts to keep Backlog to a satisfactory level. It should, therefore, be negative and it is. This all adds to one's confidence that the influence diagram is correct. This process of building one's confidence in the model by seeing it pass more and more tests is the validation which was mentioned in Chapter 1.

Whether, of course, these loops are successful in achieving what they are supposed to achieve is an entirely different matter, and we shall find later that they are far from being good at their jobs, because of poor design.

Influence diagrams are a vital tool in system dynamics and they usually look very obvious after they have been built. The newcomer to system dynamics, however, faces the problem of learning how to get started on an influence diagram for a new problem, so we now turn to three methods for building diagrams.

THE LIST EXTENSION METHOD

An influence diagram is a list of factors in a problem, together with arrows and signs showing the relationships between them. List extension is based on the rather obvious idea of starting with a small list and gradually extending it until a diagram emerges. This concept is shown in Fig. 2.8 and the reader should use a piece of paper to cover up all of that diagram except the column headings and the extreme right-hand column. The paper can be moved to reveal more of the diagram as the explanation proceeds.

The column headed 'Model list' is the starting point for developing a diagram. It contains the name of a variable which it is the purpose of the model to explain or for the control of which policies are to be designed. In this case, we shall use the Temperature Discrepancy for the central heating diagram. In a practical case, one can use two or three variables, but if more than that are put into the model list, the purpose of the analysis is probably confused and one should revisit Stage 1 in Fig. 1.3.

This list of one item is now extended by writing in the column headed 'First Extension' **the names of the variables which directly and immediately influence the variable in the model list**, adding the appropriate arrows and signs. It is essential to be clear in one's mind about direct influences; in this

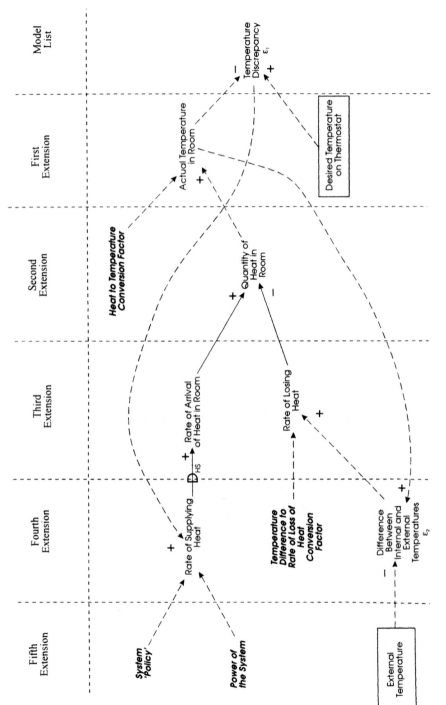

Fig. 2.8 List extension for the central heating problem.

case, the temperature discrepancy is, by definition, the difference between target and actual temperatures in the room. One might be tempted to write 'External Temperature' in the first extension, but that is an indirect influence.

The second extension continues the process by writing down the direct and immediate influences on the variables in the first extension. A chain of influence terminates when a parameter or an external force is encountered. As each extension unfolds, a check has to be made to previous extensions for links from earlier variables. Thus, in Fig. 2.8 we see links running from right to left which emerge as extensions are built. Notice, however, that Fig. 2.8 is nothing like as neat and orderly as Fig. 2.1; in particular, one of the loops is in the shape of a figure of eight. Diagrams built using list extension almost always have to be drawn again to minimize crossing links and portray a smooth flow of influences and clearly visible feedback loops.

The reader should now repeat the exercise for the central heating problem, starting from any other variable, and repeat it again for the production management problem. In each case, the process should lead automatically to the same solution each time. It is one of the very attractive aspects of list extension that two analysts tackling the same problem should come up with recognizably similar solutions even if they started with different ideas as to what were the key variables. This is a solution to the 'two-analyst problem', and it is vital to good modelling that one should get the same end view of the system, regardless of one's initial prejudices about significant variables.

List extension is a good method for starting work on a problem, especially in one's early stages of developing skill in influence diagramming. It is usually not necessary to go past four or five extensions as the pattern of the diagram will probably have emerged by then and one can use one of the other two methods to proceed. It is, however, essential to have the patience to redraw the diagram to make it look neat.

As a rough guide, the first stage of modelling a new problem should not involve more than about 15 or 20 variables and parameters. It is very easy to expand a model in later stages of a project and it is better to start with something simple. The key to successful modelling is to keep one's understanding of the model and what it says about the problem ahead of its size. A model which is too large to start with restricts that development of understanding and insight which was emphasized in Fig. 1.4.

THE ENTITY/STATE/TRANSITION METHOD

The emphasis on the **flows** of something in a system and the analogy of someone attempting to regulate the flow of water in a set of pipes and tanks by turning valves makes the idea of identifying all the **entities** in a system and mapping their **transitions** from one **state** to another a very effective approach to problems in which the physics are more complicated than in the central heating or production management cases. We shall first describe a very simple problem in manpower planning and then show how this method can lead very easily to an influence diagram.

Consider the following narrative account.

> A firm of management consultants recruits new staff to fill any shortfall between the number of qualified consultants available for projects and the number needed to cope with the demand for the firm's services. Recruits undergo a protracted training period before they become fully qualified consultants. When they complete their training, they join the consultancy team. Qualified consultants tend to stay with the firm for a period of years before they leave to go into industry as managers. Some of the qualified consultants employed by the firm have to be used to supervise the trainees, on the scale of 1 supervisor to 10 trainees, and these supervisors are not available for fee-earning assignments. The trainees are not able to work on assignments. The total number of qualified working consultants the firm needs is driven by the demand for consultancy services.

The first step is for the reader to parse this statement, as was done earlier for the washing machine problem. Notice that the last sentence repeats information from the first sentence. Such duplication is very common in practical modelling and the analyst has to be careful to avoid double-counting the same factors.

Having identified the factors in the problem by parsing, we move to the entity/state/transition method, the steps in which are listed in Fig. 2.9.

In this problem, the entities are *Trainees* and *Consultants*. Although individual trainees become consultants in due course, two separate groups of people can be identified at any one time.

A state is a condition in which an entity remains until something causes a change. By the water flow analogy, water in a tank would remain in that state until a valve is turned to allow the water to flow to another tank. For this problem, trainees can only be in the *state* of **undergoing training** and consultants can only be in the *state* of **being available** either to work on projects or train new recruits.

The *flows* cause the numbers of a particular entity in a given state to increase or decrease. For the trainees, the direct and immediate influences are the inflow of new recruits and the outflow at the end of the training period. For the consultants, the inflow is the rate of completion of training, at which point individual people move from one state to another, and the outflow is the rate at which qualified consultants leave the company. Thus, as step 4 enjoins, we have identified a flow which interconnects the two entities. Step 4 also stipulates that one should be alert for delays, and the parsed narrative makes it clear that there are two delays: the training of recruits and the period for which qualified consultants stay with the firm. Adding these delays to the picture gives the diagram shown in Fig. 2.10.

1. Identify all the separate entities or *actors* in the problem.

2. For each entity identify all the possible *states* in which members of that entity can be.

3. For each state, identify the *flows* which can cause the state to increase or decrease.

4. Check for *connections between flows.* Does the outflow from one state feed another?

   Ensure that any *delays in flows* are represented, especially when relating the outflow from one state to the inflow to another.

5. Identify the *controlling* flow rates which drive the system. In general, these will have arrows coming out of them, showing that they influence something, but no arrows going in to show what influences them.

6. Identify and represent the *information and action influences* on the controlling flow rates. This is usually done from the parsing of the narrative account.

Fig. 2.9 The steps in the entity/state transition method.

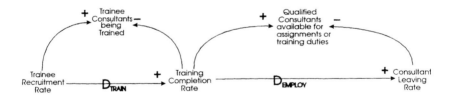

Fig. 2.10 The entities, states and transitions in the consultant problem.

It should be clear that the entity/state/transition method is geared to the injunction to *think physics* and the notion of identifying the direct and immediate influences is that used in list extension. It is not surprising that the various methods of building influence diagrams have much in common; they are directed towards the same end.

In Step 5, we identify the controlling flow rate as the trainee recruitment rate and must now move on to the information and action links to complete the diagram. This is usually rather more difficult and it is often useful to do a minor list extension exercise to get one's ideas straight.

The narrative states that the number of consultants required for work depends on the demand for the firm's services. It also states that some of the qualified consultants have to be used to train recruits and we shall assume for the moment, because the narrative does *not* tell us to assume anything else, that qualified consultants can be moved instantly between working on assignments and training recruits. This leads to the final picture, shown in Fig. 2.11. Note the parameter, 'Time to Correct Consultant Discrepancy', which was not mentioned in the narrative but which, for reasons to be discussed later, has to be introduced into the model. The reader should study Fig. 2.11 carefully, relating it to the narrative for this problem and comparing it to the previous examples, noting and understanding the similarities and differences.

The point made in the previous paragraph about not making an unsupported assumption is rather important. System dynamics models, as we shall see when we come to write equations, are *very* easy to write. That means that they are also easy to extend as more information comes to hand and as understanding of the problem deepens. However, the ease of writing also tempts one to make the model larger than it needs to be in the early stages of a project. It is essential to keep one's understanding of a model ahead of its size and that means that it is good practice to keep the model simple to start with and only to make it more complicated in due time. Making a model more complicated is not the same as improving the sophistication with which it represents a problem, and subtlety is always to be preferred to mere size.

THE COMMON MODULES METHOD

Figures 2.1 and 2.7 use the same unit of structure in that they both contain a flow which, after a delay, leads to an effect. In one case the flow starts with the supply of heat, in the other with the ordering of raw materials, but, apart from these differences in names, the two portions of the influence diagrams are identical. Figures 2.7 and 2.11 both contain information and action components in which target and actual states are compared and action is taken to correct a discrepancy over a time constant. Figure 2.11 uses the same format for the delay in training recruits and for the duration of employment of qualified people. In short, managed systems contain common **modules** of structure which recur in utterly different problems. Once one has learned to recognize these modules, building influence diagrams becomes quite easy. Caution must be exercised, however, against simply throwing modules into a diagram without taking sufficient care to ensure that they are the right ones for the particular case.

Some of the standard modules are shown in Fig. 2.12, which uses 'level' to suggest the analogy of the contents of a water tank. The reader should not bother to remember the names of the modules, but should learn to recognize the phenomena they represent.

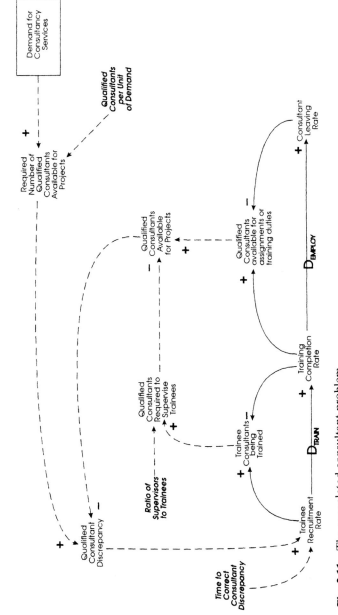

Fig. 2.11 The completed consultant problem.

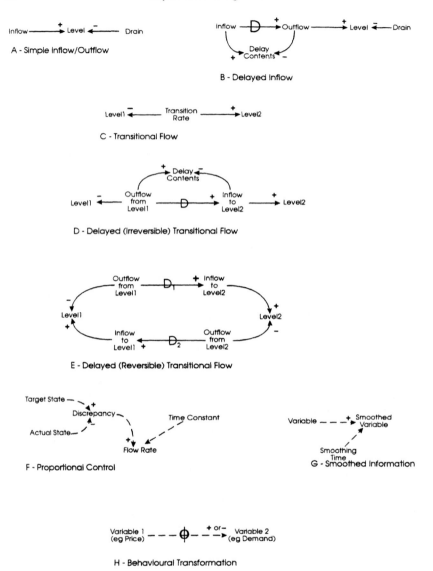

Fig. 2.12 Some common modules.

Case A, simple inflow/outflow, occurs in Fig. 2.6 when orders flow into backlog and production depletes it. Case B, delayed inflow, represents the ordering of raw materials in Fig. 2.6, except that the contents of the delay, the amount of raw material ordered but not yet delivered, is not shown in that illustration, the reason being that the narrative account did not mention it. Clearly, the contents of a delay are present in a system and it is typical, in practical work, that narrative accounts do not mention everything. The value of the common modules is that they help one to recognize items that may have been omitted from a narrative. The reader should now study the earlier diagrams in this chapter and find examples of cases C, F and G, noting how modules sometimes overlap. Figure 2.7 uses cases F and G combined to represent the influences on production rate.

Case C implies that the transition can be in either direction, such as swings of allegiance between one political party and its rival; a positive transition rate flowing from Level 1 to Level 2, and a negative rate flowing the other way. In other cases, such as the flow through a pipe, or the ordering of raw materials in Fig. 2.6, the flow cannot be reversed, and this is shown in case D. If, for example, the qualified consultants required a period of time to adapt from working on projects to training recruits, Case D would be an appropriate module, and the delay contents would be those consultants who were neither available for work nor to supervise training. Unfortunately, case D would not allow them to move back to project work as needs changed, and for that we should have to use Case E, which provides for a reversible flow, the two delay contents not being shown; readers should practice drawing them for themselves[2].

Case H represents a common phenomenon in managed systems whereby one variable is transformed into another, sometimes by managerial choice, but also, as shown here, by an ungovernable force of the economy in which price is transformed into demand. The Greek letter ø, phi, denotes that there is a function or formula which models the transformation of one variable into another. The effect may be $+$ or $-$ or even $+/-$, meaning that in some circumstances it is $+$ while in others it is $-$. This is an example of **non-linear** behaviour in managed systems, which we shall study in much more detail later in the book.

WHICH IS THE RIGHT METHOD TO USE?

We have studied three methods of building influence diagrams so it is a natural question to ask which method to use in different circumstances. In practice, there is no simple answer, though we shall use this section to suggest some broad guidelines. The reader should try to apply them to the influence diagram case studies in the next chapter, as that will help you to see how those models came about.

It should be realized that an experienced system dynamics modeller will often use none of the methods *formally*, in the sense of actually writing out a list extension or the entities, states and transitions. The modeller will,

---

[2] In quantitative modelling, it would be necessary to ensure that the flows were not working in both directions at the same time, a problem we shall address in a later chapter.

however, be using these methods, especially the common modules, though perhaps without recognizing their names, mentally as the diagram proceeds. The diagram seems to emerge on paper without effort, though there is a good deal of thought behind it.

For the less experienced modeller, list extension is usually a good way to start, mainly because it requires one to stipulate the model variables, which is usually a good way of being clear about the purpose of the model. The list extension should, however, be stopped after about four or five extensions and the diagram drawn again to be as clear as possible.

When a diagram drawn using list extension is reviewed it typically leaves the analyst with a feeling that he is being rather vague about some aspects of the problem. This is where the entity-based method comes in to resolve ambiguities about what is flowing and how it flows. Being clear about the flows helps immeasurably when it comes to mapping the information/action elements which control the flows.

The common modules method is a useful check on the correctness of the representation of the flows.

SOME GENERAL GUIDELINES FOR DRAWING INFLUENCE DIAGRAMS

Having studied some examples of influence diagrams and considered three approaches to their construction it is time to offer some general comments on influence diagramming in practice, as shown in Fig. 2.13.

These guidelines are not absolutes, and, as we are about to see, the strict rules of influence diagrams, shown in Figs. 2.2 and 2.3, should not be made into a handicap to thought. Readers should, however, review Fig. 2.13 from time to time and develop their own guidelines as practice is gained and skill increases.

DIAGRAMS AT A MORE AGGREGATED LEVEL

A fully detailed influence diagram can sometimes be an obstacle to communication with a sponsor and it is often useful to draw a diagram to show less detail while still capturing the main features of the problem. For instance, Fig. 2.14 is a rather generalized portrayal of the relationships between an international mining company, 'Anglo-Consolidated Zinc' (ACZ), and its operating subsidiaries around the world.

ACZ derives cash flow from the distributed profits of the mines of which it is part owner. Money retained by operating companies is invested either to maintain an existing mine in production or to develop new mines. ACZ invest their cash flow either in exploration to discover new reserves or to develop new mines. The diagram shows these as the key features of the

1. Always be clear about the *purpose* for which the model is being built. Is it a top-level, strategic view or a more detailed, tactical, problem? Who is the customer for the model – a fee-paying client or the academic community?

2. Always be alert for the difference between the information/action/consequences parts of the paradigm.

3. Choose variables which are defined so that they can, in principle, be measured and which can vary over time. 'Attitude of Customers' would a poor name, 'Customer Satisfaction' would be a good one.

4. Whenever a variable such as 'actual backlog' is seen, look for the corresponding 'desired backlog' and vice versa. As we have seen, without the comparison of an actual and a desired state, negative, goal-seeking, feedback does not exist and control cannot be achieved.

5. Distinguish carefully between actual values and perceived values. For example, actual 'Customer Satisfaction' can almost certainly not be known. The best that could be achieved would be a perception of it, so 'Perceived Customer Satisfaction' would enter the diagram in addition to the actual value. Perceptions usually lag behind reality, so the time delay would need to be included.

6. Watch out for unintended effects. An increase in the production rate might lead to a reduction in quality, with consequences, perhaps, for real and perceived Customer Satisfaction.

7. If a link between two variables cannot be easily explained to others it may be because an intermediate factor needs to be included. A direct link from production rate to customer satisfaction might be hard to understand without including the effect of production on product quality.

8. Above all, keep the diagram simple to start with.

Fig. 2.13 Some guidelines for drawing influence diagrams.

problem, though without the use of solid and dotted lines, and without identifying parameters. The exogenous driving force of metal prices, which even ACZ is not big enough to influence, is shown with the usual box. In short, this is a top-level view of the problem. It is, in fact, deliberately drawn to be a rather poor top-level view, because it says little about the **policies** by

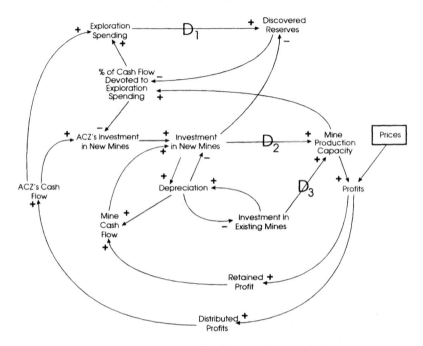

Fig. 2.14   Salient features of 'Anglo-Consolidated Zinc' model.

which ACZ decides to split its cash flow between the two streams other than to imply that, when ACZ has sufficient discovered reserves, more money will be put into developing new mines and vice versa. The ideas of 'target' or 'acceptable' reserves and mine capacity are not present. The diagram says nothing at all about the operating companies' policies.

Other apparent faults are that the diagram includes delayed flow modules of type D from Fig. 2.12, but it shows only the inflow to the delay, Exploration spending measured in $/year, no outflow, such as Exploration spending becoming effective, and Discovered reserves must be measured in tons of metal, so there is a dimensional inconsistency, caused by a missing parameter.

In short, the diagram seems to break all the rules of influence diagrams which has been done to engender in the reader the habit of looking *critically* at an influence diagram: a most essential skill to develop. The aggregated diagram may, nevertheless, be a more useful diagram for discussion with senior managers in ACZ. Rules are often said to be for the observance of fools and the guidance of wise men. That is putting it a little too strongly, but the reader should break the rules when it helps, especially when some practice has been gained in following them.

It would probably be good practice for the reader to redraw Fig. 2.14 in

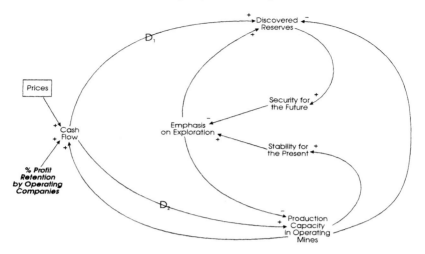

Fig. 2.15 A very aggregated view of 'Anglo-Consolidated Zinc'.

more detail, observing the modules, introducing parameters and distinguishing between information, action and consequences by using solid and broken lines.

Figure 2.15 is an even more aggregated view of the same problem. It introduces two new variables, Security for the Future and Stability for the Present, to capture what is really conferred by the development of reserves and investment in mine production capacity. The effect of prices is shown as an exogenous factor, with a + sign, and the percentage retention of profits by the operating companies, which would form part of the financial agreement with the nation where the mine is located, is strongly emphasized. Figure 2.15 is, perhaps, the most appropriate simplification of all to convey ideas to the most senior management of ACZ, not because they are stupid, but because it corresponds most closely to the problems with which they must deal at a strategic level (this broad view is closest to the idea of a 'causal loop diagram').

Readers should study these two diagrams quite closely and form their own views on how well, or badly, they conform to the guidelines discussed above. Neither diagram is perfect.

THE CONE OF INFLUENCE DIAGRAMS

We have drawn two different diagrams for the same problem and this leads to the ideas shown in Fig. 2.16, namely that it is not only legitimate, but it can also be desirable to draw many diagrams of the same problem.

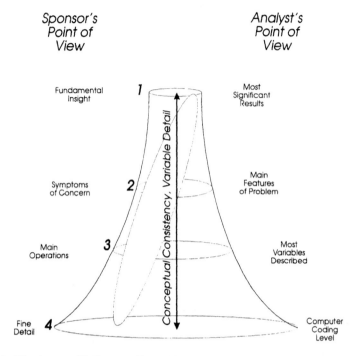

Fig. 2.16 The 'cone of influence diagrams.

Each of the ellipses in Fig. 2.16 symbolizes an influence diagram, the size of the ellipse suggesting the amount of detail the diagram contains. The diagonal ellipse implies a diagram in which some parts of the problem are shown in great detail, while other parts are shown in broad outline. Notice the division between the sponsor's and the analyst's points of view. Clearly, there is no magic about having four levels, and it is usually fruitless to argue about whether a given diagram is at a particular level. The true point is that influence diagrams are flexible tools and should be adapted to the purpose.

Notice, though, that the expression 'Conceptual consistency, variable detail' in Fig. 2.16 does not mean that the same variable names have to be used in diagrams at different levels. As we saw in Fig. 2.15, new variables representing security and stability were introduced because they were appropriate for the level 1 view. These variables do not exist as such in Fig. 2.14 because they would not be appropriate at that level.

In a practical project, the problem is usually first encountered at level 2. Further analysis leads one to level 3 and perhaps to level 4 if a simulation model is constructed. The emphasis in Fig. 1.4 on insight and understanding implies that one might then work back up the levels to convey the most significant results. The level 1 diagram is often drawn only at the end of the study.

LIMITATIONS OF INFLUENCE DIAGRAMS

As we shall see in later chapters when we deal with quantitative models, even the level 4 diagram does not show every single detail of the equations. Attempts have been made over the years to adapt the influence diagram conventions to include details of such factors as choices between two alternatives, non-linear effects and so on. These attempts have not succeeded because they destroy the essential value of the influence diagram, namely its simplicity and its value as tool for communication which does not require an intimate knowledge of the simulation techniques of system dynamics.

LEVEL/RATE DIAGRAMS

Before discussing what constitutes a 'good' influence diagram and summarizing this chapter it is necessary to mention another form of diagram which is used in system dynamics. This uses standard symbols for the variables which correspond, on the water flow analogy, to the levels in the tanks and the flows in the pipes. A third symbol is used for so-called **auxiliary variables**, which are neither levels nor flows. Figure 2.17 shows part of Fig. 2.7 drawn in this way.

These diagrams make very clear the distinction between what flows and where it accumulates. The 'cloud' symbols at the ends of the line represent sources and sinks from which flows arise and into which they vanish when they are no longer of interest, though if Fig. 2.17 had been drawn in full in

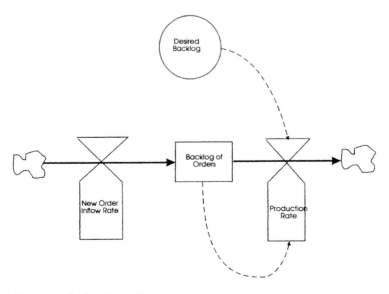

Fig. 2.17   A partial level/rate diagram.

1. Have the purpose and the target audience for the diagram been carefully chosen? It would be no good using a level 4 diagram for a level 1 audience.

2. Are the factors which it includes consistent with the purpose?

3. The objective of system dynamics is policy analysis, so are the policies clearly shown in the diagram?

4. System dynamics also aims to produce policies which are robust against a range of circumstances, so are the exogenous factors which might present the system with setbacks or opportunities clearly identified?

5. Are the variables capable of being easily explained to the target audience, are they capable in principle of being measured and can they vary over time? A variable such as 'Security for the Future' in Fig. 2.15 might be measured in tons of metal in known reserves, though what those tons provide is security, in which a top-level audience might be more interested.

6. If it is a level 1 diagram, does it capture the most significant insights and was it drawn *after* careful analysis?

7. Can the diagram be redrawn to be one level lower or higher without losing conceptual consistency?

8. Has the diagram been constrained by too slavish an adherence to the conventions? If it is too complicated by solid and broken lines it may fail to communicate. On the other hand, the conventions identify important aspects of a system. A difficult choice may be needed.

9. Is the diagram neat and tidy, with a minimum number of lines crossing? Can feedback loops be clearly seen?

10. Do I understand the diagram myself before I try to explain it to someone else or to write about it?

Fig. 2.18   Some criteria for assessing influence diagrams.

this style, the arrow would have led to another box for Raw Material Stock. The circle denotes the auxiliary variable and other symbols exist for delays.

Obviously, such diagrams are incredibly tedious to draw by hand or with a graphics package, but they are the basis on which two of the software packages for system dynamics work. In those packages, symbols are drawn on the screen using standard icons, which reduces the labour involved. Once the symbols have been drawn, *some* of the equations for the simulation model can be generated automatically, though the user has to create the rest. Equivalent facilities exist in other packages which use influence diagrams, rather than level/rate diagrams. The topic of software is explored in more detail in Appendix A.

WHAT IS A 'GOOD' DIAGRAM?

Simply drawing diagrams is not enough: one must know how to recognize a good diagram from a poor one. There are no simple answers, but some criteria are listed in Fig. 2.18. These should be reviewed carefully and applied to the diagrams which appear in this book.

SUMMARY

This has necessarily been quite a long chapter, as it sets out to explain the foundation on which system dynamics analysis is based. We have studied the development of diagrams and suggested conventions. From that point, a number of diagrams were developed.

Towards the end of the chapter, we introduced the important notion of drawing different diagrams for the same problem and put forward some criteria for a 'good' diagram.

The fundamental point is that the diagram should portray the interaction of cause and effect within a system, because it is that interplay which will govern its ability to generate its own future behaviour.

In the next chapter, we shall study some more diagrams, some of which will be the basis for the quantitative models developed later in the book.

CHAPTER 3

Influence diagram case studies

INTRODUCTION

In the last chapter we studied some simple examples of influence diagrams and developed conventions for drawing them. We now need to develop skill in handling rather more realistic problems.

The cases discussed in this chapter are intended to show the practically limitless range of problems to which influence diagrams, and system dynamics in general, can be applied. These cases deal with manpower planning, product development, a predator/prey system, military combat and the decline and fall of a civilization. This is a wide range of subjects, but many years of experience suggest that analysts are restricted only by their ability to grasp a problem and see how it can be framed in system dynamics terms, and it is that insight which these cases are intended to foster.

Each problem is introduced and analysed in turn. Most of them are revisited at the end of the chapter in what amount to exercises for the reader to extend the diagram from a further narrative description.

These problems are much more complicated than those in the previous chapter and there is, unfortunately, no substitute for carefully going through them and trying to solve them oneself before looking at the 'answer'. Bear in mind that, when studying managed systems, there is rarely an unambiguously 'right' answer. Two reasonably competent analysts should, however, come to recognizably similar answers, though there might be differences of interpretation. Do not, therefore, look upon the solutions in this chapter as being definitive.

The solution to the first problem will be explained in a good deal of detail to give the reader some ideas on how to approach the other case studies. First, we must consider the vital question of the purpose of a model.

THE PURPOSE OF A MODEL

The essential first step in studying any problem is to be clear about what the model is for, as was mentioned in Fig. 2.18. A model's purpose has a strong influence on what it includes, so being clear about the purpose is an important step in good analysis.

There are three main aspects of purpose:

- What has happened to the system to make its managers worry about it? This is usually expressed as a pattern of behaviour, experienced in the past or expected in the future, which is in some way unacceptable. The pattern is often referred to as the **reference mode.**
- What is the time period of concern? If the problem concerns corporate growth over a period of years there is little point in representing processes involving delays of days.
- What are the main policy levers which managers can use to improve behaviour? These must be the focus of the policy design effort, but the levers that have usually been used may not be those which produce the best behaviour, so the analyst may be looking for policy *redesign* opportunities.

These points will be briefly mentioned in each case study, but the reader should bear them continually in mind.

THE CONSULTING FIRM'S PROBLEM

In Fig. 2.11 we studied the flows of trainees and qualified consultants in a very simple version of a management consultancy firm. We shall now develop a more realistic portrayal of such a firm, starting with a narrative description.

Narrative description

The management of a consultancy firm has been concerned about the firm's seeming inability over the past few years to grow with the market for consultancy services. The firm also seems never to have the right numbers of qualified consultants and trainees to offer a balanced service on existing jobs and to obtain new business.

The firm recruits new staff who require two years of training before they are considered to be qualified. At the end of the training, a proportion of the trainees leave, for various reasons, the rest becoming full-fledged consultants. Qualified consultants stay with the firm for about 10 years on average before they leave to pursue other careers. For every 10 trainees, a qualified consultant acts as tutor, and tutors and trainees are able to work on projects, though they are only about half as productive as qualified people who are not acting as tutors. Qualified consultants can move freely between acting as tutors or working on projects.

Projects tend to be similar in size and, on average, take about 5 months to complete once work has started, but the actual time depends on the effort the firm can deploy. On average, each project requires a certain number of qualified consultants, or their equivalent in trainees and tutors, but, if the actual number of equivalent consultants is less than the number required for the projects being undertaken, jobs tend to take longer than normal, and vice versa.

There is an industry standard practice that each project requires a certain number of qualified consultants who are not also acting as tutors. If the number of projects the firm is undertaking at a given time is higher than this standard, the firm tends to be less successful at obtaining new projects, and conversely. Otherwise, the rate of obtaining new projects is governed by an exogenous demand for consultancy services which grows slowly and tends to rise and fall roughly in line with the business cycle.

Once a project has been obtained, there is a delay of, on average, two months before work is started, due to the preparation of contracts etc.

The number of projects currently being worked on governs the number of qualified consultants and equivalents the firm should have. The discrepancy between that and the actual number is a component of the desired recruitment rate. The firm wishes, however, to keep up a reasonably steady stream of recruitment so that the average current recruitment rate is also a component of the desired recruitment rate. The firm also has a policy of not dismissing staff on the grounds of redundancy, so trainees are taken on if the desired recruitment rate is positive, but people are not dismissed if the desired rate is negative.

Getting started on the problem

The first step is to parse the narrative, being alert for the same concept being expressed in different words (a typical phenomenon in practical work), detecting the entities, and trying to spot common modules. It may be a good idea to make notes by drawing parts of the diagram. These will help when it comes to applying list extension, and a good variable for the model list might be the number of projects currently being undertaken.

It is also necessary to look for the policy areas of the model. In fact, there is only one, the recruitment of new trainees.

In two parts of the narrative, the consequences in the paradigm produce ungovernable behaviour, stemming from the characteristics of the market for consultancy services. The first is the extent to which the number of qualified consultants who are not acting as tutors and the number of projects currently on hand act to influence the rate at which the firm obtains new projects. The second arises from the influence that the total workforce exerts on how long the average project takes to complete.

Both arise because of a mismatch between required and actual resources and, as the discrepancy between available and required staff was mentioned as a driver of the recruitment process, it is easy to fall into the trap of thinking that a discrepancy is the right concept to apply in these two instances. In fact, when dealing with behavioural effects in managed systems, it is usually more effective to think in terms of the *ratio* of desired and actual. To see that, think in terms of a desired value of 100 and an actual value of 99. For the recruitment sector, *which controls a physical flow*, that would lead to a slow rate of recruitment which would be exactly the same rate if the two values were 2 and 1, the discrepancy being the same. For the *behavioural effects*, the difference between 99 and 100 is obviously far less serious than the difference between 1 and 2; this clearly seen when we think of ratios of 99% and 50% respectively.

A suggested solution to the problem is shown in Fig. 3.1, but readers will essentially be wasting their time if they look at that without first having had a try at formulating a diagram. That will be a bit of a struggle, but worth the effort. It is suggested that readers spend about an hour trying before looking at the solution.

A suggested solution

Notice, first, that Fig. 3.1 contains only one place at which links cross. It is, indeed, impossible to draw the diagram with no crossing links. It is very unlikely that the reader's solution is as neat, and time must be taken to redraw diagrams so that they are clear and contain as few crossing links as possible. If this is not done, the diagram is likely to confuse rather than illuminate and a diagram which resembles a plate of spaghetti in a bad temper is poor system dynamics practice.

As before, solid lines show physical flows and broken lines show information, actions and behavioural effects. Parameters are in italics and without signs.

Figure 3.1 is drawn in terms of the physical flows of three entities (trainees, qualified staff and projects), although the physical flows of trainees and consultants are measured in a common unit; they are said to have the same **dimensions**.

Both projects and people have essentially the same structure of two delayed modules in sequence, though the number of projects obtained but not yet started is not shown, *because it was not mentioned in the narrative*. As was stated in Chapter 2, such omissions from narratives are very common. In this problem, the number of projects being worked on drives the recruitment process, but the number obtained but not yet started is ignored in that decision. That may have a significant effect on the ability of the firm to grow with the market while surviving the oscillations of the business cycle, so perhaps the number of projects pending should be included in the recruitment decision; a good example of the qualitative analysis which was mentioned in Fig. 1.3. The reader should test this, and other policy options, when we undertake policy analysis for this problem in Chapter 6.

Most of the delays in the problem are shown by a simple D on the link with an identifying subscript, as was done earlier. This implies that the

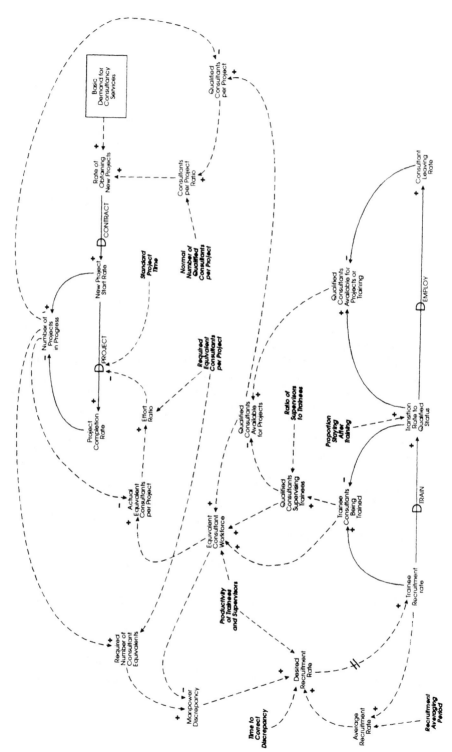

Fig. 3.1 The consulting firm's problem.

delay's duration is constant; in terms of the water flow analogy, the length of the pipe is fixed.

The narrative states, however, that the delay between project starts and project completions is *not* constant; the pipe is, as it were, elastic and can change its length. That is shown by the arrow and sign leading to $D_{PROJECT}$ from Effort Ratio[1] and by the arrow from Standard Project Time[2]. These express the narrative statement that if the Equivalent Consultants per Project, to allow for the lower productivity of trainees and supervisors, are not sufficient for the projects being undertaken then projects take longer than normal. This is really only common sense, and painstaking thought and common sense goes a long way in system dynamics.

The Effort Ratio derives from the workforce available per project relative to that required per project, and the Equivalent Consultant Workforce makes allowance for the lower productivity of trainees and tutors. Notice an example of the limitations of influence diagrams (and of level/rate diagrams). The Equivalent Consultant Workforce is shown as being influenced by three variables and one parameter. The productivity parameter is necessary because it is equivalent consultants the firm wants, but trainees which it recruits. However, there is no indication of how these four influences determine the variable.

It is, however, evident that the relationship must be:

EWF = QCAFP + (QCST + TCBT)*POTAS

where EWF means Equivalent Consultant Workforce, QCAFP means Qualified Consultant Available for Projects, QCST means Qualified Consultants Supervising Trainees, TCBT means Trainee Consultants Being Trained and POTAS means Productivity of Trainees and Supervisors, but this formula cannot be included in the diagram without losing the clarity of its portrayal of the forces at work in a system. The more sophisticated software packages for system dynamics do allow one to browse through portions of the influence diagram and the corresponding equations and definitions.

As with the Effort Ratio, a Consultants per Project Ratio is defined as a means of modelling the behavioural influence on the Rate of Obtaining New Projects. Notice that Basic Demand for Consultancy Services drives the system with a behavioural link, not a physical flow, the reason being that the physical flow only starts when projects are obtained.

The link from Desired Recruitment Rate to Trainee Recruitment Rate has a ‖ symbol to show that there is a **non-negativity constraint**: trainees are only recruited when the Desired Recruitment Rate is positive and they are

---

[1] We shall adopt the convention of using initial capital letters for the names of variables which appear on influence diagrams.

[2] As before, the reader may well have used different names for variables and parameters, but that does not matter.

not dismissed, or negatively recruited, when there are too many staff for the projects on hand.

With these comments in mind, the reader should carefully study this solution and his own version and ensure that he fully understands what has been done.

The level of the diagram

How does Fig. 3.1 relate to the cone of influence diagrams displayed in Fig. 2.16?

Its level of detail, showing many parameters, is such that it is practically at Level 4. As we shall see in Chapter 6, it is only necessary to give numerical values to the parameters, to specify the rate of increase in demand and the severity of its business cycle oscillation and to define in detail the effects of the two behavioural links to have a complete simulation model. In short, it is a very detailed depiction of the problem and, as such, may be too detailed for some purposes. The reader should certainly attempt to redraw the diagram to show less detail but to capture essential features for, say, the senior partner in this firm. To do so, it may help to study Figs. 2.14 and 2.15.

Qualitative analysis of the diagram

Qualitative analysis was mentioned in Fig. 1.3 as an important component of the system dynamics methodology. In essence, it involves studying the diagram to see what can be learned about its adequacy as a model and its qualities as a control system. These are very broad ideas which are hard to pin down until one has had some practice.

In this instance, there is only one policy area by which the system can be managed, the recruitment of trainees. As the model stands[3], recruitment is governed only by the number of projects in progress and the size of the workforce, including trainees and tutors. An average project takes 5 months, but there is also a 2 month delay in signing contracts. No account is taken of the projects in the contract pipeline, so about 2/7 of the firm's order book is ignored in determining the manning requirements. When trainees are recruited, qualified people are immediately assigned to supervise them, which has an effect on the availability of qualified consultants not acting as tutors, but they are the people who have an influence on the ability to generate new projects. This line of thought suggests that there are other policy options which the firm might use in controlling recruitment. After studying Chapters 6 and 7, the reader should experiment to find policies which have a beneficial effect on corporate performance.

---

[3] As with the production planning model in Chapter 2, this is based on a real problem in which the policies were as described in the narrative.

The feedback structure of the model

Loops are important to the dynamic behaviour of models and Fig. 3.2 highlights some of the loops in Fig. 3.1 by drawing them with different patterns of line. The reader should follow the loops by eye, noting that Loop 4 has a common path with Loop 3 between Qualified Consultants Available for Projects and Qualified Consultants per Project. Similarly, Loop 3 is common with Loop 2 from Qualified Consultants per Project *via* Projects per Consultant Ratio and thence to Number of Projects in Progress and is common with Loop 1 from manpower Discrepancy to Trainee Consultants Being Trained. *In tracing these loops, take care to follow them in the direction of the arrows.*

Loops 1 to 3 are negative, goal-seeking, loops. Loop 1 controls recruitment to make sure the firm has enough staff. Loop 2, however, acts to reduce business when there are not enough qualified staff available to generate it, or generates business to absorb those staff. Loop 3 acts in a similar way to Loop 2. Loop 4, which is positive or growth-producing, should act to increase the firm's business in that the more qualified consultants there are, the more projects should be obtained, which should support more staff. The reader should think round the loops, confirming these ideas.

However, there are some snags in this system of loops which cast doubt on how well they will function.

Although Loop 1 should ensure that recruitment does not go too far it competes with a loop, not marked in Fig. 3.2, which runs from Trainee Recruitment Rate, through Average Recruitment Rate and back to the start. That appears to be a positive loop which would drive recruitment ever higher, or lower. The unmarked loop exerts an 'inertia' effect which ought to be counteracted by Loop 1. Notice that the non-negativity constraint, which operates in Loops 1, 3 and 4 means that these act only when there is a shortage of manpower. However, as soon as trainees are recruited, some qualified consultants are immediately switched to supervision, reducing the resources available to Loop 4 to generate business. Further, the long delay, D_{TRAIN}, also means that Loop 4 has little chance to get started.

This kind of qualitative thinking about a system helps to generate ideas for policy design which are tested in Phase 5 of the system dynamics approach which was discussed in Figs. 1.3 and, especially, 1.4.

Other applications of this model

System dynamics models are so easy to build that it is nearly always very bad practice to take a model built for one purpose and adapt it to deal with another problem. It is usually just as quick, and more satisfactory, to build a new model for the new problem.

Having said that, it is evident that there is a large class of manpower planning problems to which the ideas used in this model could be applied.

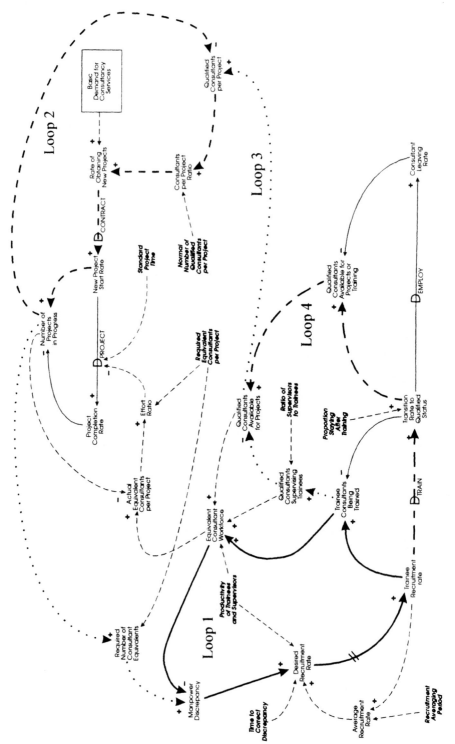

Fig. 3.2 The consulting firm's feedback structure.

The training of airline pilots is an obvious example. Similarly, many large firms, and especially the civil and uniformed services, have quite complex career structures in which one moves up through several ranks, promotion sometimes being more or less automatic in the early stages but becoming much more competitive at the higher levels.

These problems can quickly become quite complicated. For example, there is usually a requirement that there must be some ratio of people at one rank to those at the next lower rank, that new recruits should have some known chance of being promoted during their careers and the capacity of the training system may be restricted. At any given rank, it is usually necessary to serve for some time at that level before becoming eligible for consideration for further promotion. The reader may wish to try drawing an outline influence diagram for such a problem. The trick would be to avoid trying to represent every single rank category and to think instead in broader terms, such as junior, middle and senior grades.

In these systems, the proportion of people moving to the next higher level is a management control influenced by circumstances, rather than a parameter as in Fig. 3.1. That, of course, suggests the idea that, while the consulting firm may well not be prepared to make people redundant, there is no reason why the proportion staying after training should depend only on the preferences of those individuals. Maybe management should have a say in the matter, in which case the proportion would cease to be a parameter and become a variable, influenced by, perhaps, the manpower discrepancy.

This style of thinking is a very important characteristic of good policy design, and the reader should practice developing it as more examples unfold throughout this book.

THE PHARMACEUTICAL COMPANY'S PROBLEM[4]

In the previous problem we examined a model in which the system was heavily influenced by an exogenous force, the Basic Demand for Consultancy Services, which it could do little to influence. The best the firm could hope for would be to expand its market share, the size of the market being beyond its control. Such a system is said to be **driven** or **open**. In this case study we shall examine a **closed** or **undriven** system, that is, one in which the behaviour arises entirely within the system itself.

Narrative description

A pharmaceutical company both develops new products and launches them on to the market. Its products are technically good and the demand for health care is practically limitless, so the company should

---

[4] Again, based on a real problem, but with the product changed.

be experiencing steady growth. However, there seem to be instabilities in its behaviour and products which have been developed in the research laboratories are quite frequently abandoned before they can be launched on the market.

The flow of disposable revenue (revenue after direct manufacturing and selling costs and overheads) into the firm depends only on the number of products it has on the market at any one time and on the disposable revenue the average product generates. The disposable revenue for the average product depends on its 'age'; the time for which it has been on the market. Novel products command much higher revenues than older ones, partly because of patent protection and partly because, as products get older, another company will have time to bring out a younger and perhaps better rival. At the end of their market lives, products are withdrawn from sale.

The company's research chemists are competent and, given money to spend on research, they will, after a considerable time lag, develop products that are viable in the market place. The company must, however, incur considerable extra expense and endure further substantial delay before a product which it has decided to introduce can actually be offered in the market. Products which have been developed but not introduced can be thought of as having a certain 'shelf life'. If they have not been placed on the market, they will have to be abandoned because other companies will have introduced comparable products.

The company is very marketing-orientated and devotes an increased fraction of its disposable revenue flow to product introductions as the average age of its products increases. Whatever is left from the disposable revenue flow is the research budget.

Getting started on the problem

As always, the first step is to parse the problem, and detect entities as a precursor to list extension or the use of common modules. The reader will probably find it helpful to do this problem twice using different variables for the model list. The injunction to *think physics* still applies!

The tricky area in the model is the research and development sector. Product research starts lead inevitably to completions, but completed products are not inevitably introduced to the market. Instead, they accumulate into a pool from which they may be drawn by management to be introduced to the market, or they may be eliminated as they become obsolete.

A suggested solution

As with the consulting firm's problem, the suggested solution in Fig. 3.3 has been carefully drawn, and in this case there are no crossing lines.

The physical flow for products being introduced and at the market is two delayed modules. In Fig. 3.3 the products in the D_{RESEARCH} and D_{INTROD} delays are shown, even though they were not mentioned in the narrative. Recall the discussion of this point in the consulting problem. There is no law about whether the contents of delays should, or should not, be shown in an influence diagram. The issue is whether doing so clarifies or confuses the issue and, more importantly, being aware that the contents of delays might offer options for policy design.

The physics for product research is more subtle. There is, indeed, a standard delayed flow module for the research phase, after which products accumulate into a pool of products available for the market. There are, however, two flows from that pool; one a matter of management action, the other a consequence of actions.

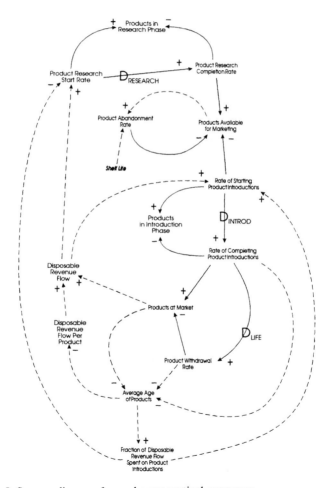

Fig. 3.3 Influence diagram for a pharmaceutical company.

The influences on Average Product Age are all negative. At this stage, that is intuitively correct; introducing new products should reduce age, as should withdrawing old ones. Similarly, the more products there are in the market, the lower should be the average age.

The rest of the model is straightforward. The older the products, the greater the proportion of available money spent on introductions. The greater the proportion and the greater the flow, the larger will be the Rate of Starting Product Introductions and the lower the Product Research Start Rate. It is essential that the reader does not simply glance at these explanations, but follows them closely on the diagram.

The level of the diagram

While it is usually not worth debating in detail the level of any particular diagram, it is useful to be aware of roughly where it lies on the cone of Fig. 2.16. In this case, the diagram is not at level 4.

There are at least two missing parameters: the respective costs of Product Research and Product Introductions. The reader should have noticed that while, for example, Disposable Revenue Flow must be measured in $/month, the Product Research Start Rate must be in Products/Month. A parameter of $/Product is missing. If the reader did notice this and did include these parameters in the model, the reader's diagram is better than the suggested solution, a good example of not regarding these solutions as definitive.

Qualitative analysis

There is much else that is missing from this diagram, and the purpose of giving the reader an incomplete solution is to develop a critical habit of looking at narratives and models. It is implicitly assumed that the research capacity and the ability to introduce new products are both unlimited and completely flexible; the research laboratories can be switched on and off at will. In Fig. 2.13, point 4, the variable (actual) Products in Research Phase should make one think of a Desirable or Feasible equivalent, to represent the ideas of keeping the research team reasonably busy or their physical capacity, respectively. Similar ideas apply to product introductions.

The feedback structure of the model

As with the consultancy company's model, it is worth taking a closer look at some of the feedback mechanisms in the pharmaceutical company as shown in differing patterns of lines in Fig. 3.4. Bear in mind that these are not the only loops the model contains and the reader should search for, and think about, more.

Loop 1 is clearly positive and capable of producing growth in the size of the firm. Since there is no exogenous demand, as there was with the consultancy company, there might be no immediate limit to the growth this loop could produce, bearing in mind the narrative comment that the demand for health care products is virtually limitless. In Loop 1[5], Products at Market produce Disposable Revenue, so more products can be introduced, and so on. Loop 3 shares a common path with Loop 2 between Rate of Starting Product Introductions and Average Age of Products, and is also positive. As Average Product Age Falls, more revenue is generated per product and

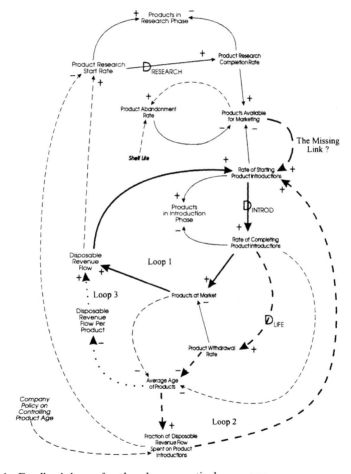

Fig. 3.4 Feedback loops for the pharmaceutical company.

---

[5] The order in which the loops are numbered is deliberately arbitrary and does not imply that Loop 1 is more important then the other loops mentioned here.

hence more can be spent on introducing new products, driving product age further down. On the other hand, Loop 2, which is negative, acts to inhibit Loop 3 in that, whenever products are relatively young, the firm cuts back on product introductions to leave more money for product research. Further, the link from Product Withdrawal Rate to Products at Market connects Loop 2 with Loop 1 and may inhibit Loop 1 from generating its full growth power.

It is clear that reasoning about loops in this fashion rapidly becomes cumbersome and it is best seen as a source of insight into aspects of the model which might repay closer study. In this case, it suggests that the corporate policy which varies Fraction of Disposable Revenue Flow Spent on Product Introductions, which was implied in Fig. 3.3 but is explicitly shown in Fig. 3.4, will need to be worked out very carefully if smooth growth is to be achieved.

Similarly, following the links up the left-hand side of the diagram leads to an impasse in that there are no links coming back down and hence there is no feedback to harmonize the activities of research and marketing. This is emphasized by the link added from Products Available for Marketing to Rate of Starting Product Introductions, a link which was missing from the original system but the addition of which might have a major effect on the corporate behaviour. We shall test some of these possibilities in a quantitative model in Appendix B.

A PREDATOR/PREY SYSTEM

The dynamics of the relationship between two populations of animals, one of which preys upon the other, is of great interest in ecological thinking and has been a fruitful area for analysis. One topic has been the academic need to understand why populations of animals undergo extreme fluctuations, another has been to suggest policies by which populations might be managed for economic reasons or to preserve threatened species. In this case study we shall model the mutual influences of populations of rabbits and foxes, extending the discussion later to other animal species. We commence with a very simple narrative. Later in the chapter, the reader will be invited to improve the model.

Narrative description

A fairly large area of land is inhabited by populations of foxes and rabbits, but the sizes of the two populations fluctuate quite severely over a period of years.

The rabbits eat only grass and the rate at which they breed depends on the size of the population and the inherent fecundity of the species.

Rabbits are killed by foxes and rabbit deaths from natural causes are very rare.

Foxes eat only rabbits. The rate at which rabbits are killed depends on the numbers of both species and the efficiency with which foxes hunt. The rate at which foxes are conceived depends on the numbers of foxes and their natural fecundity. Since the foxes have no natural enemies, their life span is determined by the inherent life span of the species, but life span is reduced if the average food intake of foxes (from killed rabbits) is less than the food a fox requires for a normal life span.

Comment on the narrative

This narrative is fairly typical of what might be gleaned from an elementary textbook on animal biology; that is, various parameters are implied but not stated and some broad statements are made such as that the rate of rabbit kills 'depends on' three factors. Part of the skill of influence diagramming lies in using painstaking common sense to work out whether those dependencies are positive or negative.

The narrative also raises questions such as whether it is necessary to distinguish between male and female foxes, whether one should allow for immature foxes which are incapable of hunting but which require to be fed by their parents, as also might pregnant vixens by their partners.

Much else is omitted or implied. For example, it is implied that there is a limitless supply of grass. The statement that rabbit deaths from natural causes are very rare excludes both extreme weather conditions and diseases such as myxomatosis, which is highly infectious and almost inevitably fatal to rabbits, with obvious effects on fox food supply.

In all such cases, the best advice is to ignore as much detail as possible in the first attempt at system description via an influence diagram. The model can always be refined as understanding develops. Refining a model is not the same thing as making it more complicated, and a model which starts by being complicated is unlikely ever to become sophisticated.

A suggested solution

An initial and, as we have just pointed out, very simple diagram for this problem appears in Fig. 3.5. This differs from previous diagrams in that abbreviations for the variable names are given under the text for each. This is often done as the basis for the connection between the influence diagram and a subsequent simulation model. The reader should decide whether it adds to or detracts from the diagram.

The rabbit population sector consists of a simple flow module, with rabbit births depending on Rabbit Conception Rate delayed by a gestation period. The physics of fox birth and death are a little more complex as we need to

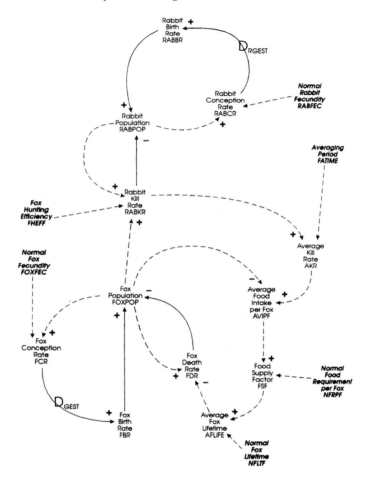

Fig. 3.5 A predator/prey system.

allow for the fact that foxes only die from old age, accelerated, perhaps by shortage of food.

The Fox Death Rate problem introduces a new type of delay. Normally, an inflow leads to an outflow with a delay between the two. The delay might be constant or variable but, if the variation arises *from the contents of the delay*, we need to show the influences involved explicitly. If we do not, we shall get the wrong feedback structure. Thus, Fox Death Rate is shown as depending on the Fox Population and the Average Fox Lifetime, with the latter deriving from the Fox Population mediated through the food intake effects. The effects of food intake are that, the greater the Average Food Intake per Fox, the greater the Food Supply Factor, hence the longer the Average Fox Lifetime and the lower the Fox Death Rate.

At this point, readers should pause and check that they understand the reasons for solid lines and dotted lines in the population segments of the diagram.

An invaluable skill in modelling is to think in terms of the dimensions of the variables and parameters. In this case, Rabbit Population must be measured in [RABBITS], where [ ] enclose the units in which the variable is measured. Similarly, Rabbit Conception Rate must be measured in [RABBITS/YEAR]. Normal Rabbit Fecundity must, therefore, have dimensions of [(RABBITS/YEAR)/RABBIT], in which we use ( ) within the [ ] to draw attention to that which flows in the system.

Using a similar line of thought, the Fox Hunting Efficiency must have dimensions of [1/(FOXES*YEAR)], in other words, the number of rabbits killed by a fox in a year of normal hunting. The reason is that the narrative implies that

RABKR = FOXES*RABBITS*FHEFF

Since RABKR must be [RABBITS/YEAR], and foxes and rabbits have dimensions of [FOXES] and [RABBITS], ordinary algebra gives the dimensions for FHEFF when the variable names in the equation above are replaced by the dimensions of the known variables and the resulting **dimensional equation** is solved for the unknown dimensions of FHEFF. As we shall see in later chapters, dimensional validity is an essential aspect of the correctness of a model and it is as well to get into the habit of thinking dimensionally at an early stage.

Average Kill Rate is needed because it is very unlikely that foxes start to die early simply because today's hunt was not very successful. There must be a period after which hunger starts to have an effect. Note, however, that the Averaging Period is a parameter of fox physiology and not, as was the case with Averaging Period for Order Rate in Fig. 2.7, a matter of management choice.

While simple diagrams are generally preferable, this one is too simple for comfort. Later in this chapter the reader will be asked to extend it to take account of the food available to rabbits.

The feedback structure of the model

This model contains six loops, as shown in Fig. 3.6. Some of the loops have links in common with others, so the reader should be careful in following them. Loops R1 and F1 are positive and are capable of producing unlimited growth in the two populations in the absence of any limiting factors. Loop R2 provides some limitation in that, the more rabbits there are, the more will be killed by foxes. Loop F2 increases fox deaths as fox population increases, thus slowing down the power of F1. Loops F3 and F4 are negative and positive, respectively and they function in rather a complex way. F3's effect is that, the more foxes there are, the less food there will be for each and hence the shorter their lives, thus decreasing the population. This is

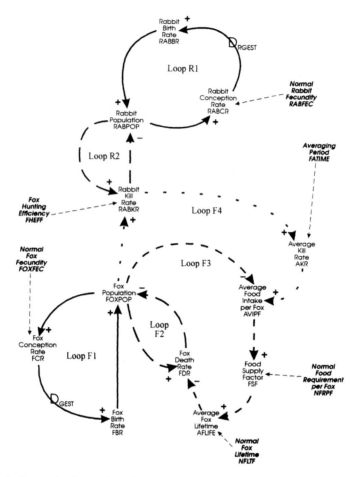

Fig. 3.6   Loops in the predator/prey system.

counteracted by F4 which ensures that more foxes will kill more rabbits producing food to keep foxes alive for longer to kill yet more rabbits. Obviously, this conflicts with loop R2, which ensures that the more rabbits are killed the fewer there will be to kill.

Reasoning about loops in a system as complex as this is clearly of limited value and we shall have to rely on simulation, in Chapter 6, to understand it fully. The seeds of complex behaviour are, however, clearly present.

Other applications of this model

Problems similar to this arise in cases such as fishing policy. How many fish can safely be harvested and what the consequences are for both fish and people of fishing quotas versus fishing industry subsidies are obvious policy areas. A complication arises if the fish have other predators, such as sharks.

A SIMPLE COMBAT MODEL

System dynamics has proved to be rather good for representing the dynamics of combat. This case study indicates a line of approach to the development of an extremely simple combat model which will be considerably extended later in the chapter.

Narrative description

Two nations, traditionally referred to as Blue, for one's own side, and Red, for the potential enemy, are hostile, and it is possible that Red may attack Blue. Blue wishes to improve its combat power with the aim of deterring Red, but has a limited defence budget. Blue's problem is to detect the '**pressure points**' in the system at which defence expenditure might have the greatest effect, or defence cuts the least effect, on combat power. Blue approaches this problem by modelling a potential combat, with the aim of assessing the broad effects of defence improvements[6], not of predicting the exact outcomes of battle.

The scenario is that both Blue and Red have forces on the frontier and others in reserve, some distance away. For both sides, if reserves are ordered into battle, there is a delay before they can arrive and, because of limitations in the road network, there is a limit to the rate at which reserves can be despatched.

Each side inflicts combat losses on the other depending on the rate at which each man fires, the number of men in action and the proportion of shots which hit a target.

The Red commander wishes to achieve a speedy victory and therefore commits his reserves as fast as possible. The Blue commander hopes to wait for Allied forces to arrive from a neighbouring country and then to achieve an overwhelming victory. He therefore commits reserves only when he needs to. His criterion of need is that he does not wish to be outnumbered by more than 2 to 1 at the point of combat[7].

A suggested solution

Parsing the narrative and consideration of Fig. 2.12 should show that the physics for each side consists of a simple Type A module, the inflow to

---

[6] Indicating the broad effects of system changes is exactly what system dynamics is good at in all manner of problems.

[7] This is a *very* simplified version of the Battle of Waterloo.

which is the outflow of a Type B delayed flow module. This is shown by the solid lines in Fig. 3.7.

The Red commander's policy on reserve commitments is to do so as fast as possible, so the transport limitation is the only true influence shown on Reserve Commitment Rate. The link from Red Reserves Remaining is marked with the two lines for a non-negativity constraint to show that Red will have to cease committing reserves when there are none left to commit.

The transport capacity and non-negativity constraints also apply to Blue with the addition of the need to commit reserves when Blue is outnumbered on the battlefield by more than an acceptable ratio. This is expressed by comparing the actual and acceptable ratios and triggering the movement of reserves when the actual ratio exceeds the desired ratio. The link from Force Ratio Trigger is also marked with two lines to show that there is a non-negativity constraint, as was the case with the link from Desired Recruitment Rate to Trainee Recruitment Rate in Fig. 3.1. The two non-negativity constraints mentioned in this paragraph represent slightly different things. In the first case, Reserve Commitment Rate stops completely when there are no more reserves left to commit. In the second, commitment takes place spasmodically as and when the need arises. Obviously, the first constraint will take precedence over the second.

The loss rates for the two sides are driven by the total firing rate, measured in [shots/hour], which is the number of men multiplied by the rate at which each man fires [(shots/man)/hour], and the casualties inflicted by the average shot [men/shot]. As mentioned earlier, the sooner the reader develops the habit of thinking in terms of the dimensions of the variables, the better. In this case, it is a common error to regard a parameter such as Red/Blue Shot as a kill probability: a probability has no dimensions, which would lead to a dimensionally invalid equation.

Pressure points in the system

The purpose of the analysis is to guide Blue's defence planning by identifying the pressure points where expenditure might be beneficial. These points are shown in ellipses in Fig. 3.8. In particular, Blue might increase its combat power by spending money on any of the following measures.

- Increasing its own transport capacity by building roads.
- Decreasing its own movement delay by better intelligence on enemy plans so that the reserves can be positioned more appropriately.
- Reducing Red's transport capacity and increasing his movement delay by using sabotage teams, building obstacles or buying long range artillery.
- Providing Blue forces with rapid-fire weapons.
- Protective armour to reduce Blue/Red Shot.

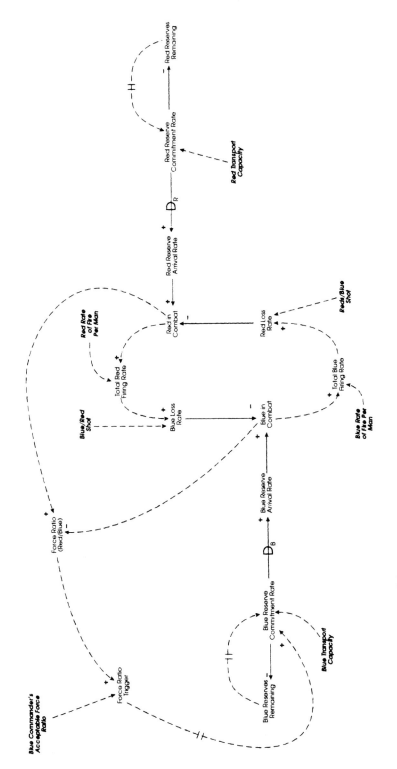

Fig. 3.7 A simple combat model.

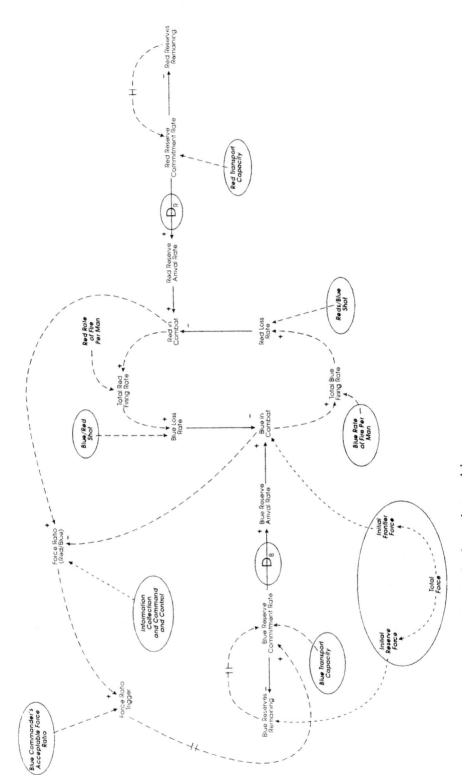

Fig. 3.8 Pressure points in the simple combat model.

- More training to improve the marksmanship of Blue soldiers.
- Having a larger total force and/or dividing the total force between the frontier force and the reserves in different ways.
- Better information collection and command and control to monitor Red in Combat more closely. At present, Blue waits until he is outnumbered by more than 2 to 1 before starting to commit reserves. Predicting Red's movements in advance might enable better results to be achieved.

All the foregoing involve spending money on assets. There may, however, be great benefit from changing the Blue commander's policy on reserve commitment, as shown by the ellipse round Blue Commander's Acceptable Force Ratio. In all system dynamics, not just military models, the key is to find the policy which makes the best use of increased assets, not just to find the best mix of assets for a preordained policy. As we shall see in later chapters, optimization provides a powerful means of dealing with this difficult problem.

The idea of pressure points is a useful way of analysing models, and the method of this book is to bring in new ideas in a steady stream. The reader should, therefore, look back at some of the earlier influence diagrams and check them for pressure points. The predator/prey case might appear to have none, but that would be very rare, as all systems are to some extent managed. In that example, people might not wish to see rabbits wiped out and might shoot foxes to prevent that occurring. On the other hand, rabbits can be immensely destructive and reducing excessive populations by releasing into the wild foxes bred in captivity might be a more satisfying policy than gassing rabbits or spreading diseases among them.

The feedback structure of the model

This model has an interesting mixture of loops, as shown in Fig. 3.9.

Loop 1 is positive and produces a runaway victory for whichever side can gain the upper hand. For example, an increase in Red in Combat produces a greater firing rate, more Blue losses, fewer Red losses and hence, in effect, produces a snowball effect in Red's favour (or in Blue's, depending on the situation).

The negative Loop 2 represent Blue's efforts to prevent this happening by committing reserves. Loop 3, which has a common path with Loop 2 from Force Ratio to Blue in Combat and with Loop 1 from Blue in Combat to Red in Combat, is also negative. It acts to supplement Loop 1 in that the additional Blue troops committed by Loop 1 inflict increased losses on Red, thereby helping to bring the Force Ratio under control.

Two loops are marked as false loops. This means that they do not exert a continuous effect, as does Loop 1, or even the spasmodic effect of Loops 2 and 3. Instead, they simply reflect that a flow must stop when its source reservoir is empty.

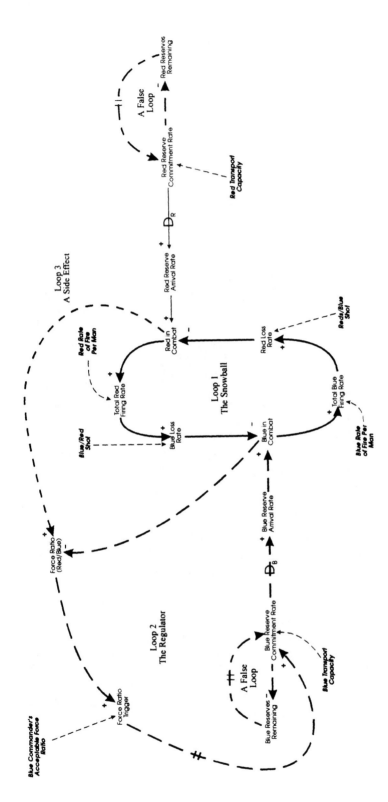

Fig. 3.9 Loops in the simple combat model.

THE DECLINE OF THE MAYAN CIVILIZATION

The Mayan civilization flourished in the southern part of Central America for a few hundred years from about 400 AD. It produced magnificent art and great monuments in its numerous cities and their temples and palaces. Over a short period around 800 AD the civilization collapsed and the cities were abandoned, though recent research suggests that the Maya people continued to live as farmers in the area for a few centuries afterwards[8].

Simplifying considerably, the view of archaeologists (or, at any rate, some of them) is that Maya society consisted of a ruling elite and a mass of commoners[9].

The prestige of the elite was reflected by the magnificence and profusion of the national monuments, and, when their prestige was adversely affected, perhaps by loss of trade with neighbouring societies due, apparently, to warfare and pressure from the north, or by declining agricultural output due to population pressure, they reacted by dragooning commoners into building more monuments. Inevitably, fewer people were available to produce food and the consequent poverty seems to have produced yet further monument building, possibly in an effort to placate the gods. Whether the commoners obligingly starved to death while continuing to build, though working ever more slowly as hunger took its effect, or whether they decamped or rebelled is not known, but the Maya society certainly collapsed.

In this case study we have not presented a formal narrative, but the reader may still wish to draw an influence diagram. Hosler's version is shown, with some amendments, in Fig. 3.10. The Prestige of the Elite is shown as driving the process of monument building, together with one of Hosler's parameters, the Elite's Averaging Time for Monument Construction, which represents their 'policy' for reacting to events. The flow modules for population and monument construction provide the skeleton around which the rest of the model revolves.

The summary can be taken further when we consider the feedback loops shown in Fig. 3.11. Ideally, the reduced prestige should have been adjusted by the Control Loop building more monuments. Unfortunately, an excessive attempt to apply that control so reduced the people available to produce food that the Collapse Loop came into play as people, in effect, starved more quickly than they could build, and then attempted to build more to placate the gods (or so runs the archaeological theory, according to Hosler).

The point of this case study is not whether the theory or the interpreta-

---

[8] The discussion in this section, and Fig. 2.10, are based on Hosler *et al.* (1977). One of Hosler's co-authors was a system dynamicist, D. Runge.
[9] For a fascinating account of similar social structures among the Mexica (Aztecs) see Thomas (1993).

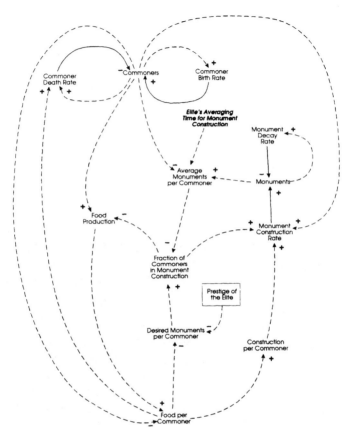

Fig. 3.10 The decline of the Maya.

tion are correct, but to illustrate the use of influence diagrams and simulation models to provide a framework of understanding another discipline.
The collapse of the Maya is a large topic in American archaeology and
many books have been written on it. This diagram puts onto one piece of
paper a summary of a very complex argument. In their simulation model,
Hosler *et al.* claim to show that a less severe reaction by the elite to the
decline in their prestige would have averted the collapse.

EXTENDING THE CASE STUDIES

The pharmaceutical company revisited

Having studied Fig. 3.3 the management of the pharmaceutical company
realizes that the narrative description it gave to the consultant was woefully
inadequate. This is an exceedingly common occurrence in practical system

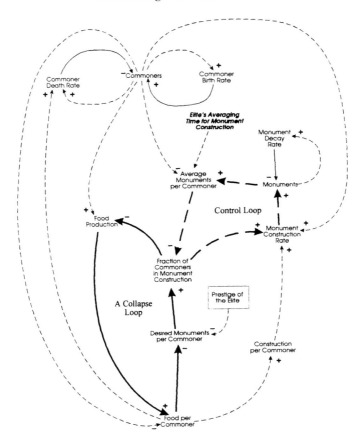

Fig. 3.11 Loops in the decline of the Maya.

dynamics and does not mean that the managers or the consultant were incompetent. It almost invariably comes about because the clarity of a well-drawn influence diagram helps people to think more clearly about what they do and enables them to describe it more accurately to someone else; a very useful property for a diagram to possess.

In this case, the management explains that it does not change the flow of money into product introductions simply because of the age of the products, even though the Marketing Director might want to do that. Instead, it attempts to keep spending in the research labs and in the product launch department on a reasonably even keel, to avoid wasting money by developing products which are then abandoned because they have passed their shelf lives and to keep a reasonably steady pool of Products at Market. The spending problem is approached on the following lines:

• Management takes an average of the product withdrawal rate and bases the required rate of spending on product introductions on that, in an attempt to replace old products with new ones. However, it reduces the required rate as the number of products on the market increases.

- Similarly, management arrives at an idea of the rate of spending on product research by smoothing out the rate of starting product introductions, to try to provide for newly-researched products to replace those being introduced. However, it takes account of products available for marketing to try to avoid wasting money.
- The total of these two required spending rates may be more or less than the disposable revenue, but the two required rates enable management to calculate the fraction of the revenue flow which ought to be spent on product introductions.
- Having done these calculations, the company is still very market-orientated, so the average product age is used to sway the ratio in favour of product introductions as the age rises and falls.

This revised narrative is represented in Fig. 3.12, but readers should attempt the revision themselves before studying the answer. They should, however, study Fig. 3.12 rather carefully to see exactly what has been changed and how those changes conform to the new narrative.

Fig. 3.12 is, perhaps, at level 3 in the cone because are still at least six missing parameters. These are the respective unit costs of developing and introducing an average product, which were missing from the earlier diagram, the two parameters for the times over which the averages are taken, and two more which reflect the time constant over which Products Available for Marketing and Products at Market will be spread in assessing the two required spending rates. This is clearly seen by thinking in terms of the dimensions of the variables. For example, if RRSPI is Required Rate of Spending on Product Introductions, UCPI means Unit Cost of Product Introductions, APWR stands for Average Product Withdrawal rate and PAM symbolizes Products at Market we can write the equation implied by the diagram and then write the equation again, replacing variable names with the respective dimensions in [ ]:

$$RRSPI = UCPI*(APWR - PAM)$$
$$[\$/month] = [\$/Product]*([Products/Month] - [Products])$$

We can now see that the second equation is not dimensionally consistent as it involves subtracting [Products] from [Products/Month], which is not legitimate.

Although these parameters would have to be given in a simulation model, the purpose of an influence diagram is to promote communication and thinking, and Fig. 3.12 does that.

For instance, the reader should trace the new loops and see that they provide the *equivalent* of the missing link which was identified in Fig. 3.4: Products Available for Marketing does now influence, indirectly, Rate of Starting Product Introductions. However, there is still no concept of targets

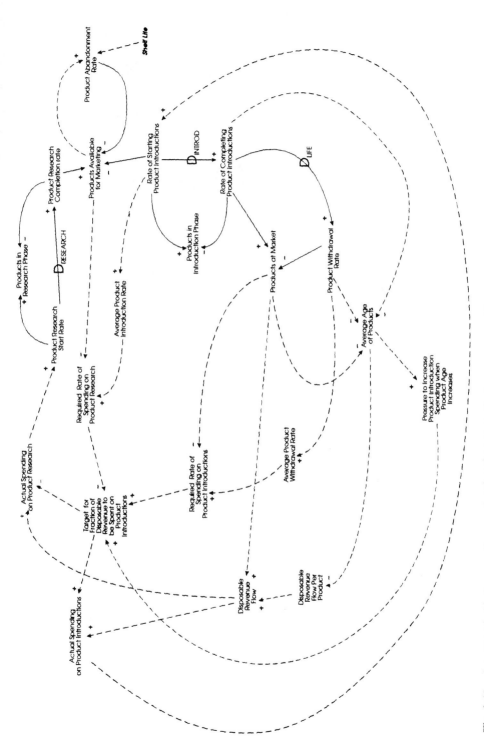

Fig. 3.12 An extended influence diagram for the pharmaceutical company problem.

for Products Available for Marketing or Products at Market, so the negative feedback loops are still rather ill-designed.

Indeed, the very idea of using Products at Market to *reduce* the Desired Rate of Spending on Product Introductions rather conflicts with the company's aim of corporate growth, which would require positive feedback, not negative. This is an example of revisiting the statement of the problem (Step 1 in Fig. 1.3), when study of the system description reveals apparent conflicts between what the system is said to be supposed to do and what its structure shows that it is trying to do. Of course, it may also be the case that the aim of growth is correct and that the Marketing Department is hindering that by adopting poor policies. This sort of ambiguity is very typical of practical problem solving.

The food supply of rabbits

The predator/prey model was rather too simple, so a useful exercise is to extend it to allow for the grass which the rabbits eat, taking account of the following facts.

At any one time there is a certain amount of grass available in the area. The rate at which grass grows depends on the weather, and the rate at which it is eaten depends both on the number of rabbits and the grass available per rabbit. If there is abundant grass the rabbits gorge themselves, but if there is little they spend time searching for it and fighting over it.

To achieve its normal life span, a rabbit needs a certain average rate of food consumption. If it eats less, it will be likely to die earlier, unless, of course, a fox gets it first.

The suggested solution shown in Fig. 3.13 should present few surprises. A third flow module has appeared for the new entity of grass. The effect of food intake on rabbit life span is modelled in very much the same way as for foxes.

The main difficulty is in correctly relating Rabbit Death Rate and Average Life Span. In the case of the foxes, we could model Fox Death Rate as depending only on Fox Birth Rate, though the length of life might vary. In effect, foxes die only as result of having been born, since they have no predators. Rabbits, however, die either as a result of having been born or by being killed by foxes. The flow of water goes in to the pipe at one end, but there is a leak for those killed by foxes, before the rest comes out at the other end. For influence diagram purposes the form shown in Fig. 3.13 is correct and, as we shall see in Chapter 8, it is also consistent with the techniques for writing equations for a leaky delay. Such effects are quite common in managed systems (the rejection of low-quality items during manufacture is one example), so it is important that the reader studies the solution and learns to recognize when it should be used.

Perhaps the food intake also has an effect on rabbit fecundity. If such is

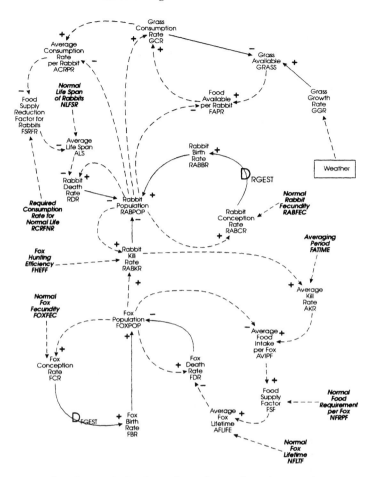

Fig. 3.13 Adding the prey's food supply to the predator/prey system.

the case, a variable for Actual Rabbit Fecundity would be needed, with appropriate influences from food intake. However, food intake might not have the same effect on fecundity as it does on life span. As influence diagram practice, the reader should experiment with altering the diagram. In practical modelling one always has to be careful about extending a diagram too much, as data for the parameters and effects might be quite hard to come by.

In the discussion of pressure points reference was made to the possibility of culling foxes to keep the rabbit population up or releasing them into the wild to keep the rabbit population down. Add these extra flows to the diagram, bearing in mind that foxes will now no longer die only as a result of having been born. Do not forget to allow for a Desired Fox Population, dependent, perhaps, on a Desired Rabbit Population.

Before leaving this model we shall use it to illustrate a wider problem in modelling. Adding fox culls based on a comparison of actual and desired fox populations leads to a diagram which implies that the fox population is instantaneously and precisely known. That is clearly not the case, and one should refer to perceived actual population, drawing a link from actual to perceived population, including a delay to imply a time lag in perceiving what is going on and adding a new factor to represent distortion and error in observing populations. Similar ideas would apply to the combat model: it is very unlikely that the Blue commander knows the actual strength of his own troops in combat, let alone Red's. The reader should study how the effect of delay and distortion affects the information parts of the system dynamics paradigm and how to write equations for such aspects.

Extending the combat model

The combat model of Fig. 3.8 is very simple and the reader should attempt to incorporate the following factors:

- Blue's soldiers go into battle carrying negligible amounts of ammunition and the commander of Blue's support services is responsible for bringing up fresh supplies. He has a target for the amount of ammunition which should be available for each man in combat, from which he can calculate the discrepancy between the total requirement and the total availability. The reserve ammunition is held some distance from the combat area and requires the same time to move forward as do troops, and it has to compete for the same transport capacity, a certain number of rounds taking up the same space as one man.
- The support services commander and the tactical commander therefore have to resolve the demands for transport capacity if the one wants to move ammunition at the same time as the other wishes to commit reserves.
- The ammunition available per man at the front governs the rate at which each man fires. If there is plenty of ammunition the individuals fire rapidly, if there is little they fire more slowly.
- Blue also has long range artillery which cannot be attacked by Red but which can fire at Red's supply route. The artillery is not accurate enough to hit Red troops on the march, but the holes it makes in the road have the effect of reducing Red's transport capacity and of increasing his movement delay.
- Red has a certain number of engineer troops who can repair holes in the road but in doing so, use up engineer supplies.

In parsing this narrative, be careful to distinguish clearly between total quantities and quantities per man.

A suggested solution is shown in Fig. 3.14. Study it carefully noting that:

- Four new flow modules, for Ammunition, Long Range Ammunition, Current Amount of Damage (holes in the road), and Engineer Supplies. Total Blue Rate of Fire obviously depletes the front line ammunition stock.
- Red Transport Capacity is now a variable.
- Blue Transport Capacity is still a parameter, but now affecting two flows.
- There is no non-negativity constraint from Blue Ammunition Reserve Remaining to Ammunition Despatch Rate, implying that there is a huge stock of reserve ammunition. The narrative was silent on that point and that is, again, typical of real SD problems, not just military ones. One of the roles of the influence diagram is to make one question the information available. If the reserve ammunition supply is so large that it has no effect, it can be omitted from the problem.
- Similarly, there is a link from Long Range Ammunition Stocks/Gun to Rate of Fire per Gun. That was not specified in the narrative and illustrates the risks of jumping to conclusions, in this case by assuming that what was said about the behaviour of soldiers in combat is also true of long range artillery.
- The same concept is used for the Red engineer troops and their supplies.
- Non-negativity constraints are not needed from the two ammunition stocks and the engineer supplies, as the reduction in the individual rate of fire or use will ensure that fire ceases when there is no ammunition left. Do not complicate an influence diagram unnecessarily.
- The variable called Percentage of Transport Capacity Allocated to *Troop* Movement does not imply that troops always take precedence. Rephrase that variable to refer to *ammunition* movement, make changes to signs as required and note that there is no change to the signs of the loops which pass through that point. As long as a diagram is carefully thought out, the types of the loops do not depend on the names chosen for variables.
- Even though the diagram is now much more complicated, there is only one point at which links cross. Redrawing a diagram to make it clear is always worth the effort.
- The diagram is still nearly at level 4, but a few parameters are not shown. Practice dimensional thinking to work out what they are.

SUMMARY

The illustrations in this chapter are the foundation on which you will build skill in system description and the knowledge of that skill will add much to your confidence when building simulation models.

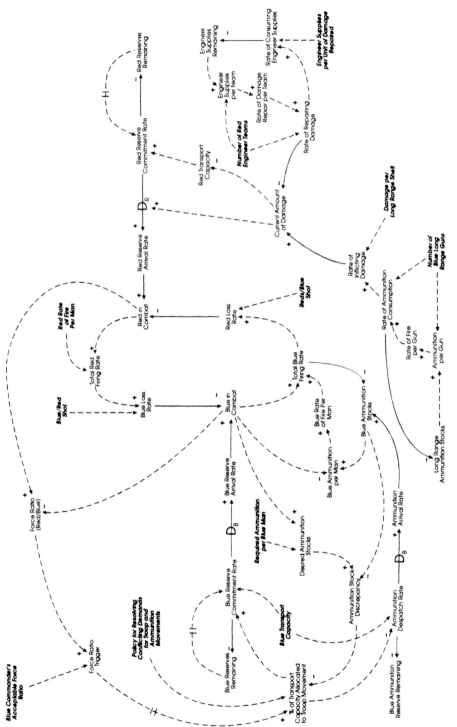

Fig. 3.14 Combat model extended to include ammunition supplies and long-range fire effects.

The main theme of the chapter is that influence diagrams can be applied to just about any kind of problem in which dynamic behaviour might occur, whether it be the dynamics of the firm, the ecology, the battlefield or a much wider socio-economic system. Careful study of the diagrams will reveal a good deal of commonality between them, even though the topics and the names of the variables are so different. That commonality is the idea behind the common modules studied earlier, and the basis of system dynamics is that managed systems all work in pretty much the same way, that is they have similar units in their structure, almost regardless of the problem domain. Recognizing commonalities, without jumping to conclusions, is an essential part of the art of influence diagramming and system dynamics.

REFERENCES

Hosler, D., Sabloff, J.A. and Runge, D. (1977) Simulation model development: a case study of classic Maya collapse, in *Social Process in Maya Prehistory* (ed N. Hammond), Academic Press, London.
Thomas, H. (1993) *The Conquest of Mexico*, Hutchinson, London.

CHAPTER 4

Introduction to simulation

FUNDAMENTAL IDEAS

Having seen how influence diagrams are developed and used for qualitative analysis, it is time to cross the dotted line which separates Stages 3 and 4 in Fig. 1.3 and move on to quantitative models.

Influence diagrams are models, in the sense that they represent a real system by using variable names connected by signed links. A quantitative model, in contrast, represents a system by using variables in equations. In practice, there is no hope of 'solving' those equations by normal algebra or calculus; managed systems are too complicated for that. We therefore make use of **simulation**, meaning that we will create a set of equations to represent the system and then allow the equations to run forward in simulated time to attempt to mirror the behaviour of the real system as it runs forward in real time. In principle, one could run the equations forward by pencil and paper calculation; in practice, a computer and specialized software are used. The key to quantitative simulation is that the equations must do the same things that the real system would have done, *and for the same reasons*, if the model is to be regarded as satisfactory.

In system dynamics, simulation is governed entirely by the passage of time and is referred to as 'time-step' simulation. The other type of simulation in management science is called 'event-based' and is discussed in the standard management science textbooks. This chapter introduces the essential ideas of the simulation process used in system dynamics and some of the basic techniques which we shall need to build models. In a later chapter we shall show that there is no limit to the complexity which can be represented in system dynamics, and it is, in fact, possible for a model to be both time-step and event-based.

The essential idea in time-step simulation is that the model takes a number of steps along the time axis; as many, in fact, as are required to simulate reasonably accurately the total period which the modeller wishes to investigate. Each step is quite short, so that there might be many steps in total. The step length is a critical factor in good simulation and we shall discuss in due course how a satisfactory step length is determined. The step length is always denoted by DT.

At the end of each step, two things happen (automatically in the specialist system dynamics simulation packages). The *first* is that the variables which represent the system's state are brought up to date to represent the *consequences* which have ensued during the previous time step; special measures being needed to get the simulation to start the first time step. *Next*, the variables which represent the flow of *information* and the initiation of *action* are evaluated and the necessary actions are set in train. What actions, if any, are to take place will have been stipulated by the modeller when the equations for the system were written.

To bring this about requires two types of variable. The *first* represents the **states** of the system. In control engineering these are called 'state variables' while in economics they are termed 'stocks', but system dynamics calls them '**levels**'. The *second* represents the physical flows in the system which arise as a result of actions and produce the consequences in the information/ action/consequences paradigm. In system dynamics terminology they are called '**rates**'; they have no special name in control theory, but economists recognize them as 'flows'. These variable types are shown in Fig. 4.1.

Levels and rates are the mathematical basis of system dynamics, but, in practice, a third type of variable is also used: the **auxiliary**. These reflect the detailed steps by which information about the current state, the levels, is transformed into rates to bring about future change. This is shown in the inner loop of Fig. 4.1, the right-hand side of which is a dashed line to show the information/action component of the information/action/consequences paradigm. The left-hand side of the inner loop is shown by a solid line to signify physical flow, and it also has an integration sign to show that, the longer the rate continues to flow and the larger it is, the greater will be the consequence, as felt by the level. The perfect analogy for this is the quantity of water in a bath when water flows in or out.

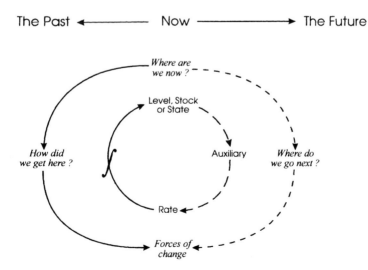

Fig. 4.1 The fundamental variable types.

The outer loop in Fig. 4.1 and the text and arrows at the top remind us of the connection with time. The levels are the current state of the system, shown by the word 'Now'. The right-hand side is movement into the future and the left-hand side is movement from the past. Past changes produced present conditions; present conditions will produce future change. Rates are, therefore, the movements across time steps, levels are the thresholds between time steps.

What of auxiliaries? These are intermediate stages by which the levels determine the rates and they therefore also exist at the thresholds between time steps and are 'now' variables.

It is important to realize that every variable in a system dynamics model is calculated at every DT. Some of the calculations bring the system up to date, others move it forward in time.

THE FUNDAMENTAL EQUATIONS OF SYSTEM DYNAMICS

What we have said about variables which exist at points in time and those which exist across time steps suggests a notation for variables to emphasize their time dependencies. Figure 4.2 shows a time axis with two of its time steps drawn to a very exaggerated scale.

The three **points** are conventionally labelled J, K and L. The **step** from J to K is thus referred to as JK and that from K to L as KL. The time length between successive points, DT, is very small, perhaps only a fraction of a week. With these time labels, a level, to which we shall give the name LEVEL, as though it were an ordinary variable, would be referred to as LEVEL.J or LEVEL.K, depending on which time point was involved. System dynamics variable names have, therefore, two parts: the name itself and the **time label**. The name for a rate would be RATE.JK or RATE.KL, again depending on the time step in question. Since auxiliaries are 'now' variables, existing at the thresholds, they are labelled AUXILIARY.K.

Figure 4.2 shows that LEVEL has increased from J to K because RATE flowed into it between J and K, DT being so small that RATE can validly be treated as constant over the time-step. Common sense now leads to the first of the fundamental equations:

L LEVEL.K=LEVEL.J+DT*RATE.JK

In words, what is in the level at the present time, K, is what was there at the previous time, J, plus what flowed in during the interval from J to K. The **amount** that flowed in was the rate of flow, RATE, multiplied by the time for which it flowed, DT. To emphasize that this is a level equation, it starts with the **type label** L.

This seems very obvious, but it is reinforced when we consider the vital matter of **dimensions**, or units of measurement. That can be done by writing

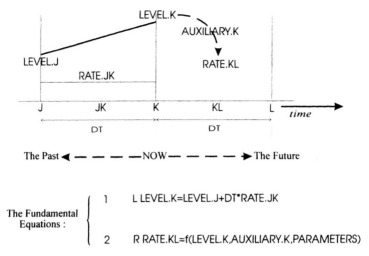

Fig. 4.2   The fundamental equations of system dynamics.

the equation again with, underneath each term, the appropriate dimensions. In this case, we will use the bath water analogy, LEVEL being not the depth but the amount of water in the bath. The amount must be measured in litres, and the rate of flow from the tap (faucet) will be in litres/minute. DT, being time, must be measured in minutes, though we shall write 'minute' to be consistent with the time unit for RATE. Omitting the type label and spacing out to show the match between the level equation and its corresponding **dimensional equation**, this leads to:

LEVEL.K=LEVEL.J+   DT*RATE.JK
litres =   litres   + minute*litres/minute

By ordinary algebra, the minute for DT cancels out the minute for RATE, so the dimensional equation becomes

litres=litres+litres

This is called a **dimensional identity** because the units of measurement are the same on both sides of the = sign, and it demonstrates that our first fundamental equation is dimensionally consistent. Before we move to the second fundamental equation we must explain why dimensional consistency matters.

Those levels in a system which represent aspects such as the amount of stock, the number of people, the quantity of water or the balance of cash, are **conserved quantities**; what the levels contain cannot be lost or created. Even if water flows out of the bath, it has not been lost, merely transferred elsewhere, so there must always be a mass balance in the system and hence in the model of that system. The rates in a model are the forces of change and must be measured in units which are consistent with the levels they feed. If we use [ ] to denote the dimensions of a variable, then:

[RATE]=[LEVEL]/[TIME]

and

[LEVEL]=[RATE]*[TIME]

For conservation, a rate must have units of the level it feeds divided by the units in which TIME is measured. The units of a conserved level must be those of its inflowing rate multiplied by those of TIME[1]. That is how a real physical system works, and a model must work in the same way if it is to do the same thing as the real system *and for the same reasons.*

Again, what of auxiliaries? They represent the steps by which a level causes a rate to flow and must, therefore, have whatever dimensions are needed to ensure dimensional consistency between rates and levels.

We shall have much more to say about the details of dimensional analysis when we come to formulate equations for the case models.

The first fundamental equation copes with the *consequences* of what happened during the JK time-step. The second is the *information/action* phase which will take place at K and produce rates for the impending step KL. Figure 4.1 emphasized that levels and auxiliaries, together with any parameters in the model, determine the rates as forces of change, so the second fundamental equation is much more general in form than the first:

R RATE.KL=f(LEVEL.K,AUXILIARY.K,PARAMETERS)

where the f implies that there is a function or equation which connects LEVEL, AUXILIARY and PARAMETERS to RATE, and the type label is R to emphasize that this is a rate. That equation will model the ways in which the system makes decisions, given the values of LEVEL and AUXILIARY. (There may, of course, be more than one of each in the equation.)

Notice that in the level equation, the rate has the time label **.JK** on the right-hand side of the =, because past history is being used to update present conditions. In the rate equation, the time label is **.KL** on the left-hand side of the = to show that future changes are being generated by present conditions.

The significance of the two fundamental equations is that the RATE equation can be as complex or as simple as is required by the problem; provided the modeller can understand what is happening, there are no constraints on what can be modelled. The LEVEL equations, on the other hand, ensure that the physical quantities are conserved so that the model cannot get out of balance. That property of the LEVEL equation contrib-

---

[1] Later in this chapter we shall develop a special type of level equation to represent the smoothing of information which was shown in Fig. 2.12G. The dimensional rule given above does not apply to smoothed levels.

utes much to ensuring that the model does the same things as the real system and for the same reasons, to repeat a slogan which the modeller should learn by heart and repeat whenever a new equation is written. In summary, the RATEs are the entrepreneurs driving the system forward and the LEVELs are the accountants, keeping it in balance.

The LEVEL equation should also be seen as representing the **memory** of the system. The size of the cash balance level at time point K is the net effect of all the cash outflows and inflows which have ever taken place. Later in this chapter we shall discuss a special form of level, the **smoothed level**, which gives a 'memory' for information just as the standard, or 'pure', level which we have discussed so far is a memory for quantities.

TIME SHIFT AND RELABELLING

Figure 4.2 seems to imply that the sequence of time points in a simulation would have to be labelled N, O, P and so on, and, even using the Greek and Hebrew alphabets, one would soon run out of labels. The problem is solved in the system dynamics languages by realizing that J, K and L are simply labels for a window in time. As simulated time moves forward, the window is advanced and can be relabelled as a new J, K and L rather than L, M and N. This is shown in Fig. 4.3; the current position of the window on the time axis is the nth step, so the next will be $n+1$, at which point the relabelling automatically happens.

It is important to understand that this can only happen because there are two main types of variable, levels and rates. The rates are forces for change into the future, governed by the current state of affairs, the levels. As the time windows move forward, what was the future for one time point becomes the past at the next; in Fig. 4.1, we got where we are now because of forces for change during the last time step and where we want to go next determines the forces for change in the impending time step.

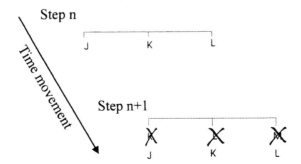

Fig. 4.3 Time shift and relabelling in system dynamics simulation.

THE SIMULATION PROCESS

Level equations always come in pairs: the equation itself and an additional equation called the **initial condition**. Thus our level equation would be:

L LEVEL.K=LEVEL.J+DT*RATE.JK

as before, together with

N LEVEL=100

The initial condition has the type label N and it has no time label because this is the value at TIME=0 when the simulation starts. In this case the initial condition, usually called an N equation, states that there are 100 units in the level to start with.

The simulation process thus starts by loading up all levels with their initial values and then proceeds as shown in Fig. 4.4.

In step 1, all the equations are sorted into the correct sequence for calculation. This is something that programmers must do for themselves in an ordinary language such as BASIC, but it is done automatically by the

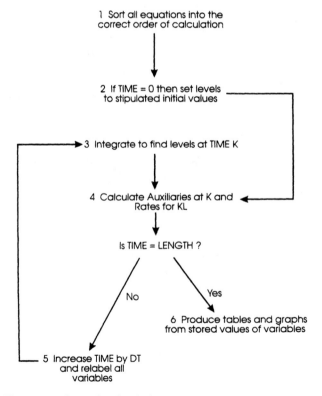

Fig. 4.4   The system dynamics simulation process.

system dynamics languages, saving much time and effort and allowing the modeller to concentrate on the model, rather than having also to worry about the details of calculations. At this stage the model is thoroughly checked for errors and, if serious ones are found, the process stops. All being well, step 2 is performed to load the levels with their correct values.

At stage 2 a variable for time is automatically created. This will be in the form of a level:

L TIME.K=TIME.J+DT
N TIME=0

and provides a clock for the simulation.

Having loaded the levels, attention moves to step 4 to get the rates moving for the first DT in the simulation. After that, the process loops round through steps 5, 3 and 4 until TIME has grown to the total period which the modeller wishes to simulate, which, reasonably enough, is called LENGTH. Throughout the simulation, values of variables selected by the modeller will have been saved for output at the end of the simulation. The languages differ somewhat in the facilities they provide for output. Such, in essence, is the process, and we shall see much of it later.

A SIMULATION EXAMPLE

Before considering some of the special aspects of simulation, such as delays and information smoothing, it will reinforce the ideas of this chapter if we construct a *very* simple simulation model. This will also serve as another example of influence diagramming, and the reader might find it helpful to draw the diagram for the following problem before looking at the solution. The example is deliberately chosen to be rather unrealistic so that the reader can concentrate on the modelling lessons and not be distracted by the details of the problem.

> The benevolent ruler of a city obtains food from other places, but he simply has to take what amounts he can get. As food flows in, it is placed in a stock from which the ruler issues free supplies to his people. His policy is to release food at the rate of 10% of the stock; if there are 100 tons in stock, he will release food at a rate of 10 tons/day and so on. In other words, he tries to spread the available food over the next 10 days. If that is not sufficient for the people, they have to buy their own food elsewhere.

We shall write a simple model of this problem, assuming that initially there is no food in stock and that none is arriving. On the tenth day, food starts to arrive at a rate of 100 tons per day. What happens?

The ruler may be able to arrange for deliveries of an additional 50 tons/day to start in 120 days time. What would be the effect of that?

Figure 4.5 shows the influence diagram for this simple model. The 'Food Stock Time' corresponds to the 10 day period for which current stocks are made to last. It is shown in ***bold italic*** because it is a parameter, representing the ruler's policy.

The first step in modelling is to decide on the types of the variables in the influence diagram. Are they levels, rates or auxiliaries? This process is called **type assignment**. It is always a good idea to **start by identifying levels**, because all the rest will follow from that. In this case, the Food in Stock is a level because it is the accumulation of quantity in the system. The description of the system tells us so. As Fig. 4.2 showed, the variables which fill or drain levels are rates, so Food Inflow Rate and Food Outflow Rate are both rates. Each variable needs a name, so we will call them FOOD, FINRA and FOUTR, respectively. The system dynamics software packages differ in their conventions for variable names, so readers should consult the user manual for whichever package they are using.

The equation for food in stock will be

L FOOD.K=FOOD.J+DT*(FINRA.JK−FOUTR.JK)
N FOOD=0

It is *essential* to notice that, in this equation, FINRA has a positive effect on FOOD and FOUTR has a negative one. These are exactly the signs on the corresponding links in Fig. 4.5.

The equation for FOUTR is very simple. The current stock is to last for 10 days, so the outflow at any time is the stock at that time divided by 10:

R FOUTR.KL=FOOD.K/FST
C FST=10

FST, the food stock time, is the ruler's policy and it is defined as a constant, constants having the type label C.

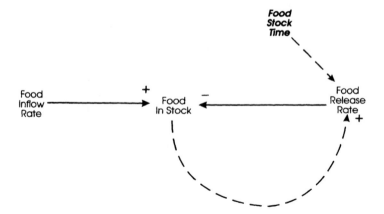

Fig. 4.5 Influence diagram for food problem.

Finally, the food inflow is 0 until TIME=10, at which point it takes a sudden step to 100 tons/day. Again, the software packages differ slightly in the facilities they offer for common dynamic processes, but most of them include a standard function, STEP, to produce step changes of the type we require. In this case:

R FINRA.KL=STEP(100,10)

would ensure that FINRA would increase from its previous value to 100 when TIME=10. Since there is nothing else in the equation, the previous value is 0, which will have the required effect. In practice it would be better to put:

R FINRA.KL=STEP(100,10)+STEP(STEPHT,STEPTM)
C STEPTM=120
C STEPHT=0

to allow for the possibility of the second delivery flow starting at TIME=120. At this stage, STEPHT is set to 0, but we can turn it to 50 later to see the effect of the second inflow.

It was stated earlier that the modeller can write the equations in any order which is convenient to understand and explain, and the packages will do the chore of sorting them into the correct order. That is exactly what we have done here, but you might like to type in the equations in a different order just to satisfy yourself that the results are not changed.

The complete model is shown in Fig. 4.6 and appears on the disk as FIG4-7.COS. This model happens to be in the COSMIC conventions, but it would look very similar in most of the other languages. Some points about Fig. 4.6 should be studied.

- The line numbers, starting from 0, are not part of the model, but are added by COSMIC when the model is printed out. They have been retained for ease of explanation.
- Lines 0 and 47 are the start and end of the model. The text in line 0 will appear as the model title on all output.
- All the model is in lower case type, as it is usually easier to read on the computer screen.
- The lines starting with 'note' are comments and section headings. It is good practice to group notes in threes to improve model legibility. As a rough guide, a well-documented model will devote about 20% of its lines to explanatory comments. A badly documented model is useless.
- Lines 22–29 set a value for DT, a topic we discuss later, make the model run for a LENGTH of 300 days and produce printed tables and plotted graphs of behaviour. Different languages have slightly different conventions for achieving the corresponding effects. It is *essential* that all the

```
 0    * Figure 4.7 Simple Simulation Model
 1    note
 2    note    file named fig4-7.cos
 3    note
 4    note    food inflow rate
 5    note
 6    r finra.kl=step(100,10)+step(stepht,steptm)
 7    c stepht=0
 8    c steptm=120
 9    note
10    note    food stocks
11    note
12    l food.k=food.j+dt*(finra.jk-foutr.jk)
13    n food=0
14    note
15    note    food outflow
16    note
17    r foutr.kl=food.k/fst
18    c fst=10
19    note
20    note    output and control
21    note
22    c dt=0.25
23    c length=300
24    c pltper=1
25    c prtper=10
26    print 1)finra
27    print 2)food
28    print 3)foutr
29    plot food=a(0,2000)/finra=b,foutr=c(0,150)
30    run steady inflow of food
31    c stepht=50
32    run 50% increase in food inflow
33    note
34    note    definitions of variables
35    note
36    d dt=(day) time step in model
37    d finra=(ton/day) food inflow rate
38    d food=(ton) stock of food
39    d foutr=(ton/day) food outflow rate
40    d fst=(day) food stock time
41    d length=(day) simulated duration
42    d pltper=(day) output plotting interval
43    d prtper=(day) output printing interval
44    d stepht=(ton/day) step height in food supply
45    d steptm=(day) time of step increase in food supply
46    d time=(day) simulated time
47    +
No syntax errors detected
===============
```

Fig. 4.6 A model for the food problem. 'Figure 4.7' appears in line 0 because this line is used to generate the caption for Fig. 4.7.

variables in a model be printed out and the values are examined closely. **Simply looking at graphs of a few variables is an excellent way to leave mistakes in a model.**

- Line 30 commands the model to run and line 31 sets STEPHT to 50 to test the effect of the extra supply, after which another run takes place. This ability to change the parameters and rerun the model is the basis for the experimentation phase, Step 5A in Figs. 1.3 and 1.4. All the languages, including COSMIC, allow this to be done interactively on a PC, but that would not be convenient in a book, so we shall use the traditional method of presetting the experiments to be performed.
- Lines 36–46 are the definitions of the variables, and COSMIC and one of the other languages differ considerably from the others in this respect. The type label for a definition is, obviously, D, which is followed by the variable name, an = sign and, in ( ), the dimensions of the variable or constant.
- After the dimensions, the meaning of the variable is defined in words. Notice that definitions must be given for DT, TIME, LENGTH, and the plotting and printing controls, PRTPER and PLTPER. The definitions, 'd statements' as they are called, fulfil two essential purposes. The first is the documentation of the model so that it can be explained to oneself and to the sponsor. All system dynamics languages provide this in some form. The second, which is only available in some of the languages, is automatic checking of the dimensional consistency of the equations. Dimensional analysis was briefly mentioned above, but it is *essential* for a model to be dimensionally valid, otherwise it may not be doing the same things as the real system *and for the same reasons*, and no confidence can be placed in the output. It is possible to check dimensions by hand calculation, but software does it far more thoroughly.
- Finally, after line 47 is the encouraging message that no syntax errors were discovered when the model was checked by the simulation package.

The results of running the model for the base case and with the extra food inflow are shown, respectively, in the upper and lower halves of Fig. 4.7. Again, all the packages produce similar graphs and all should give the same results for this model. As we shall see many more graphs in the book it is necessary to spend a moment on understanding the conventions.

- The captions on the graphs are the name of the model, which appeared in line 0 of Fig. 4.6, and the text comment in the run commands in lines 30 and 32.
- The horizontal axis is TIME, in this case from 0 to 300. The d statement for TIME is drawn immediately under the axis.
- Three variables were plotted, each of which has its own pattern: a solid line is used for FOOD, for example. The same pattern is used to draw the vertical axis and to point below the vertical axis to the name of the

variable. In some places, FINRA and FOUTR have the same values, so their lines merge.

- In the base case, the upper graph, food starts to arrive at TIME=10; the STEP effect is clearly visible. FOOD starts to rise smoothly, as does FOUTR. The behaviour does not settle down until TIME=80, approximately 70 days after the food started to arrive. This is called the **settling time** of the system and the settling times of managed systems are usually very long. FOUTR stabilizes at 100 tons/day and FOOD at 1000 tons. It is clear that this must happen. The system can only be stable when the inflow and outflow are equal and that can only happen when FOOD is 10 times FOUTR; a consequence of the ruler's policy.
- The **mode**, or general form of dynamic behaviour, does not change when the extra food arrives, though the numerical values at stability do. The settling time is unchanged.

The purpose of this very simple model was to introduce some of the ideas of simulation. We shall revisit it later to see that it can become very interesting indeed when some more factors are included.

TESTS OF CONFIDENCE IN A MODEL

How do we know that the behaviour in Fig. 4.7 is right? This is an essential component in building up one's confidence in the model, and there are a number of steps which apply to all models and which can be illustrated from this one.

- The influence diagram must correspond to the statement of the problem.
- The equations must correspond to the influence diagram; in particular the + or − signs in the equations must match the signs in the influence diagram.
- The model must be dimensionally valid. This one has been checked by the software, but readers ought to take each equation and write it out with the dimensions from the d statements underneath, as was done earlier with the first fundamental equation, and satisfy themselves that this is indeed a dimensionally consistent model.
- The model does not produce any ridiculous values, such as a negative food stock.
- The behaviour of the model is plausible – what it does is what we expect it to do – and the values at which it stabilizes can be confirmed by simple arithmetic. These tests are harder to apply to a more complex model, but they are still necessary.
- The model's masses should balance. This means that the total quantity of, in this case, food which has entered and left the system, together with what is still there, should be accounted for. This is achieved by making

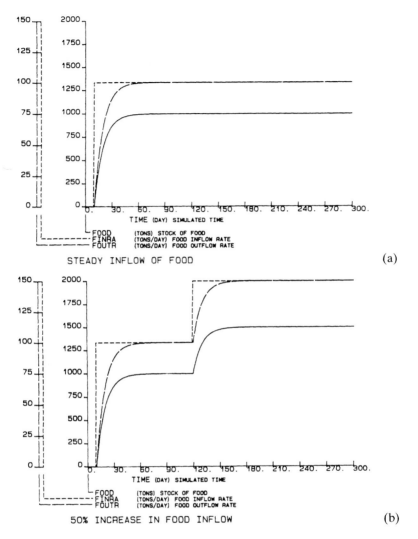

Fig. 4.7 A simple simulation model.

some additions to the model, which appear in FIG4-7A.COS on the disk. One adds two new equations for the cumulative inflow and outflow of food:

L CUMIN.K=CUMIN.J+DT*FINRA.JK
N CUMIN=0

- Even simple equation-writing practice is useful, so readers should write their own equations for CUMOUT. The initial condition for FOOD is worth rewriting as:

    N FOOD=IFOOD
    C IFOOD=0

At first sight, this makes no difference to the model, but it allows the initial amount of food to be changed in a later run. With these additions, a check variable can be written:

    A MCHECK.K=CUMIN.K−(FOOD.K−IFOOD)−CUMOUT.K

In words, MCHECK compares the total which came in with the change in the food stock and the total amount which was distributed. Ideally, MCHECK will be 0; in practice, because of rounding error in the computer, it will be some small number, such as -12.45×10^{-3} (why does it not matter that this is a negative number?). Notice in FIG4-7A.COS on the disk that *all* the new variables have been added to the printed tables so that MCHECK can be seen to check, and that they have also been added to the model documentation so that dimensional analysis can be done again to make sure that the additions have not undermined the model's dimensional consistency.

- We mentioned earlier that the rates in a model can be seen as the entrepreneurs and the levels as the accountants. On that analogy, *mass balance equations are the auditors, making sure that the accountants have done their job properly.* These precautions may seem unnecessary for such a small model, but it is as well to get into good modelling habits from the start.

Having examined a small model to familiarize ourselves with some of the procedures, we shall now look at some important components of system dynamics models.

REPRESENTING DELAYS IN MODELS

We have seen in Chapter 1 that delays are an extremely significant feature of managed systems, and we examined, in Chapters 2 and 3, how delays can be included in influence diagrams. We must now develop a method of representing them in quantitative models.

In the earlier influence diagrams we showed a delay by:

INFLOW − D → OUTFLOW

with a + sign on the link. This is simple, but does not draw attention closely enough to the fact that between INFLOW and OUTFLOW there will be a quantity of material in the pipeline. If INFLOW is the rate of recruitment of workers and OUTFLOW is the rate at which they complete their training, then the 'contents' of the delay are the number of people in the process of being trained, which will rise and fall as INFLOW and OUTFLOW vary.

CONTENTS is clearly a level variable and is, obviously, fed by INFLOW and depleted by OUTFLOW.

Figure 4.1, the fundamental variable types, stated that levels control rates, and, if this is true for dynamic systems as a whole, it must also be true for delays as important components of such systems. This is shown in Fig. 4.8 for three increasingly complicated cases.

In the first case, the outflow at any time is simply CONTENTS divided by the magnitude of the delay. Thus, if the delay in training is, on average, 10 days, then the daily rate of completion of training will be 10% of the number of trainees. This is, of course, also the city ruler's policy for issuing food; on average there will be a 10-day delay between food arriving and it being issued.

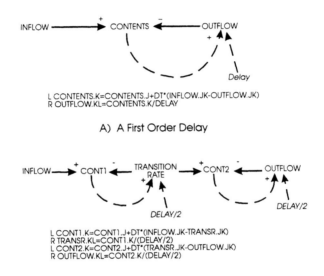

L CONTENTS.K=CONTENTS.J+DT*(INFLOW.JK-OUTFLOW.JK)
R OUTFLOW.KL=CONTENTS.K/DELAY

A) A First Order Delay

L CONT1.K=CONT1.J+DT*(INFLOW.JK-TRANSR.JK)
R TRANSR.KL=CONT1.K/(DELAY/2)
L CONT2.K=CONT2.J+DT*(TRANSR.JK-OUTFLOW.JK)
R OUTFLOW.KL=CONT2.K/(DELAY/2)

B) Two First Order Delays 'Cascaded' to Give a Second Order Delay

L CONT1.K=CONT1.J+DT*(INFLOW.JK-TRANSR1.JK)
R TRANSR1.KL=CONT1.K/(DELAY/3)
L CONT2.K=CONT2.J+DT*(TRANSR1.JK-TRANSR2.JK)
R TRANSR2.KL=CONT2.K/(DELAY/3)
L CONT3.K=CONT3.J+DT*(TRANSR2.JK-OUTFLOW.JK)
R OUTFLOW.KL=CONT3.K/(DELAY/3)

C) Three First Order Delays Give a Third Order Delay

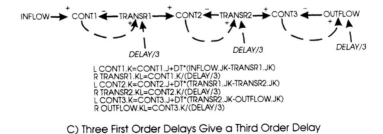

Fig. 4.8 Representing delays in models.

Introduction to simulation

This type of delay is called '**first-order**' because only one level is used to store the contents[2]. Its behaviour is shown in Fig. 4.9[3], in which DELAY has been set to 20 days, in response to a step in INFLOW. OUT1, the response of the first-order delay is initially immediate, but is eventually sluggish and, theoretically, never reaches the value of INFLOW; some trainees never learn.

The first-order delay, for all its simplicity, can be quite a good model for certain processes. For example, a bus company recruits drivers, all of whom have the necessary licence to drive a bus and all of whom know the town fairly well. Some of them will become fully productive quite quickly; others will take longer to learn their route, the fares and so on.

For other problems, the first-order delay is too simple. The training of scientists, for example, must take at least three years to complete a first degree, after which some will be productive quickly, while others will take longer. Such delays are modelled by 'cascading' as many first-order delays as required to produce a higher order of delay. Figure 4.8(b) shows a second-order delay. There are now two internal levels, CONT1 and

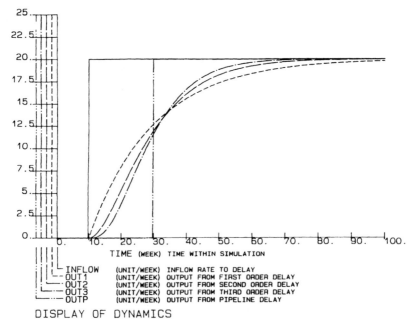

TIME (WEEK) TIME WITHIN SIMULATION

```
INFLOW   (UNIT/WEEK)  INFLOW RATE TO DELAY
OUT1     (UNIT/WEEK)  OUTPUT FROM FIRST ORDER DELAY
OUT2     (UNIT/WEEK)  OUTPUT FROM SECOND ORDER DELAY
OUT3     (UNIT/WEEK)  OUTPUT FROM THIRD ORDER DELAY
OUTP     (UNIT/WEEK)  OUTPUT FROM PIPELINE DELAY
```

DISPLAY OF DYNAMICS

Fig. 4.9 The dynamics of delays.

[2] A first-order delay has the same output shape as an Erlang Type 1 distribution, and correspondingly for the higher order delays discussed below.
[3] Notice that the model LENGTH in Fig. 4.7 was 300 days, to allow for the effects of the second step to be seen. In the rest of this chapter, LENGTH is reduced to 100, which is sufficient to show the dynamics.

CONT2, between which there is a transition rate, TRANSR. Notice that each of the components of CONTENTS controls its successor rate *and* that the total delay is split equally between the two stages. This produces the behaviour shown for OUT2 in Fig. 4.9. After the step in INFLOW there is a short dead time before OUT2 starts to respond, after which the rise in OUT2 towards INFLOW is swifter than OUT1's was, though OUT2 is only just over half way to INFLOW's level after the 20 days have passed.

Cascading three delays gives the '**third-order**' response shown for OUT3.

Figure 4.9 suggests that, as the order is increased by cascading more delays and splitting the total delay across more transition rates, the shape of the response comes closer to a step delayed by 20 days from INFLOW's step. This is, indeed, the case and, theoretically, one can create a delay of infinite order in which OUTFLOW has exactly the shape of INFLOW and lags behind by exactly the amount of the delay. In system dynamics, an infinite order delay is called a '**pipeline delay**', though some packages use slightly different terms, and this is shown in Fig. 4.9 by OUTP, which exactly matches INFLOW exactly 20 days later.

It would obviously be incredibly tedious to write the equations shown in Fig. 4.9 every time one wished to model a delay, and all the packages include some standard delay functions. A third-order delay is usually written as:

R OUT.KL=DELAY3(IN.JK,DEL)
C DEL=10

The time label .JK within the DELAY3 does not mean that OUT.KL depends on IN in the preceding time step; it simply reminds one that the current value of OUT arises from previous values of IN, approximately DEL time units ago. COSMIC has standard functions for all the delay types.

The total amount within the delay is modelled as:

L TAMOUNT.K=TAMOUNT.J+DT*(IN.JK−OUT.JK)

but what should be the initial value of TAMOUNT? To answer that we recall that a DELAY3, for example, has three internal levels. Like all levels they have to have initial values and these are generated *automatically* by the software package. In most cases, the three internal levels of DELAY3 are each set to an initial value of:

N CONT1=IN*(DEL/3)

and so on. The total amount in the delay will, therefore, be IN*DEL, so the correct initial condition for TAMOUNT will be:

N TAMOUNT=IN*DEL

otherwise the external level, TAMOUNT, which the user has created, will not be consistent with the total amount in the three inaccessible levels within the DELAY3 function. This equation for the initial value of TAMOUNT would be correct for any order of delay. Initializing delays is a most important concept to which we shall return when we consider the modelling of more than one output from a delay.

Although Fig. 4.9 was produced using a step for INFLOW, for simplicity, the corresponding behaviour patterns are produced for any other shape of INFLOW. In particular, the output of a pipeline delay will always exactly match the inflow and will lag behind the inflow by exactly the magnitude of the delay. The program disk includes FIG4-9.COS, and the reader should run this program, experimenting with the size of delay and, when we have studied how to write the equations, with different patterns for INFLOW.

We have shown that different types of delay can be modelled using the standard functions available in the various software packages, but which type of delay should one use in any particular case? We shall deal with that in Chapter 6 when we build models for the case problems.

As we have remarked, delays are a vital component of managed systems and often have a significant effect on the dynamics, so it is as well to be quite clear about what is being modelled in a delay. The important point is that the delays do *not* represent individual entities, such as particular bus drivers or bags of food. They *do* represent the overall flow in the system. This can lead to apparently strange results in that, while an issue rate of 151.279 tons of food per day does have meaning, what meaning could be attached to 151.279 drivers completing training per day? At first glance, 0.279 drivers is nonsense, but the point to grasp is that system dynamics concerns itself, usually, with the broad behaviour of the system, rather than the fine detail. When necessary, however, one can model exact integer quantities, as we shall see later, providing one has thought through whether the extra effort and detail are really worthwhile, given that the purpose of system dynamics is strategic policy design for a system.

Before we leave the question of delays, notice that the influence diagram in Fig. 4.5 is the same as Fig. 4.8(a), the dynamics in Fig. 4.7 are those of OUT1 in Fig. 4.9, and recall that in Chapter 3 we modelled the elimination of a discrepancy between desired and actual stocks over a period of time. It should be clear, therefore, that the ruler's and the production manager's policies both have the same effect as a first-order delay and that the time constants in those policies are delays.

SMOOTHING OF INFORMATION

Suppose that the city ruler decides that making his people wait 10 days for food is too harsh and instead he would like to issue food at a rate of half the INFLOW and 10% of the stock. He realizes that, if he issued at the same rate as INFLOW there would be no stock to tide people over any irregularities in INFLOW. We are tempted to write an equation:

R FOUTR.KL=FOOD.K/FST+FINRA.KL*0.5

but this would constitute a **most fundamental error of system dynamics theory**. The reason is that values with the time label .KL are *yet to happen and cannot be used in decisions about what is yet to happen*. In common sense terms, the ruler does not know what food will come in today, so he cannot base today's decision on something unknown. Even if someone rode ahead of the food convoy yesterday and told him what was expected to arrive today that would still not be the same as what does arrive today. It would be information at time K, not flow during KL.

In short, to use a rate variable on the right-hand side of another rate or an auxiliary **is to make the serious error of confusing the future with the present**. Review Fig. 4.1.

The best the ruler could do, therefore, would be to take account of the average rate at which food had arrived over the past few days, AVFINRA, and use a policy of:

R FOUTR.KL=FOOD.K/FST+AVFINRA.K*0.5

Notice that the time label on AVINFRA is **.K** because it is something the ruler knows now.

This process is called **smoothing of information** and, like delays, it is a very important component of dynamic systems. It has a standard equation form which is shown in Fig. 4.10(a).

The smoothing equation is a **level** with a definite structure. The difference between the variable being smoothed and the smoothed variable is weighted by the effect of the time constant and added to the smoothed variable. The effect of this is best understood by imagining that FINRA has been constant for a long time. Clearly, AVINFRA, FINRA's average value, will be the same as FINRA and will have been the same for a long time. That corresponds to the system's 'memory' that FINRA has been stable. Suppose that FINRA now steps up to a new value and stays there. In the first few DTs, the memory will not change much, as the system cannot abandon all its previous history of FINRA and suddenly switch to the new value. As time passes and FINRA stays at its new value, the system will come to accept the change as permanent and AVINFRA will move closer and closer to the new FINRA.

This behaviour is shown in Fig. 4.11, in which FINRA starts at zero and steps to 100 tons/day, as it did in Fig. 4.7. Now, however, it becomes very noisy and unpredictable after day 30, because of storms in the mountains. When the step happens, the smoothed value tracks up to the new value and then follows behind the variations, more sharply in the upper graph, in which the smoothing time constant is 4 weeks, than in the lower graph

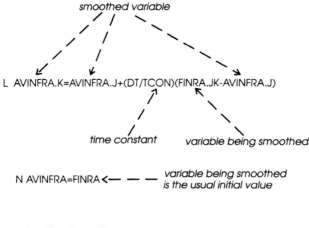

smoothed variable

L AVINFRA.K=AVINFRA.J+(DT/TCON)(FINRA.JK-AVINFRA.J)

time constant   variable being smoothed

N AVINFRA=FINRA ◄─ ─ ─ variable being smoothed
                          is the usual initial value

A) The Equation

L AVINFRA.K=AVINFRA.J+(DT/TCON)(FINRA.JK-AVINFRA.J)

$$\frac{(tons/}{day)} = \frac{(tons/}{day)} + \left((day) \middle| (day) \right) \left(\frac{(tons/}{day)} - \frac{(tons/}{day)} \right)$$

B) The Dimensions

Fig. 4.10   The smoothing equation.

where it is 10 weeks. Choosing a suitable time constant is a matter of balance between accepting a genuine change reasonably quickly while avoiding chasing the random noise. The model which produces Fig. 4.11 is included on the model disk, but it employs some modelling techniques which we shall not study until later. The reader should, however, run the model with various values of TCON and get a 'feel' for the way in which the behaviour of AVINFRA changes.

The dimensions of the smoothed level are shown in Fig. 4.10(b). FINRA has dimensions of [tons/day] and DT and TCON both have dimensions of days. The effect is that the **level**, AVINFRA has the same dimensions as the **rate**, FINRA. At first sight, this is strange, especially as we stated earlier that a conserved level must have the dimension of its inflowing rate multiplied by those of time. The reason is that the smoothed level is not a conserved level. In an equation such as

L TAMOUNT.K=TAMOUNT.J+DT*(IN.JK−OUT.JK)

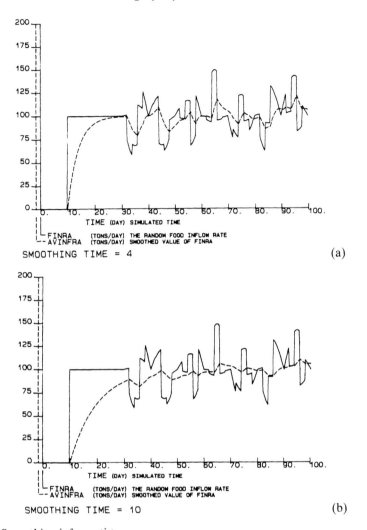

Fig. 4.11 Smoothing information.

if IN and OUT suddenly become zero, TAMOUNT will not change from its previous value but will continue to act as a memory for all the flows which have taken place. In the smoothed equation:

L AVINFRA.K=
AVINFRA.J+(DT/TCON)(INFRA.JK–AVINFRA.JK)

if INFRA becomes zero, then AVINFRA will eventually decay to zero, thus acting as a memory of INFRA's **behaviour**, while FOOD will act as a memory of INFRA's consequences. In short, the variable name Average Food Inflow **Rate** does not mean that the variable is rate type.

Let us also note that the proposed policy equation:

R FOUTR.KL=FOOD.K/FST+AVINFRA.K*0.5

would be bad modelling practice. The value of 0.5 could not be changed in a rerun to experiment with different policies and it has no stated dimensions, which may well lead to errors when dimensional consistency is evaluated for the model. It would be far better practice to put:

R FOUTR.KL=FOOD.K/FST+AVINFRA.K*FRAC
C FRAC=0.5

together with

D FRAC=(1) fraction of average inflow which will be issued

in which (1) means that FRAC is a dimensionless constant.

POLICY EXPERIMENTS WITH THE FOOD MODEL

Having discussed the equations for smoothing, let us now move into the policy analysis phase of system dynamics and see what effect this change of policy has. Model FIG4-12.COS on the disk is set up to provide this by adding a '**policy switch**' to the equation for FOUTR:

R FOUTR.KL=STOCK.K/FST+PSWITCH*(AVINFRA.K*FRAC)

The policy switch is a constant which can be set to 0 or 1. If it is 0 the effect of AVINFRA.K*FRAC is suppressed, whereas, when PSWITCH=1, the new policy can have its effects. Obviously, one could achieve the same effect by setting FRAC to 0 or to any other value, but the switch method is useful in more complicated cases, which is why we mention it now.

The consequences of the change in policy are shown in Fig. 4.12. Fig. 4.12(a) and (b) show two graphs with the policy off and on. The general result is that implementing the new policy will lead to generally lower stocks, which might have been expected. The food outflow rises more quickly in the first 20 days but, after then, there is a generally less smooth pattern of food outflow. The new policy seems to have made matters worse overall.

This can be seen more clearly in Fig. 4.12(c), which gives comparative results for the two policies. In these **comparative plots** the caption states which variable is being compared and the patterns correspond to the different runs – the scale line leads to the variable name followed by, not its definition, but the text on the run statement.

These policy comparisons, even for such a simple model, lead to some important general comments.

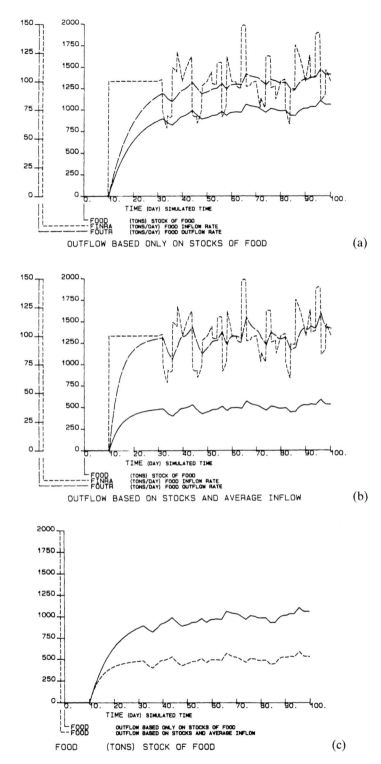

Fig. 4.12 Comparison of ruler's policies.

Policy usually has a very marked effect on the behaviour of a system. We shall see this many times, but it is important for the reader to grasp this now. It is one reason why it is usually not worthwhile to complicate a model so that it always deals with integer numbers of, say, bus drivers. Concentrating on the policy issues is usually a better strategy for bringing about an improvement in managed systems.

Systems often exhibit **counter-intuitive behaviour**, that is, what they do is often not what one would have expected. It seems obvious that issuing food as it comes in is going to improve the lot of the people. It turns out that it does not. Food reserves are smaller and the rate of issue is more variable.

Policies need not be simple choices. Is there a policy that gets the benefit of a rapid rise in issue rate as food starts to come in, as the new policy does, but gives the higher stocks as protection against the future and smoother behaviour on a daily basis? To test that, the reader should change FIG4-12.COS as follows:

- Replace FRAC by an auxiliary variable, FRAC.K, in the equation for FOUTR.KL.
- Replace C FRAC=0.5 by C BFRAC=0.5 to give a base value for the fraction issued. This will allow BFRAC to be changed in re-runs to allow one to test alternative values for it.
- Add a new group of equations:

    A FRAC.K=BFRAC−STEP(BFRAC,CTIME)
    C CTIME=200

As LENGTH is only 100, the STEP will not have any effect, so the two polices are unchanged. If, however, CTIME is set to, say, 30 in a run in which PSWITCH=1, one would be testing the new policy being implemented when the advantage of the more rapid rise in food issue rate had been enjoyed, after which the policy would revert to the original, smoother case, or so one would expect. Does it?

It is essential that you should make these experiments, consulting the user manual for the package you happen to be using for guidance on exactly how to do it. You will gain much more from carrying out experiments for yourself than from any amount of reading about experiments in a book. It is also essential to *document all variables, work out their dimensions and test the changed model to ensure that it is still dimensionally consistent.* This is such a fundamental of good modelling practice that it should always be done whenever a model is changed.

Before summarizing this chapter we must consider some aspects of DT, the step size.

SELECTING A VALUE FOR DT

If DT is too small time will be wasted waiting for models to run. If it is too large, numerical instability may occur, which means that dynamics in the *model* may be due to errors of calculation, not to the dynamics of the *system*. A good compromise is to ensure that

DT≤DEL/(4*ORDER)

for the smallest delay in the model. Thus, if a model contained a DELAY3 with a delay magnitude of 10 and also contained a time constant of 4 in a smoothing equation, the candidate values for DT would be 10/(4×3)=0.833 and 4/(4×1)=1, because a smoothing process is a first-order delay, as is the ruler's policy of FST=10. However, DT must also be set to one of several standard values in order to avoid rounding error in the calculation of TIME affecting the times at which time-dependent actions such as STEP take place. The standard values are 0.5, 0.25, 0.125 and so on, so the correct value in this case would be 0.5. In the model in this chapter we have used 0.25, which is smaller than strictly necessary, but gives smooth graphs.

It is not wrong to use very small values, such as 0.015625, but it would be unusual, and such small values should only be used when one has had a good deal of experience of modelling.

THE SIGNIFICANCE OF DT

The equations in a model are supposed to represent the way in which the real system works. However, in order to get the equations to run on a computer, an additional factor, DT, has to be invented by the analyst. DT has nothing at all to do with the real world; it is a figment of the calculation. It is, therefore, a *fundamental error* to use DT on the right-hand side of a rate or auxiliary equation. Those equations describe information and actions in the real system; the real system does not contain DT, so the equations describing the system's workings must not contain DT. There are two exceptions to this rule, which we shall encounter in Chapter 5.

SUMMARY

This chapter has necessarily been rather long in order to cover the essential foundations for much further work. We have explained the fundamental ideas about how system dynamics simulation works and showed the main equation types, such as delays and smoothing. These concepts were illustrated in a simple model. That model was also used to test the effects of different policies on the behaviour of the system and we emphasized that policy can have significant effects on system performance; the effects sometimes being contrary to intuition and common sense. Finally, we have discussed the selection of a suitable value of DT and considered its significance.

CHAPTER 5

System dynamics modelling techniques

INTRODUCTION

In the previous chapter we built a very simple model to demonstrate the principles of simulation, to show the type of output which a system dynamics model produces and to introduce the idea of counter-intuitive behaviour, which is very important in policy design. In Chapter 3, however, we saw that influence diagrams can be much more complicated than the simple diagram for the city ruler in Fig. 4.5 so, if system dynamics is to be useful for quantitative modelling, it must be capable of representing the rather complex problems which can arise in real systems. This chapter therefore covers some standard techniques and tips for modelling to act as a bridge from Chapter 3's influence diagrams to the models we shall develop for some of those problems in the next chapter. For example, we shall find out how to write the equation which produces the noisy behaviour of FINRA in Fig. 4.11.

At this stage, you should go through this chapter with the aim of finding out what it contains so that, when in a later chapter you see a CLIP function for instance, you will know where to find out more about it. In a sense, this chapter is a very abbreviated user manual for system dynamics software. In most cases, the forms for functions are stipulated without using system dynamics conventions, and then illustrated with the conventions. The chapter covers basic features which are widely used; more advanced topics are treated in Chapter 10.

It was stated earlier, but bears repetition, that the secrets of success in system dynamics are first to build a good influence diagram and then to make careful, painstaking and imaginative use of whichever of the several system dynamics software packages is available. Success does not really lie in using any particular package though they differ considerably in their capabilities.

The equation forms used in this chapter are as close as possible to traditional system dynamics formats, deriving mainly from the DYNAMO compiler, which was the first to be developed. Some packages differ quite considerably from that 'standard' and you must consult the manual for the package you are using. This chapter focuses on ideas and good practice, not on the use of a particular package. As mentioned earlier, this book was written using the COSMIC and COSMOS packages but it must be emphasized that the book is *not* about COSMIC modelling; it is about system dynamics modelling.

Throughout this chapter, features of system dynamics modelling will be illustrated by a few equations. The reader should practise setting up these examples as complete models and running them to demonstrate the behaviour discussed. Bearing in mind the requirements of your own software, use Fig. 4.6 as a guide to good practice, and follow the advice to print out all variables and study the results, as well as looking at the graphs of dynamic behaviour on the screen. A value of DT=0.25 will be suitable for all these examples. Do not neglect these simple exercises. The practice is vital in gaining confidence in using your software and even more so in developing the trick of thinking in terms of the passage of time.

TIME

As was mentioned in Chapter 4, all the system dynamics packages contain an inbuilt variable called TIME. An extra equation is created automatically which, if it were written as a level, would be:

L TIME.K=TIME.J+DT

This is, therefore, a perfectly ordinary level equation which acts as a memory for time in the model. The equation is hidden from the user who cannot amend it in any way. TIME's meaning is defined as shown in line 46 of Fig. 4.6.

The units in which TIME is measured have to be chosen to suit the problem. For the predator/prey model, months might be appropriate; for the decline of the Maya, certainly years, perhaps decades; for the combat model, probably hours. DT, the time-step, must have the same units as TIME. In most of the illustrations in the chapter, TIME will be expressed in weeks.

Like all levels, TIME requires an initial value which is automatically set to

N TIME=0

though the user can override that and initialize TIME by including in his model an equation such as

N TIME=1990

to simulate time starting on a given date.

The duration for which the model is to run is the maximum value for TIME and is set by, for example,

C LENGTH=100

which would simulate 100 weeks starting from 0 if TIME has not been initialized.

Although the equation for TIME cannot be altered, TIME can be used in equations just like any other variable. It has a time label of .K unless it is used on the right-hand side of a level equation, in which case its label is .J.

A TIME-DEPENDENT PROCESS – THE RAMP FUNCTION

In the model for Fig. 4.7 and 4.11 we used the STEP function to generate a pattern of food inflow which depended on the passage of time. Another useful time-dependent process is the RAMP function, which takes the form:

R FLOW.KL=RAMP(SLP,START)
C SLP=2
C START=5
D FLOW=(UNIT/WEEK) A FLOW VARIABLE IN A MODEL
D SLP=(UNIT/WEEK) A SLOPE FOR FLOW
D START=(WEEK) THE TIME POINT AT WHICH THE SLOPE
X   STARTS

The effect of RAMP is that the rate variable, FLOW, will start at 0 at TIME=0. When the inbuilt clock of TIME reaches START, FLOW will increase at a steady rate of SLP per week, so that, by TIME=20, the value of FLOW will be 30. X is the label for a continuation line.

One could also write

R FLOW.KL=BASE+RAMP(SLP,START)

Run a model to find the effects of

R FLOW.KL=RAMP(SLP1,START1)+RAMP(SLP2,STAR2)
C SLP1=2
C START1=5
C SLP2=−4
C START2=20

Also, find what happens if SLP2=−2.

Predict in your head the effects of

R FLOW.KL=STEP(STEPHT,STPTM)+RAMP(SLP,START)
C STEPHT=20
C STEPTM=10
C SLP=3
C START=30

and test your predictions with a model. What happens if START=5?

The point of these questions is to show that all the standard functions, not just STEP and RAMP, can be combined in any sequence and to any degree of complexity. Complexity for its own sake is not a virtue in a model, but users are restricted only by their own understanding of a problem. There are no limits in the software on what can be modelled.

In fact, there is no reason why factors such as SLP, or any of the constants used later in this chapter, have to be constant. They can be variables and might themselves be calculated from system dynamics functions, as we shall see later.

In the examples above, FLOW is a rate variable, but we could equally well have modelled an auxiliary variable, VAR, by the same method. However, the dimensional arguments and the form of the first fundamental equation of system dynamics explain why

L LEV.K=RAMP(SLP,START)

would not be legitimate.

RAMP could, however, be used directly for the rates in a level equation, perhaps to show the effects of rabbits immigrating into an area:

L RABPOP.K=RABPOP.J+DT*(RAMP(IMIG,IMTIME)+ ...
D RABPOP=(RABBITS) ACTUAL POPULATION OF RABBITS
X   IN THE AREA
D IMIG=(RABBITS/MONTH) RATE AT WHICH RABBITS
X   IMMIGRATE INTO THE AREA

though it would be better practice to define a separate variable which could be printed and plotted to make it easier to check that the model is working correctly:

L RABPOP.K=RABPOP.J+DT*(IMRATE.JK+ ...
R IMRATE.KL=RAMP(IMIG,IMTIME)

MORE TIME-DEPENDENT PROCESSES –
THE SINE AND COSINE FUNCTIONS

The standard trigonometric functions can be used to represent periodic time-dependent behaviour, usually in a model's exogenous input. Thus

R NO.KL=BASE+AMP*SIN(6.283*TIME.K/PERD)
C BASE=100
C AMP=30
C PERD=25
D NO=(WMS/WEEK) INFLOW OF NEW ORDERS TO WASHING
X   MACHINE PROBLEM OF FIGURE 2.7

```
D BASE=(WMS/WEEK) BASE LEVEL FOR NEW ORDER
X                INFLOW
D AMP=(WMS/WEEK) AMPLITUDE OF OSCILLATION IN
X                NEW ORDER INFLOW
D PERD=(WEEK) PERIOD OF OSCILLATION IN NEW ORDER
X                INFLOW
```

The number 6.283 is the value of 2π so this equation would make NO start at 100 and oscillate between 130 and 70 with a period of 25 weeks. If some number other than 6.283 is entered, the sine function will still work, but, in this case, the period will not be 25 weeks. Note the use of TIME.K within an equation.

SUDDEN CHANGES – THE PULSE FUNCTION[1]

What would happen to the washing machine firm if its usual pattern of orders fluctuated as above, but, in addition, it suddenly received a huge order for 500 washing machines? This requires the use of the PULSE function, which takes the form

V=PULSE(HEIGHT,FRST,INTVL)

where V is a variable or a term in an equation, HEIGHT is the height of the PULSE, which lasts for 1 DT, FRST is the value of TIME at which the first pulse occurs and INTVL is the interval between successive pulses; FRST = 10 and INTVL=20 would make pulses happen at TIME=10, 30, 50 and so on until LENGTH is reached. If we wrote

```
R NO.KL= . . . +PULSE(OSIZE,FRST,LENGTH)
C OSIZE=500
C FRST=50
D OSIZE=(WMS) SIZE OF LARGE ORDER
```

where . . . represents the previous equation for NO, there would be one pulse at TIME=50 and then no more, as the second would be called for at TIME=50+LENGTH and the model would have stopped at TIME=LENGTH.

However, this would not produce the required effect, because the PULSE lasts only for 1 DT and, as DT=0.25, the effect would be a height of 500 multiplied by 0.25 weeks, or 125 in total. The correct form would be

R NO.KL= . . . +PULSE(OSIZE/**DT**,FRST,LENGTH)

which makes OSIZE, which is measured in washing machines because it is a single large order, appear as a rate of inflow of 2000 washing machines per week for 0.25 weeks, or 500 in all, as required[2]. This is an excellent example

---

[1] Be very careful when using the PULSE function. Some packages treat it in different ways, so check the user manual for the package.
[2] Bold type cannot, of course, be used when entering this equation into a model.

of the use of dimensional analysis to get the correct equation. The first attempt required washing machines per week, which are the dimensions of the . . . part of the equation, to be added to washing machines, which are the units for OSIZE.

This use of DT on the right-hand side of a rate or auxiliary equation is one of the two exceptions mentioned on p. 109 to the rule prohibiting that. The other is to handle non-negativity constraints, which are discussed below.

As well as being used to inject quantities into a model, PULSE can also be used to make sudden changes in endogenous variables. Suppose that one wished to run a model with time measured in months, for a LENGTH of 10 years with DT set at 0.25. Some processes in the system, such as financial budgeting, take place annually, so it is necessary to record the time within a year and set that record back to zero at the end of each year.

The following equations would be appropriate:

L TYEAR.K=TYEAR.J+DT*(1−PULSE(YLEN/DT,YLEN,YLEN))
N TYEAR=0
C YLEN=12
D TYEAR=(MONTH) TIME WITHIN A YEAR
D YLEN=(MONTH) NUMBER OF MONTHS IN A YEAR

TYEAR will grow from 0 to 12 in the first year and then the pulse will subtract a year from TYEAR during the first DT of year 2. During that DT, however, the first DT of year 2 will be accumulated so that TYEAR will fall to the value of DT at the end of the first DT of each successive year. In short, TYEAR will only be zero when TIME=0. To make TYEAR be zero at the end of the year the first equation would have to be

L TYEAR.K=TYEAR.J+DT*(1−PULSE(YLEN/DT,
X   YLEN−DT,YLEN))

The reader should experiment!

INTERMITTENT INFORMATION – THE SAMPLE FUNCTION

It is often necessary to represent information being of interest or available only at intermittent times. For example, whether or not to cull foxes might be decided at intervals of 6 months, with the decision then being in force for the next 6 months. This intermittent measurement of the fox population can be modelled by:

A OBFOX.K=SAMPLE(FOXPOP.K,INTVL,INIT)

where OBFOX means the number of foxes observed when the sample was taken, INTVL is the interval between samples, which are taken when TIME=INTVL, TIME=2*INTVL and so on, and INIT is the value of

OBFOX until the first sample is taken, in other words the value which OBFOX will have from TIME=0 to TIME=INTVL. In order to get the initial value right it is usually possible to write:

A OBFOX.K=SAMPLE(FOXPOP.K,INTVL,FOXPOP.K)

in which case the initial value of OBFOX will be set to the initial value of FOXPOP. Note the use of a variable as one of the inputs to this function. Obviously, OBFOX, FOXPOP and INIT must all have dimensions of [FOXES] and INTVL must have the same dimensions as TIME.

RANDOM EFFECTS – THE NOISE AND NORMD FUNCTIONS

Random effects are often included in system dynamics models. There are two functions: NOISE, which produces values uniformly distributed between −0.5 and 0.5, and NORMD, which samples from a Normal (Gaussian) distribution of given mean and standard deviation (some packages use different names for this function). The principles for using both are the same, so we shall discuss only the NOISE function.

It is tempting to write and, indeed, one frequently sees in system dynamics models, an equation such as

A VAR.K=NOISE(SEED)
C SEED=99 (for instance)
A VAR=(UNIT) A VARIABLE WHICH VARIES UNIFORMLY
X   BETWEEN 0.5 AND −0.5
D SEED=(1) A NUMBER TO SEED THE RANDOM SAMPLING

in which VAR is a random variable and the significance of SEED is that, every time it is set to a given value, the same sequence of random numbers will be generated. SEED is a simple number which is regarded as dimensionless, so its dimensions, or lack of them, are denoted by (1).

The equation above is, however, *fundamentally wrong*. The reason is that, in system dynamics models, every variable is calculated every DT, so the equation says that a new random sample is to be taken every DT. If DT was halved, random samples would be taken twice as frequently. However, as we discussed in the last chapter, DT is an invention of the modeller in order to get the model to run on the computer, and it has nothing whatever to do with the nature of the real system. In this case, we must take samples at the same frequency that nature in the real world generates randomness, and we must ensure that the randomness is of the same magnitude as nature's. This requires us to combine two of the system dynamics functions and put

A VAR.K=SAMPLE(SCALE*NOISE(SEED),NFREQ,
X   SCALE*NOISE(SEED))
C SCALE=0.1 (for example)

```
D SCALE=(1) A MULTIPLIER TO MAKE THE SAMPLES HAVE
X            THE SAME RANGE AS NATURAL VARIATION
D NFREQ=(WEEK) THE FREQUENCY WITH WHICH
X              RANDOMNESS OCCURS IN THE REAL
X              WORLD
```

This will produce a fresh sample every FREQ weeks and SCALE will force the randomness to lie in nature's range, whatever it is, not in NOISE's of 0.5 to −0.5. In this example, the random variable, VAR, will be between 0.1 × 0.5 and 0.1 × −0.5, or 0.05 to −0.05. The use of SCALE and NOISE in the third argument of SAMPLE ensures that a suitable random value is used between 0 and TIME=FREQ.

To illustrate NOISE more specifically, suppose one wished to model the random effects of minor breakdowns on production capacity. Clear thinking is required because, when the sampling of NOISE produced positive values, one would be close to saying that production capacity could be *increased* by random effects. A possible approach would write

```
A PCAP.K=NCAP*RANEFF.K
A RANEFF.K=(1−SCALE*0.5+SAMPLE(SCALE*NOISE(SEED),
X           FREQ,SCALE*NOISE(SEED))
D PCAP=(UNIT/WEEK) CURRENT PRODUCTION CAPACITY
D NCAP=(UNIT/WEEK) NOMINAL OR MAXIMUM
X                  PRODUCTION CAPACITY
D RANEFF=(1) A MULTIPLIER TO REDUCE CAPACITY TO
X            REFLECT THE RANDOM EFFECTS OF MINOR
X            BREAKDOWNS
```

The effect of these equations is that a sampling which produces the result 0.5 will be exactly cancelled by the SCALE*0.5 from which the random values are subtracted, and there will be no increase in production above the maximum capacity. If the sampling produces any other value, production capacity will be correspondingly reduced.

Note that SCALE is used both within the SAMPLE and outside it to ensure that, even if this parameter is subsequently changed to test the robustness of the system policies against an increased magnitude of randomness, the equation will still be correct. Note also the use of RANEFF as a separate variable which can be printed or plotted to ensure that it never exceeds 1, a useful check that the model is working correctly.

It would be bad practice to write

```
A PCAP.K=NCAP*(1−SCALE*0.5+SAMPLE(SCALE*NOISE(SEED),
X          FREQ,SCALE*NOISE(SEED))
```

precisely because it would require more effort to see that what is happening in the model is correct. The greater the visibility of a model's workings the more one can be confident in it.

LOGICAL OPERATIONS – THE CLIP FUNCTION

The example for the SAMPLE function implies that culling of foxes might take place if OBFOX was greater than some target value but not otherwise. This is an example of a **logical choice** which, in the original system dynamics packages, requires the use of a function with the strange name of CLIP. This takes the form:

V=CLIP(A,B,C,D)

in which V=A if C⩾D and V=B if C<D. In dimensional terms, [V]=[A] =[B] and [C]=[D], in other words, C and D must be alike in order to be validly compared and A and B must be consistent with the result, V. Notice carefully that the answer A is produced when C is greater than *or equal to* D.

To apply this to the fox problem we could define a 'culling switch' which is 1 when culling can take place and 0 when it is forbidden. That could be done by:

A CSWTCH.K=CLIP(1,0,OBFOX.K,CRITF.K)
D CSWTCH=(1) A SWITCH TO INDICATE WHETHER OR NOT
X   CULLING OF FOXES CAN TAKE PLACE
D OBFOX=(FOXES) OBSERVED NUMBER OF FOXES
D CRITF=(FOXES) CRITICAL NUMBER OF FOXES AT OR
X   ABOVE WHICH CULLING IS REQUIRED

The foregoing equation could be linked with another:

R CULLR.KL=DCULLR.K*CSWTCH.K
D CULLR=(FOXES/MONTH) ACTUAL RATE AT WHICH
X   FOXES ARE CULLED
D DCULLR=(FOXES/MONTH) DESIRED RATE OF CULLING
X   FOXES

The CLIP function produces the result A if C⩾D so how does one handle the case that V=A if C>D, a so-called strict inequality? One approach would be to write:

A CSWTCH.K=CLIP(0,1,CRITF.K,OBFOX.K)

which will give the answer 0 if CRITF.K⩾OBFOX.K and, therefore, the answer 1 if CRITF.K<OBFOX.K, which is the same as OBFOX.K⩾CRITF.K. Another approach would use

A CSWTCH.K=CLIP(1,0,OBFOX.K,CRITF.K+EPS)
C EPS=0.01
D EPS=(FOXES) A SMALL NUMBER OF FOXES

which makes the CLIP produce 1 as soon as OBFOX.K equals or exceeds CRITF.K plus a small amount. The value of EPS need not be 0.01 but can be chosen to suit the problem.

It may seem confusing to have two ways of tackling a problem, but the point is to show that the system dynamics packages are almost infinitely flexible and there can be many ways of achieving the same result. Which one is used is partly a matter of personal style. In such instances, the author prefers the EPS method, as it is consistent with the standard use of CLIP and consistent practice is usually a good way of avoiding errors.

Logic requiring two conditions to be met can be modelled using two CLIPs[3]. If, for example, V=A if C⩾D AND E⩾F, to represent the foxes being culled if there are too many of them and the fox breeding season is over, then one could put:

A CSWTCH.K=CLIP(1,0,OBFOX.K,CRITF.K)*
X   CLIP(1,0,TYEAR.K,BRDTIM)
D TYEAR=(MONTH) A VARIABLE RECORDING THE TIME
X   WITHIN A YEAR, NOT THE SAME AS
X   TIME WITHIN THE MODEL
D BRDTIM=(MONTH) THE TIME WITHIN THE YEAR AFTER
X   WHICH FOXES STOP BREEDING

If both conditions are satisfied, then both CLIPs will produce 1, and the product of the two values will set CSWTCH to 1. If however, one condition or other is not met, one of the CLIPs will produce the value 0, which will negate the value being produced by the other CLIP. It would, however, be better practice to write

A CSWTCH.K=FOXCON.K*YEARCON.K
A FOXCON.K=CLIP(1,0,OBFOX.K,CRITF.K)
A YEARCON.K=CLIP(1,0,TYEAR.K,BRDTIM)
D FOXCON=(1) SWITCH TO INDICATE THAT FOX NUMBERS
X   REQUIRE CULLING
D YEARCON=(1) SWITCH TO SHOW THAT CULLING IS
X   PERMISSIBLE

This would allow FOXCON and YEARCON to be printed out so that one could check that the model was working correctly. Obviously, one could

[3] COSMIC is one of the system dynamics packages which allow for more complex logic. In this case COSMIC has single functions which are the equivalent of two CLIPs combined into one. Other packages allow logic to be entered in the form 'IF x THEN y ELSE z' etc.

also check that by looking at the values of the arguments to the CLIPs and making mental calculations, but that is more prone to error, and the less chance of error in a model the more one can be confident that it is doing the same things as the real system and for the same reasons.

Logic involving V=A if C≥D OR E≥F can also be handled using two CLIPs:

A VAR.K=CLIP(1,0,C.K,D.K)+CLIP(1,0,E.K,F.K)

or, preferably, for the reasons given above,

A VAR.K=VAR1.K+VAR2.K
A VAR1.K=CLIP(1,0,C.K,D.K)
A VAR2.K=CLIP(1,0,E.K,F.K)

This is correct unless both C≥D and E≥F in which case both CLIPs produce the value of 1 and VAR will be equal to 2. One way round that would be to write:

A VAR.K=VAR1.K+VAR2.K*(1-VAR1.K)

with VAR1 and VAR2 as before. In this case, if VAR1=1 and VAR2=0, then VAR=1. If VAR1=0 and VAR2=1, then VAR=1, but if both VAR1 and VAR2 are 1 then VAR is still 1, as required. As we shall see later, there is another way of achieving this result.

This example also illustrates an aspect of writing system dynamics models that one needs to understand at an early stage. In a normal programming language, such as BASIC or Pascal, the programmer will arrange for logical paths that are followed in some cases, with different paths being taken in other cases; parts of the program are simply not used in some circumstances. System dynamics **modelling** is different from computer **programming** in that all the system dynamics languages work by calculating all the variables every DT. One cannot bypass VAR1 and use VAR2, or vice versa; both are always 'alive' and the modeller must arrange to switch off the one not required. This is a habit which quickly becomes second nature, but it requires a little practice and careful debugging of the model in one's early stages of learning. There is no substitute for printing out all the variables in a model, as was done in Fig. 4.6, *and carefully studying the printout to check what is happening to the variables as time passes in the model.*

LIMITING FUNCTIONS – MAX AND MIN

The maximum and minimum functions have their usual form of

V=MAX(A,B) and V=MIN(A,B)

where at least one of A and B must be a variable if the use of MAX or MIN is to have any point. A triple maximum cannot be written as

V=MAX(A,B,C)

Instead one must write

V=MAX(A,MAX(B,C))

MAX and MIN can be combined, if desired, for example

V=MAX(A,MIN(B,C))

to any desired degree of complexity.

The previous example of VAR1 and VAR2 can be handled easily by writing

A VAR.K=MIN(VAR1.K+VAR2.K,1)

The purpose of showing the more complicated method was to emphasize that, in system dynamics languages, there are always several ways of solving any given problem.

MAX and MIN, like all the system dynamics functions, can be 'nested' in this way, sometimes up to 9 deep, depending on the package. Nesting is, generally, very inadvisable as it makes testing the model's operations for correctness more difficult. In proper system dynamics notation, and assuming that all the variables are auxiliaries, one should write

A VAR.K=MAX(A.K,DUMMY.K)
A DUMMY.K=MAX(B.K,C.K)

so that DUMMY can be printed or plotted to make it easier to see that what is happening in the model is happening for the right reasons.

MAX and MIN should be used only when the logic of the model requires them, and never to cover up strange behaviour. For example, one might wish to model Market Share, MS, as depending on a rate of growth, MSGR. To do so requires a level equation

L MS.K=MS.J+DT*MSGR.JK

It is evident that MS can never exceed 1, and if the model produces a value of more than 1 there is a temptation to write

L MS.K=MIN(1,MS.J+DT*MSGR.JK)

Most system dynamics packages would accept this, though they might give a message that this is an unusual format for a level equation, but the equation is actually wrong. The fault lies in the equation for MSGR, which is allowing MS to grow past its limit, and using a MIN in this way is simply covering up a fault somewhere else in the model, which is hardly good practice.

NON-NEGATIVITY CONSTRAINTS

In the management consulting firm's problem in Fig. 3.1 and in the combat model of Fig. 3.7 we encountered non-negativity constraints, though on p. 68 we make it clear that these are two rather different cases.

In the first case, the consulting firm determines the Desired Trainee Recruitment Rate, DTCRR. If this is positive, it indicates that trainees should be recruited, while a negative value would indicate that trainees should be dismissed. However, the firm has a policy of never dismissing people, so we wish to suppress the negative option. That requires

```
R TCRR.KL=MAX(0,DTCRR.K)
D TCRR=(PEOPLE/MONTH) ACTUAL TRAINEE
X                     RECRUITMENT RATE
D DTCRR=(PEOPLE/MONTH) DESIRED TRAINEE
X          RECRUITMENT RATE
```

a perfect instance of a MAX being used where the logic of the problem requires it.

In the second case, once combat starts, the Red commander commits reserves as fast as he can, as long as he has reserves left to commit. How fast he can commit them depends solely on the transport capacity available to him. That can be modelled by using the following three groups of equations:

```
L RRES.K=RRES.J+DT*(-RRCR.JK)
N RRES=IRRES
C IRRES=50000
D RRES=(RMEN) RED RESERVES REMAINING
D RRCR=(RMEN/HOUR) RATE OF COMMITTING RED
X                  RESERVES
D IRRES=(RMEN) INITIAL RED RESERVES
```

This level simply accounts for the remaining reserves as they are used up. Note the use of $(-RRCR.JK)$ to represent the drain; most packages will not allow $-DT*RRCR.JK$. It is good practice to set up initial values by using a constant as shown. The value of IRRES can be changed in a later run, which would be impossible if one had used N RRES=50000. Note also the use of RMEN, for Red men, rather than just MEN. It is always good practice to differentiate one type of unit very clearly from another, such as Blue men.

```
A WSWITCH.K=STEP(1,WTIME)
C WTIME=5
D WSWITCH=(1) A SWITCH TO DENOTE THE START OF
             COMBAT
D WTIME=(HOUR) TIME AT WHICH COMBAT STARTS
```

This is a simple use of STEP to ensure that nothing happens in the first 5 hours of the model run; a useful trick to make sure that the model is set up correctly.

R RRCR.KL=MIN(RRES.K/DT,RTCAP)*WSWITCH.K
C RTCAP=6000
D RTCAP=(RMEN/HOUR) RED TRANSPORT CAPACITY

The MIN term in the equation for RRCR is prevented by WSWITCH from having any effect until TIME=5; another instance of switching something off when it is not required. Within the MIN, the second factor, RTCAP, is obvious; it is the first factor, RRES.K/DT which represents the non-negativity constraint. To see how it works, let DT=1 for simplicity. At TIME=5 the combat starts and, since RRES is initially 50 000, RTCAP will ensure that Red can only commit reserves at the rate of 6000 men per hour. This can go on for 8 hours, by which time only 2000 men will remain. If there was no non-negativity constraint, a commitment rate of 6000 per hour for another DT would leave Red with −4000 men remaining; a clear absurdity. However, at that point, RRES.K/DT will be 2000, so the last 2000 men will be committed evenly during one more DT, at which point there will be exactly 0 men left and reserve commitment will have to stop.

In the above equations, IRRES, RTCAP and WSWITCH are all numbers, but note that IRRES and RTCAP have dimensions associated with real quantities, whereas WSWITCH is dimensionless because it represents an event occurring rather than something tangible.

Using DT on the right-hand side of a rate equation to model a non-negativity constraint is the second of the two exceptions to the prohibition against doing so. Whenever the modeller writes DT on the right-hand side of a rate or auxiliary equation it is essential to stop and work out whether one is using one of the two valid exceptions to the rule, the reasons for which are given on p. 109, or whether one is making a fundamental and catastrophic error of system dynamics logic. Notice, for instance, that DT is not used in the consulting company's non-negativity constraint.

CHOOSING A SUITABLE DELAY FUNCTION

In Figs. 4.8 and 4.9 we discussed delays and showed that they produced different dynamic responses to a STEP input. We also stated that there were other types of delay and, in particular, tested the perfect or pipeline delay which exactly mirrors the input after a time lag. How, though, does one know which delay to use in a given case and, indeed, how much difference it will make if one uses one delay type rather than another?

The best way to understand any of the system dynamics modelling techniques is to build a model and experiment, and this is no exception. Figure 5.1 is a model to show four different delay types against two different input patterns. This model is on the disk as FIG5-2.COS, in COSMIC format, though the reader should have little difficulty in converting it and, as ever, the emphasis is on the principles, not on the software.

In Fig. 5.1 the first two lines activate the model mapping which is described below and initiate automatic dimensional analysis. The driving force is in lines 8 to 18 and the reader will see a switch, SW1, to choose between a PULSE input and a noisy pattern which is very close to that used in Fig. 4.11, the equation being explained in detail towards the end of this chapter. The switch is changed in lines 84 and 85 to run the model twice. This is a traditional system dynamics technique which is useful when one wants a hard copy of the output. All the system dynamics packages also allow one to make changes like this interactively.

The output for four different cases is modelled in lines 31, 39, 47 and 56–57. The four cases are first-order, third-order (recall Fig. 4.8), pipeline and 'sixth-order' delays. The last of these is built up using two cascaded third-order delays, each of which has half the total delay. This requires the use of a dummy rate, DOUT4 in line 56 as an intermediary between the two DELAY3s. Cascading like this can produce any delay order. The reader should amend FIG5-2.COS to include a ninth-order delay.

For each delay we also calculate the contents and the cumulative output, as in lines 32 to 35 for the DELAY1. Line 48 is a special COSMIC requirement to load the pipeline initially as one wishes to have it; in this case we want it to be empty.

To check that the delays are functioning correctly, mass balances are computed in lines 65 to 68. Such balance checks are a most useful method of debugging models involving complex flows, though they are not strictly necessary here. Note that all the variables are printed and a selection are plotted.

Although this is a simple problem to help us in our study of delays, you may find it helpful to adopt the conventions of Fig. 5.1 when using your own package, as they represent much of what the author regards as good system dynamics modelling practice, in particular the heavy emphasis on generating output to assist in verifying that the model is working correctly and the full documentation of every variable and parameter.

The output for the two cases is shown in Fig. 5.2.

The PULSE happens at TIME = 10 and the graphs for INRAT and OUT3 go off the page, as both reach a value of 200/DT. However, the exact match between the two, 10 days apart, is easily seen. One can also see the immediate response and long tail of DELAY1 and the progressively sharper humps of the third- and sixth-order reactions. The contents of the respective delays in the lower half of the page show broadly similar patterns, noting especially the square form of CONT3 for the pipeline, the amount being 200 units, exactly as one would expect.

```
 0  map
 1  dim
 2  * Figure 5.2 Delay Behaviour
 3  note
 4  note    file named fig5-5.cos
 5  note
 6  note    input patterns
 7  note
 8  r inrat.kl=sw1*pulse(height/dt,frst,intvl)+
 9  x (1-sw1) *step(base+amp*sin(6.283*(time.k-15)/perd)
10  x *clip(1,0,time.k,15)*sample(noise(7),nperd,0),10)
11  c sw1=1
12  c height=200
13  c frst=10
14  n intvl=length
15  c nperd=2
16  c base=40
17  c amp=20
18  c perd=20
19  note
20  note    cumulative input for balance check
21  note
22  l cumin.k=cumin.j+dt*inrat.jk
23  n cumin=0
24  note
25  note    the delay magnitude
26  note
27  c del=10
28  note
29  note    output rate for first order delay
30  note
31  r out1.kl=delay1(inrat.jk,del)
32  l cont1.k=cont1.j+dt*(inrat.jk-out1.jk)
33  n cont1=0
34  l cumout1.k=cumout1.j+dt*out1.jk
35  n cumout1=0
36  note
37  note    output rate for second order delay
38  note
39  r out2.kl=delay3(inrat.jk,del)
40  l cont2.k=cont2.j+dt*(inrat.jk-out2.jk)
41  n cont2=0
42  l cumout2.k=cumout2.j+dt*out2.jk
```

Fig. 5.1   Model for delay demonstration. 'Figure 5.2' appears in line 2 because this line is used to generate the caption for Fig. 5.2.

```
43   n cumout2=0
44   note
45   note   output rate for pipeline delay
46   note
47   r out3.kl=dlpipe(tvals,inrat.jk,del)
48   t tvals=40*0
49   l cont3.k=cont3.j+dt*(inrat.jk-out3.jk)
50   n cont3=0
51   l cumout3.k=cumout3.j+dt*out3.jk
52   n cumout3=0
53   note
54   note   output rate for sixth order delay
55   note
56   r dout4.kl=delay3(inrat.jk,del/2)
57   r out4.kl=delay3(dout4.jk,del/2)
58   l cont4.k=cont4.j+dt*(inrat.jk-out4.jk)
59   n cont4=0
60   l cumout4.k=cumout4.j+dt*out4.jk
61   n cumout4=0
62   note
63   note   mass balance checks
64   note
65   a mchck1.k=cumin.k-cumout1.k-cont1.k
66   a mchck2.k=cumin.k-cumout2.k-cont2.k
67   a mchck3.k=cumin.k-cumout3.k-cont3.k
68   a mchck4.k=cumin.k-cumout4.k-cont4.k
69   note
70   note   model control and output
71   note
72   c dt=0.25
73   c length=50
74   c prtper=10
75   c pltper=0.25
76   print 1)inrat, cumin
77   print 2)out1,cumout1,mchck1
78   print 3)out2,cumout2,mchck2
79   print 4)out3,cumout3,mchck3
80   print 5)out4,cumout4,mchck4
81   plot inrat=a,out1=b,out2=c,out3=d,out4=e(0,60)
82   plot cont1=a,cont2=b,cont3=c,cont4=d(0,600)
83   run Delayed Response to a Pulse Input
84   c sw1=0
85   run Delayed Response to a Noisy Input
```

Fig. 5.1 *Continued*

```
 86  note
 87  note   definitions of variables
 88  note
 89  d amp=(units/day) amplitude of sine wave in input
 90  x              rate
 91  d base=(units/day) base value for input rate
 92  d cont1=(units) contents within the first order
 93  x            delay
 94  d cont2=(units) contents within the second order
 95  x            delay
 96  d cont3=(units) contents within the pipeline delay
 97  d cont4=(units) contents within the sixth order
 98  x            delay
 99  d cumin=(units) cumulative input
100  d cumout1=(units) cumulative output for first order
101  x              delay
102  d cumout2=(units) cumulative output for third order
103  x              delay
104  d cumout3=(units) cumulative output for pipeline
105  x              delay
106  d cumout4=(units) cumulative output for sixth
107  x              order delay
108  d del=(day) magnitude of delay
109  d dout4=(units/day) dummy output for first delay3
110  x                in sixth order delay
111  d dt=(day) step size in simulation
112  d frst=(day) time of first pulse in input rate
113  d height=(units) size of pulse in input rate
114  d inrat=(units/day) input rate to various delays
115  d intvl=(day) interval between pulses in input
116  d length=(day) duration of simulation
117  d mchck1=(units) mass balance check for first
118  x              order delay
119  d mchck2=(units) mass balance check for second
120  x              order delay
121  d mchck3=(units) mass balance check for pipeline
122  x              delay
123  d mchck4=(units) mass balance check for sixth
124  x              order delay
125  d nperd=(day) period of random noise in input
126  x            rate
127  d out1=(units/day) output rate from first order
128  x                  delay
```

Fig. 5.1 *Continued*

```
129  d out2=(units/day) output rate from second order
130  x                   delay
131  d out3=(units/day) output rate from pipeline
132  x                   delay
133  d out4=(units/day) output rate from sixth order
134  x                   delay
135  d perd=(day) period of underlying sine wave in
136  x                input rate
137  d pltper=(day) plotting interval
138  d prtper=(day) printing interval
139  d sw1=(1) switch to change input rate patterns
140  d time=(day) simulated time
141  d tvals=(units/day) table to load pipeline initial
142  x                   values
143  +
No syntax errors detected
================
```

Fig. 5.1 *Continued*

The conclusion from Fig. 5.2(a) and (b) is that the choice of delay order matters quite strongly when one is dealing with very sharp inputs, such as PULSEs or a STEP. The difference between DELAY3 and the cascaded sixth-order delay is, perhaps, not all that large; probably well within the uncertainties of the data in most managed systems. One can, perhaps, conclude that for practical purposes one can distinguish between three main possibilities: the output being an exact copy of the input, in which case a pipeline delay is appropriate; the case where the input has an immediate effect and then a long tail, DELAY1; and an intermediate case where there is a delayed response and then a spread of inflow, in which case DELAY3 will usually be appropriate. It is very tempting to use the pipeline under some illusion of accuracy, but such cases are rare in managed systems, dispersed delays being *far* more usual.

Fig. 5.2(c) and (d) shows a more usual case of a continual, though very unsteady, inflow. Again the pipeline matches it exactly, whereas the three distributed delays smooth out the input considerably and, indeed, have little difference between them once the effect of the STEP has been absorbed. Continuous flows are far more typical than sharp shocks in the modelling of managed systems. In such cases, the 'accuracy' of a pipeline is largely illusory, and DELAY3 is a good compromise well within the accuracy of the data usually available, which is why it is so commonly used in system dynamics modelling.

It is worth reiterating that all the system dynamics packages have a

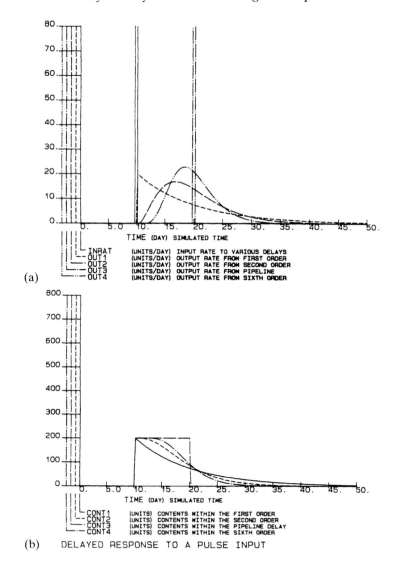

Fig. 5.2 Delay behaviour.

number of delay functions and it is useful to take the time to experiment with them. In some cases the syntax differs considerably for the forms used here, though the principles discussed here are applicable.

NON-LINEARITIES – THE TABLE FUNCTION

Much of the interesting dynamic behaviour of a managed system stems from the presence of non-linear effects within it. The influence of the

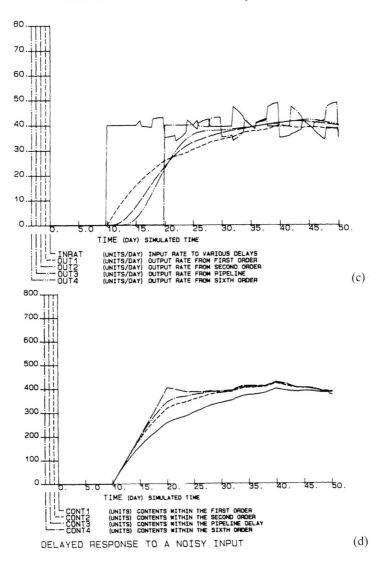

Fig. 5.2 *Continued*

Average Food Intake per Fox (AVIPF) in Fig. 5.3 on the lifetime of foxes is a good example of the difference between linear and non-linear behaviour. Suppose that AVIPF was 10% below the level required for a normal lifetime (NFRPF in Fig. 5.3), and suppose that would reduce fox lifetime to 5% below normal. If AVIPF fell to 20% below NFRPF then the influence would be **linear** if lifetime fell to 10% less than normal, because doubling the food shortfall had doubled the effect. If the relationship was truly linear, then a 100% fall in food supply, complete starvation, would lead to a 50% reduction in lifetime; clearly absurd, so the relationship must be **non-linear**.

Setting aside for the moment how one collects the numerical data for the relationship, we can approach the handling of non-linearities in system dynamics models on the following lines.

In the first place, the model for Fox Death Rate and D_{LIFE} would usually be written in system dynamics notation as

R FDR.KL=DELAY3(FBR.KL,DLIFE.K+MINLT)
A DLIFE.K=NFLTF*FSRF.K
C NFLTF=some suitable number
C MINLT=some suitable number
D FBR=(FOXES/MONTH) FOX BIRTH RATE. AN EQUATION
X   FOR THIS VARIABLE EXISTS
X   SOMEWHERE ELSE IN THE MODEL
D FDR=(FOXES/MONTH) FOX DEATH RATE
D DLIFE=(MONTH) CURRENT FOX LIFETIME
D NFLTF=(MONTH) NORMAL FOX LIFETIME IF FOOD
X   INTAKE IS AT THE LEVEL REQUIRED
X   FOR A NORMAL LIFESPAN
D FSRF=(1) A MULTIPLIER TO REPRESENT THE EFFECTS
X   OF AN INADEQUATE FOOD SUPPLY ON FOX
X   LIFESPAN.

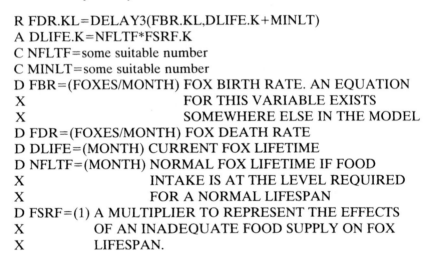
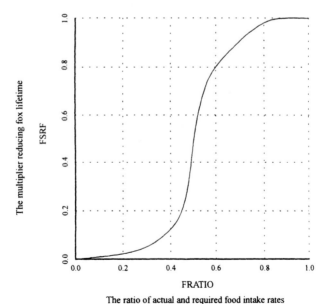

FRATIO

The ratio of actual and required food intake rates

Fig. 5.3 A non-linear relationship between food intake and the effect on fox lifetime.

D MINLT=(MONTH) THE MINIMUM LIFESPAN OF A NEW-
X   BORN FOX CUB IF THERE IS NO FOOD

MINLT is needed because, if DLIFE.K was 0, the DELAY3 function would attempt to divide by zero, as shown in the equations in Fig. 4.8. MINLT obviously act as a delay and, for the reasons given on p. 109, DT will have to be no larger than MINLT/12.

FSRF is a multiplier to scale NFLTF down according to the availability of food. Such multipliers are very common in system dynamics modelling, as they give a convenient way of modelling the effects of one variable on another, especially when the effect is non-linear. How, though, are we to model FSRF?

On the principle of introducing new variables to break what would otherwise have been complicated equations into simpler components to make it easier to see that the model is working correctly, let us define a new variable, FRATIO:

A FRATIO.K=AVIPF.K/NFRPF
D FRATIO=(1) RATIO OF ACTUAL AND IDEAL FOOD
X   SUPPLIES
D AVIPF=(RABBITS/MONTH/FOX) AVERAGE ACTUAL
X   INTAKE OF FOOD PER
X   FOX
D NFRPF=(RABBITS/MONTH/FOX) AVERAGE FOOD INTAKE
X   PER REQUIRED FOR
X   NORMAL LIFESPAN

A hypothetical relationship between FRATIO and FSRF is shown in Fig. 5.3. If FRATIO.K is 1 the foxes are fully fed and their lifespan is not affected, represented by FSRF.K=1 on the vertical scale. If FRATIO.K is zero, however, foxes will die rather quickly, corresponding to FSRF.K being zero. Similarly, the value of FSRF can be read from the graph for any value of FRATIO. Notice that the graph is highly non-linear. Small reductions below 1 in FRATIO have little effect, but the curve drops steeply until, at low values of FRATIO, fox lifespan would not be a good insurance risk. System dynamics notation converts such a non-linear curve, which we assume to be based on data about fox biology, into a set of linear segments which fit reasonably closely to the actual curve.

Such a **piece-wise linear approximation** is shown in Fig. 5.4. The dotted lines are the curve which the software will use to approximate to the true curve. It is expressed in the notation:

A FSRF.K=TABLE(TFSRF,FRATIO.K,0,1,0.2)
T TFSRF=0/0.024/0.12/0.8/0.98/1.0
D TFSRF=(1) TABLE FOR THE RELATIONSHIP BETWEEN
X   FOOD RATIO AND FOOD SUPPLY REDUCTION

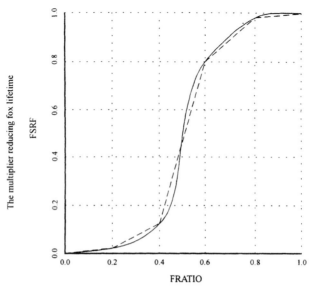

Fig. 5.4   A piecewise linear approximation to the non-linear relationship.

X   FACTOR WHICH MODELS THE EFFECTS OF
X   THE AVAILABILITY OF FOOD ON THE
X   LIFESPAN OF FOXES

The equation for FSRF means that the value of FSRF at any point in time will be calculated for a given value of FRATIO from a look-up table called TFSRF. The three numbers at the end of the equation provide that FRATIO will range from a minimum of 0 to a maximum of 1.0 in steps of 0.2. This requires a total of 6 values to be specified, and they are given in the second line, which uses the type label T for a table, the individual values being separated by /. Thus, the first value, 0.0, means that when FRATIO.K=0, FSRF.K will also be zero. The third value provides that FRATIO.K=0.4 will generate a value of 0.12 for FRSRF. If FRATIO happened to be 0.65, say, then FRSRF would be calculated by linear interpolation between the table values of 0.8 and 0.98 which are the TFSRF values for FRATIO=0.6 and FRATIO=0.8, respectively.

The 6 numbers for TFSRF are found by measuring the points at which the linear segments in Fig. 5.4 cross the dotted vertical lines. The number of linear segments is chosen by eye to get a reasonable match between the linear segments and the real curve. Figure 5.4 has been deliberately drawn to be a mediocre match, especially between 0.4 and 0.6. One might, indeed, decide to have twice as many segments to get a better fit and the reader should practice setting up such a table, which will require 11 values in total.

In the equations above, note that the definition for TFSRF extends to five lines of text. Writing very complicated equations is bad practice; it is far better to break up a complex equation into several simpler auxiliaries. Long definitions, when called for, are very good practice as they make clear to client and modeller alike exactly what the model's variables and parameters mean. This is true whether the client is a paying a fee for a service or is the academic audience for a book or paper.

The TABLE function has a strict limitation in that, in most system dynamics software packages, it will produce a fatal error if FRATIO, in this example, goes below 0.0 or above 1.0, its declared range. A variant on this is the TABHL function which has exactly the same format but which allows the argument, FRATIO, to go outside its declared range, using the first or last of the table values as a horizontal 'tail' for the output, FSRF, if that happens. Using TABHL is a temptation but is to be avoided unless the logic of the problem *really* requires it. In this example, FRATIO cannot logically be less than 0 or greater than 1, so it would be helpful for thorough model debugging if the model crashes when that happens.

If a model does crash because a table has gone outside its range, careful study of the printout of the model's variables will usually reveal the cause. There is no substitute for using printed values of model variables in debugging a model. The graphs of a few variables displayed on the screen will sometimes look right when they are not, because something is going wrong which is not being seen.

The examples used here to explain the principles of non-linearities have been expressed in the DYNAMO-type format which is used by at least two of the other packages. It is essential, however, to check the manual for the package you are using as the details of setting up TABLE and TABHL functions, and even their names, differ considerably in some of the packages. Most packages will offer additional forms for modelling non-linearities. For example, COSMIC has table types in which the output does not vary smoothly but steps up or down as the argument changes.

Tables are a most useful feature in system dynamics and can be used very flexibly. There is, of course, no reason why the output and the argument both have to lie between 0 and 1; the range of either has to be whatever is appropriate to the problem. One useful role for a table is to inject into a model an historical pattern of, say, washing machine orders. In such a case, the argument would be TIME.K and the values in the table would be the sales rates in each of the past 48 months, or whatever.

The dimensions for a non-linearity can be understood by writing the equation in abstract form:

V=TABLE(TABNAM,ARG,LO,HI,INCR)

Recalling that [ ] means 'the dimensions of', this requires that [VAR]=[TABNAM] and that [ARG]=[LO]=[HI]=[INCR]. In words, the table values have to be in the same units as the result, VAR, and the argument has to have the same dimensions as the three points which specify its range. However, there is no requirement at all for [VAR] to be the same as [ARG]. In short, the TABLE function and its relatives can validly act as a dimensional transformation between one variable and another. The curve of the relationship stands for the unknown causal mechanism which relates ARG to VAR. Clearly, if one knew that mechanism, one could write the equations for it and those equations would have to be dimensionally valid, but the dimensional transformation property allows the modeller to short-circuit that difficulty.

MODEL MAPPING AND DOCUMENTATION

Explaining a model to oneself and to a client is difficult when working from a listing such as Fig. 4.6. All the system dynamics packages therefore offer some form of printout of the equations and the definitions. The DYNAMO 'standard' is shown below for part of the model in Fig. 4.6.

The dynamic equations, levels, rates and auxiliaries are laid out as shown below, with the definitions of the variables and any constants required by the equation. The constants do not have to follow the equation in the listing: they will be searched for and printed out. This format is usually very easy to explain even to a client who is not versed in computer modelling.

Equation number 1
----

r finra.kl=step(100,10)+step(stepht,steptm)
c stepht=0
c steptm=120

FINRA=(TONS/DAY) FOOD INFLOW RATE
STEPHT=(TONS/DAY) STEP HEIGHT IN FOOD SUPPLY
STEPTM=(DAY) TIME OF STEP INCREASE IN FOOD SUPPLY

Equation number 2
----

l food.k=food.j+dt*(finra.jk−foutr.jk)
n food=0
c dt=0.25

FOOD=(TONS) STOCK OF FOOD
DT=(DAY) TIME STEP IN MODEL
FINRA=(TONS/DAY) FOOD INFLOW RATE
FOUTR=(TONS/DAY) FOOD OUTFLOW RATE

An additional form of documentation is the model map, a couple of lines of which are shown below. The variables and constants are listed in alphabetical order, DT being the first alphabetically for this model. That is followed by information on the line in the model[4] in which it occurs, the line in which it is initialized (in this case, line 0 means that DT does not require an initial value on an N statement) and the line in which it is defined. After that, the map shows the names of variables in whose equation it is used.

1 DT=(DAY) TIME STEP IN MODEL

Equation in line 22 Initialised in 0
Documented in 36
Used in equation for –

FOOD TIME

2 FINRA=(TONS/DAY) FOOD INFLOW RATE

Equation in line 6 Initialised in 0
Documented in 37
Used in equation for –

FOOD

The available packages differ enormously in the form in which this type of information is presented. Some, for example, list all the levels, followed by all the rates and then the auxiliaries.

THE COMPLEX INPUT PATTERN

Figure 4.11 used an extremely complicated pattern of food inflow to demonstrate the idea of smoothing information.

The equation which produces that pattern is shown below. It uses several of the standard functions and we shall use it to show how functions can be combined to produce a desired result.

```
r finra.kl=step(base+amp*sin(6.283*(time.k-30)/perd)*
x clip(1,0,time.k,30)*sample(noise(7),nperd,0),10)
c nperd=2
c base=100
c amp=100
c perd=20
```

---

[4] This book is about the principles of system dynamics modelling and has, therefore, been written to be as independent as possible of specific software. The line numbers given here are correct for Fig. 4.6, but users of COSMIC will find that the line numbers printed out when running this model are 2 larger than the numbers given here. This is due to the way that particular package works and does not invalidate the *principles* explained here.

The STEP is used to generate the sudden onset of food supply at TIME=10. That which is stepped is everything from base to 0), at the end of SAMPLE and the time at which the step happens is the 10 at the end which has been printed in bold to draw attention to it. The amp*sin(6.283(time.k-30) perd) is a normal sine wave, 6.283 being the value of 2*pi. However, in order to allow time for the smoothing of the STEP to be seen, we wish it to commence at TIME=30 so, instead of just using TIME.K, we use (TIME.K−30), so that when TIME=30 the sine wave will 'think' that TIME=0 and start the sine from 0. Unfortunately, system dynamics packages calculate every variable every DT so, from TIME=0 to TIME=30 the SIN function will be producing values, and those have to be switched off, as we saw earlier. The STEP switches them off until TIME=10 and the CLIP(1,0,TIME.K,30) does so between TIME=10 and TIME=30.

The effect of sample(noise(7),nperd,0) is to introduce random values lying between −0.5 and +0.5. They multiply AMP so that AMP will be between 50 and −50. These variations are added to the base level of 100, so the effect is inflow rates lying between 50 and 150. The randomness is sampled at intervals of 2 weeks to represent, perhaps, the frequency of occurrence of good and bad weather on the route taken by the food convoys.

IS THAT FUNCTION REALLY NECESSARY?

It is a great temptation to dress up a model by using lots of functions, such as CLIP, TABLE, MAX and so on. Consider, however, Fig. 5.4, which allows for the Food Ratio to range from 0 to 1. Suppose that when the model is run, FRATIO never falls below 0.8. Clearly the data in Fig. 5.4 and the equations which represent the curve are largely spurious detail and the table should have been defined to be in the range from 0.7 (to be on the safe side) and 1.0, not from 0 to 1. Similarly, a CLIP is supposed to offer a choice between two options. If it always chooses the same one, then either the detail is spurious or, more likely, there is a mistake in the model or the problem has not been correctly understood. Careful study of output is essential both to remove unnecessary detail and to detect errors.

SUMMARY

This chapter set out to be a brief introduction to the principles underlying some of the techniques required in system dynamics modelling. The formats used here are those which might broadly be regarded as traditional in system dynamics, but they are not the only ones. At the risk of repetition, the actual details of using these principles will vary from package to package and the user must consult the manual for the software being used.

The reader was encouraged to test these examples in small models because they will be used, in some cases without further explanation, in the models in the ensuing chapters.

CHAPTER 6

Case studies in modelling

INTRODUCTION

Chapter 2 explained the techniques and concepts of influence diagrams and in Chapter 3 we studied influence diagrams for some complex problems. Chapter 4 took us into the underlying theory of system dynamics modelling, including the building and running of a very simple model. Chapter 5 laid a grounding in some of the essential techniques required to build models. In this chapter, we must put those pieces of knowledge to work and build some serious models of managed systems. At this stage, the emphasis will be on building models and testing them to establish the confidence which can be placed in them; the process sometimes called 'validation', though 'confidence building' would be a far better term. The use of models in policy design will be dealt with in Chapter 7 and their optimization in Chapter 9.

We shall start with the manufacturing company, the influence diagram for which was studied in Chapter 2, because that model uses most of the standard ideas of system dynamics and it proves to be very rich in possibilities for policy design and optimization. That model will be developed in small stages to show the procedure in detail, the *process* being applicable regardless of the software package being used. Other models in this chapter will be more in the nature of exercises for the reader.

THE DOMESTIC MANUFACTURING COMPANY (DMC)

The narrative for DMC in Fig. 2.4 dealt only with how the system worked, because we were engaged in practising the techniques of influence diagramming. Before we can build a model, however, we must be clear about what it is for; that is, we must understand the problems faced by the company's management. (This problem is based on one of the author's consultancy assignments. It has been simplified, but the description of the system and its problems is essentially accurate.)

The management problem

In this case, DMC's manufacturing activities were faced with a very unstable and completely unpredictable order pattern with two peaks and

troughs in each year. The company's historical experience is that
this, the production department rode through the variations in th
pattern quite well. The raw materials department, by contrast, could
barely cope. At times, there were serious shortages of materials, to the
point that the possibility arose of having to close down production. At
others, there was such a glut of materials that space had to be rented to
accommodate them. Naturally, this led to such interpersonal tensions
that the two managers concerned were scarcely on speaking terms. Apart
from the personal factors, one of the tests for the model will be to see if it
behaves in the same general way as the real system, and for the same
reasons.

The initial stages of modelling

Some initial steps need to be performed before one can start to write
equations. They are discussed in detail for this model, although readers will
have to apply them to the others in this chapter themselves.

The **first stage** is to choose names for the variables and to decide on their
dimensions. We shall use short mnemonic names, though some modellers
prefer long ones. Variable naming is a matter of personal preference, and
you should develop your own style, because a consistent style will help to
produce better quality models. The names and dimensions for this model
are shown in Fig. 6.1.

The **second stage** is to decide whether each factor in the diagram is a
level, a rate, an auxiliary, a constant or a table for a non-linearity. This
process is called **type assignment** and probably causes more problems for
the novice modeller than anything else, though it need not if the injunction
to *think physics* has been applied, the modeller is familiar with the idea of
level variables acting as memories in the model and has studied the common modules shown in Fig. 2.12.

To assign types, the place to start is the physical flows, shown with solid
lines in Fig. 6.1. Think of the analogy of water flowing and the places in
which it accumulates. Recall the use of terms such as 'inflow', 'outflow',
'drain' and 'level' in the common modules in Fig. 2.12. Thinking in that way
makes it evident that Backlog of Orders and Raw Material Stocks in Fig. 6.1
must be level variables and that the flows in and out of them must be rates.
Module B in Fig. 2.12 suggests that Raw Material Order Rate must also be
a rate variable. Bearing in mind module G, in which a rate and a parameter
affect a smoothed variable, the Average Order Rate and Average Production Rate must also be levels, but of the smoothed variety which was shown
in Fig. 4.10.

Do not become confused by the word 'rate' in the names of the averages.
This does not mean that the variable is a rate. The key word is 'Average',
and averages in system dynamics are levels, because they are a memory for
what is averaged.

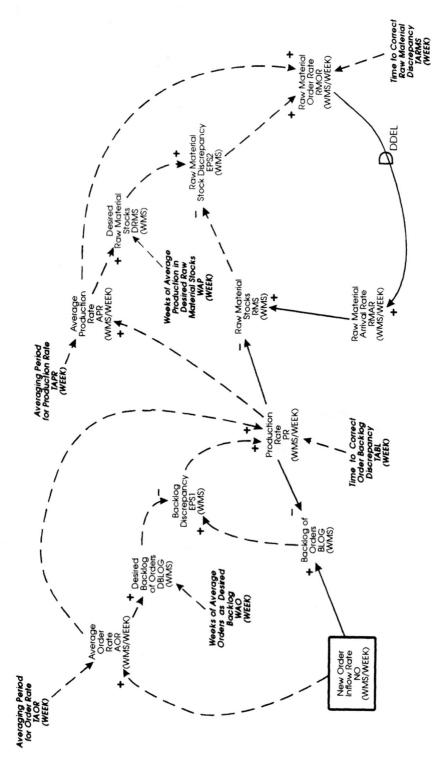

Fig. 6.1 Variable names and dimensions for the DMC problem.

To emphasize those points, Fig. 6.2 shows the level variables with a symbol denoting a bucket of water superimposed. Similarly, the rate variables have the engineering symbol for a valve superimposed[1]. The parameters are shown in italics and the other variables must be auxiliaries. There are no tables in this model.

Type assignment really is as simple as that. The main thing is to follow the physics and identify the physical and smoothed levels, and the rest will follow. Later in this chapter we shall state some rules for how variables can affect one another, but the real key is to *think physics* and to be quite clear about the principles embodied in Fig. 4.1.

Finally, one also has to identify any additional parameters which will be needed and to choose whether or not to make the diagram more complicated in order to show them explicitly. In this instance, extra parameters are required to specify the pattern of New Order Inflow Rate, NO, but we choose not to show these on the diagram as they will probably confuse the client.

Some system dynamics software packages are based on drawing on the screen, using icons, the level/rate type of diagram, a small example of which was shown in Fig. 2.17. Using that type of software still requires the modeller to make type assignments and also requires *every* detail to be shown on the diagram. Whether this is helpful or laborious is a matter of personal preference. The author's view is that models can be built just as effectively and, perhaps, more quickly, with an influence diagram at about level 3 in Fig. 2.16. The connections between the diagram and the model are clear from the equations and the diagram is kept in its proper role as a tool for communication with the client. These are, however, matters of opinion, and readers must make up their own minds.

The common element in all the packages is that, as explained in Chapter 4, the equations can be written in any order the analyst finds convenient, and the package will sort them into the correct computational sequence. In practice, the modeller chooses an order which is easy to understand and explain. Development of a consistent personal style will be a considerable aid to speed and quality of work.

The model equations

We now turn to the writing of the model, explaining the equations as we go. The principles are valid, though the details of entering the model into the computer will depend on the package being used. As explained in Appendix A, some packages will create *some* of the equations automatically, though the user has to type the others or generate them using a mouse. For the purposes of explanation, we shall proceed as though all the equations

---

[1] This symbology is not normal influence diagram procedure; it is used here only as an explanatory device.

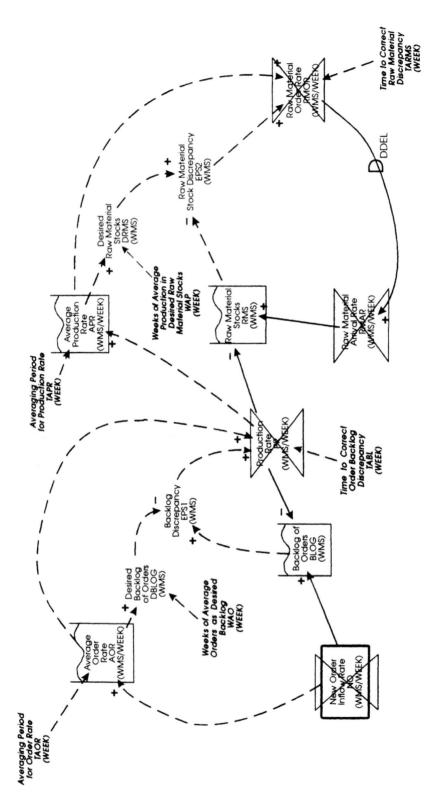

Fig. 6.2 Type assignment for the DMC problem.

are typed in by the analyst. In each case the set of equations is followed by a brief explanation. Rather than repeating definitions each time a variable is mentioned they are given in alphabetical order at the end of the explanation. The reader will have to flip back and forth between the equations and the definitions, though they are also shown in Fig. 6.1.

* Fig 6.3 The Domestic Manufacturing Company Model
note
note file fig6-3.cos

The first few lines give a title to the model.

```
note
note    current and average inflow of new orders
note
r  no.kl=blev+d1*step(20,10)+(1−d1)*amp*sin(6.283*
x  (time.k−10)/perd)*clip(1,0,time.k,10)
c  d1=1
c  blev=100
c  perd=25
c  amp=30
l  aor.k=aor.j+(dt/taor)(no.jk−aor.j)
n  aor=no
c  taor=4
```

Next, we must decide how to drive the model with a pattern of new orders. A STEP input is a standard driving force which, as we shall see, gives useful information about the model, but the company is known to have a pattern of demand with two peaks and troughs in each year. We therefore also provide for a sine wave and a 'switch', D1, to choose between them. If D1=1, the form of NO will be a constant value of 100 up to TIME=10, after which the value will be 120. If D1 is set to 0, the behaviour of NO will be a constant value of 100 until TIME=10, after which there will be a sine wave oscillating between 70 and 130. The equation for NO is rather similar to that discussed on p. 137, except that it does not allow for randomness. At this stage we do not wish to make the model behaviour too complicated to understand; the ability to cope with random behaviour will be an aspect of the system's robustness which the reader should examine in Chapter 7. Note the extra parameters for BLEV, PERD and AMP, which are not shown in Fig. 6.1. Their values are chosen to correspond to what happens in the real system. Observe the use of NOTE statements as comments to improve legibility and to identify the sectors of the model. As a rough rule of thumb, about 20% of a model's lines should be comments.

  The Average Order Rate is a standard smoothing equation, with the smoothing done over 4 weeks. The)↑(means multiplication and has the same effect as)*(. AOR is initialized to the external driving force so as to set the model into balance before it is shocked by the STEP or the sine inputs. Not all system dynamics packages will allow a variable to be used to initialize a level in this way, but COSMIC will repeat the rate equation for NO, so as to create the correct initial value for AOR. If your package will not do this, the correct initial value is 100 [WMS/WEEK], as that is the value of BLEV. That is, you will have to type in n aor=100.

```
note
note    actual and desired order backlog
note
l blog.k=blog.j+dt*(no.jk−pr.jk)
n blog=dblog
```

The influence diagram shows that the actual Backlog is fed by New Orders and depleted by Production Rate. The positive and negative signs in the influence diagram's portrayal of Backlog are mirrored exactly by the + and − signs for NO and PR. It is a good method of modelling to check off the links on the diagram as they are written into equations, making sure that the correct signs are used.

The influence diagram also shows two influences on BLOG, and the equation for BLOG has two terms within the ( ). Checks such as this, elementary and obvious though they might seem, are essential to a good model, and it is surprising how often they are not carried out.

Like all levels, BLOG requires an initial value. Rather than simply setting it to be a number it is always worth considering what the system is trying to do, where that can be worked out. In this instance, the production manager tries to keep Backlog to its desired level, DBLOG. It is, therefore, an interesting test to set Backlog initially to that target and see how well the system manages to achieve its objective. As mentioned above, COSMIC will repeat the auxiliary equation for DBLOG so as to initialize BLOG. If your package will not accept this, the correct numerical value is 600 [WMS].

```
a dblog.k=wao*aor.k
c wao=6
```

The desired backlog is the Average Order Rate multiplied by the number of weeks after which the customers will require their order to be delivered.

```
note
note    current and average production rates
note
a eps1.k=blog.k−dblog.k
r pr.kl=eps1.k/tabl+aor.k
```

```
c tabl=4
l apr.k=apr.j+(dt/tapr)(pr.jk−apr.j)
n apr=no
c tapr=4
```

The value of EPS1 is fed into the Production Rate equation to give the system dynamics format for the algebraic equation discussed on p. 25, with the addition of AOR.K, as was explained in the narrative for this problem in Fig. 2.4. The production manager's choice of policy is to correct the backlog error, EPS1, over a period of 4 weeks.

It is essential to see that the choice of PR is a model of the production manager setting the general level of activity in the factory, not of the manager dealing with individual orders of specific types of washing machines for particular customers. Although system dynamics *can* be used to model the detail, it is more useful to take this broader view of the information/action/consequences paradigm in order to get close to the *policies* for running the business as opposed to the *procedures* of its daily life. In short, one gives the production manager credit for knowing how do the job in detail and concentrates on the overall strategic aspects.

Average Production Rate is Production Rate smoothed over 4 weeks. The initial value for APR is set to the external driving force to make sure that the model is properly balanced.

```
note
note    actual and desired raw material stocks
note
l rms.k=rms.j+dt*(rmar.jk−pr.jk)
n rms=drms
a drms.k=wap*apr.k
c wap=8
note
note    ordering of raw materials
note
a eps2.k=drms.k−rms.k
r rmor.kl=eps2.k/tarms+apr.k
c tarms=4
```

The raw materials manager's sectors of the model are derived on exactly the same principles as the production manager's, which were discussed above. You are *most strongly urged* to cover up the raw materials manager's equations given above and write them again for yourself using the influence diagram and the production equations as a guide. No amount of studying models will enable one to build them; the only way is to do it oneself.

```
note
note    delayed inflow of materials
note
r rmar.kl=delay3(rmor.jk,ddel)
c ddel=6
```

This uses a standard DELAY3 to represent the inflow of raw materials in response to earlier decisions to order them.

```
note
note output and controls
note
c dt=0.25
c pltper=1
c prtper=10
c length=120
```

This section sets the parameters for controlling the simulation, the main points being the choice of DT and of LENGTH. The former is calculated by noticing that there are four first-order delay parameters, TAOR, TABL, TAPR and TARMS, all of which have the value 4. The formula given on p. 109 would require DT=1 for these parameters. However, DDEL features in a third-order delay and has the value 6. That would require DT to be no larger than 0.5, so we choose 0.25 to be on the safe side. LENGTH is chosen to allow for just over 4 cycles from the sine wave, which should be enough to show any characteristic behaviour. The parameters PRTPER and PLTPER are the DYNAMO tradition for controlling printed and plotted output, though different methods are used in other packages.

```
print 1)no,aor
print 2)pr,apr
print 3)blog,dblog,eps1
print 4)drms,rms,eps2
print 5)rmor,rmar
plot no=a,pr=b,rmor=c(0,250)
plot blog=a,dblog=b,rms=c,drms=d(0,2000)
```

The first five of these statements produce printed tables of *all* the model variables. There is absolutely no substitute for printing out tables of the variables, *and actually looking at the results*, if one is to debug a model properly.

The last two lines plot graphs of a selection of variables, using the COS-MIC techniques for flat-bed plotter output in order to produce the graphs which will be seen in Fig. 6.3.

```
note
note simulation experiments
note
```

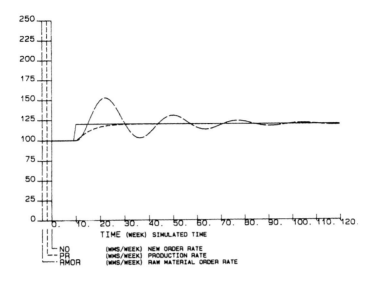

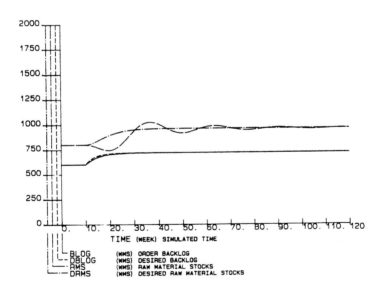

A - STEP RESPONSE OF MODEL

Fig. 6.3 The domestic manufacturing company.

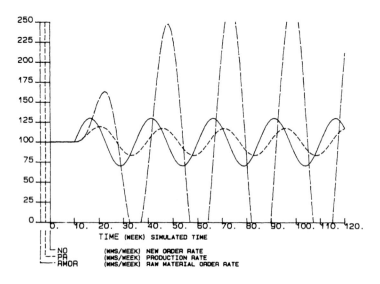

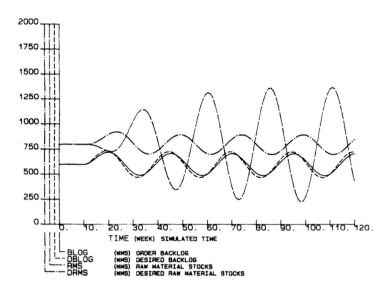

B - RESPONSE TO SINE WAVE

Fig. 6.3 *Continued*

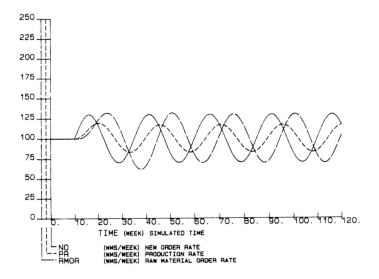

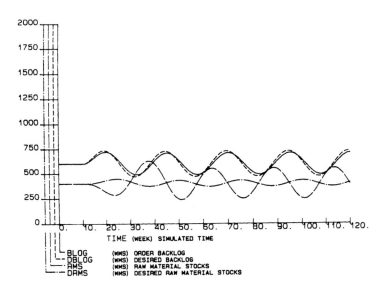

C - LOWER STOCKS AND MORE SMOOTHING

Fig. 6.3 *Continued*

run A – Step Response of Model
c d1=0
run B – Response to Sine Wave
c wap=4
c tarms=8
c tapr=8
run C – Lower Stocks and More Smoothing

Finally, we have three experiments (which could be performed interactively) which will show first the STEP response, secondly the reaction to a sine wave and, thirdly a 'what would happen if' experiment in which the company tries to hold lower stocks of four weeks average production and two of the policy reactions are slower.

note
note variable definitions
note
d amp=(wms/week) amplitude of sine wave in new orders
d aor=(wms/week) average order rate
d apr=(wms/week) average production rate
d blev=(wms/week) base level for order patterns
d blog=(wms) order backlog
d dblog=(wms) desired backlog
d d1=(1) demand pattern switch
d ddel=(week) delivery delay on materials
d drms=(wms) desired raw material stocks
d dt=(week) solution interval
d eps1=(wms) difference between backlog and desired backlog
d eps2=(wms) difference between desired raw material stock and
x   actual raw material stock
d length=(week) simulated duration
d no=(wms/week) new order rate
d perd=(week) period of sine waves on new order pattern
d pltper=(week) plotting interval
d pr=(wms/week) production rate
d prtper=(week) printing interval
d rmar=(wms/week) raw material arrival rate
d rmor=(wms/week) raw material order rate
d rms=(wms) raw material stocks
d tabl=(week) backlog adjustment time
d taor=(week) order averaging time
d tapr=(week) production rate averaging time
d tarms=(week) adjustment time on raw material stocks
d time=(week) simulated time
d wao=(week) weeks of average orders in desired backlog

d wap=(week) weeks of average production required in raw material
x   stocks

The actual format will depend on the package, but the principle of writing definitions that are as clear as possible for *every* variable and parameter is a sound one. A model which contains undefined variables is useless, and one which contains variables simply called 'X' is not much better. Some modellers prefer to put the definitions immediately after the first use of the variable to which they refer, but the author's practice is to group them in alphabetical order at the end of the model, especially with a model of any appreciable size.

Model validation

So much for developing the model's equations. But how do we know they are right? A number of tests can be applied and, the more tests are passed, the more one's confidence increases that the model is technically correct in terms of the rules of system dynamics equations, and that it corresponds to the real system well enough to be used to make policy recommendations to its managers.

On pp. 96–8 we suggested some tests of confidence in a model, and this issue is so important that it is worth restating them so that the reader can apply them to this problem.

- The influence diagram should correspond to the statement of the problem. We have already established that it does.
- The equations should correspond to the influence diagram. The reader must check that they do. Part of this test is, of course, to ensure that there are no errors of syntax, such as missing parentheses, variables which are used without having been defined or variables which are defined or documented without being used. The reader will have to run the model to verify that.
- The model must be checked for dimensional validity, either by hand or by software.
- The tests for mass balances were deliberately not included in this model to encourage you to write the extra equations yourself, not neglecting the amount of raw material in the delay contents between RMOR and RMAR.
- The tests that the model should not produce any ridiculous values and that its behaviour should be reasonable are seen in the three pages of Fig. 6.3.

Figure 6.3A[2] shows the step response, and all the plotted variables seem reasonable. That is not a sufficient test, as not all the variables are plotted,

---

[2] Recall the explanation on p. 95 for the conventions for reading these graphs. The style of graphical output varies between packages; this is COSMIC's.

and in a model even a little larger than this it would be impractical to do so. One has, therefore, to look at the printed tables. None of the printed values is implausible. However, intuitively, one would not expect to see negative values in this physical system, but EPS1 and EPS2 do sometimes go negative. Why is that acceptable?

The dynamics of the step response cannot be compared with the real system as it never experiences such a simple shock. Even so, the dynamics show some important results which we have to treat as interim results until we have tested the model against a more realistic pattern. The first is the extreme sluggishness of the response. Even by TIME=100, nearly two years after the step, the oscillations are only just dying away. The second is that the production variables, PR and BLOG, respond reasonably smoothly to the step and BLOG and DBLOG match very closely. This is consistent with the company's experience that the production department copes reasonably well with external shocks. On the other hand, the raw materials variables show a very pronounced reaction to the step, consistent with the experience that they do not cope very well. Furthermore, the time from one peak of the oscillations in RMOR and RMS to the next is about 25 weeks. This is known as the **natural period** of a system. In this case, the natural period is the same as the period of the cycles in demand, and it is usually a recipe for extreme instability to drive a system at its natural period. The film of the breakup of the Tacoma Narrows bridge is a dramatic example of that phenomenon.

The sine wave response in Fig. 6.3B is even more worrying, but, paradoxically, also more convincing.

The worry is that the response of RMOR in the top half of the graph is explosive, to the point where, at TIME=32, the graph goes off the page, thereafter going further and further apart. Similarly, the graph of RMS also shows the explosive oscillation, which was predictable from the natural period deduced from the STEP input results. RMS and DRMS only match eight times in two years, and are completely out of phase. For the production department, on the other hand, the graph of PR runs fairly smoothly through the middle of the sine wave of NO, and BLOG and DBLOG match very closely.

The paradoxical conviction is that the model is broadly on the right lines because this is *roughly* how the real system behaves. The production people can cope with the demand variations, but those in raw materials have a very hard time, which is what the model suggests. On the other hand, the explosive behaviour will eventually lead to such outrageous consequences as *negative* raw material stocks, so we are faced with the problem that, though the model may be on the right lines, it is clearly unrealistic because a real system cannot produce negative stocks.

This dilemma means that, on the face of it, we ought to make the model much more complicated to prevent these unrealistic phenomena from arising. On the other hand, we could adopt the viewpoint that the unrealistic behaviour arises because of poorly designed policies in the system and that the measures which the people in the real system would have to take to prevent negative stocks arising would be no more than desperate attempts to cope with catastrophe. It therefore seems better to spend our efforts in designing the catastrophic behaviour out of the system rather than trying to model it.

This is a fairly fundamental point about the attitude to 'validation'. A valid model is one which is, as was stated on p. 12,

well suited to its purpose and soundly constructed

We have satisfied ourselves that the model is technically sound by making sure that its equations agree with the influence diagram, that they are dimensionally valid and that the model's behaviour is, in important respects, 'like' that of the real system. Its suitability for its purpose is largely a matter of its simplicity, clarity, acceptability to the client *and* whether it contains the policy options which will lead to improvement. In other words, *does the model contain the solution to the problem?*

Testing the model

We can easily test this by the 'what would happen if' experiment which was set up within the model and which is shown in Fig. 6.3C. This shows a system which is well under control and suggests that the explosive behaviour arises from the exaggerated response which the original policies made to the unpredictable external driving force.

Although Fig. 6.3C is a considerable improvement over Fig. 6.3B, the behaviour it shows is very far from perfect. The oscillations in Raw Material Order Rate are as large as those in New Orders and much larger than those in Production Rate. In the lower half of the graph, the match between BLOG and DBLOG is less close than in Fig. 6.3B, and actual and desired Raw Material Stock still match only four times a year. RMS is worryingly small at some stages of the run. There is scope for improvement, which we shall study in Chapter 7, but the model does contain solutions to the problem of bringing this system under control in the face of unpredictable external demand.

For the present, we shall leave the Domestic Manufacturing Company, which is on the model disk as FIG6-3.COS, though we shall use it again in Chapter 7 for policy design and in Chapter 9 for optimization.

THE MANAGEMENT CONSULTANT'S PROBLEM

The influence diagram for the consultancy company's problem discussed in Chapter 3 is reproduced for convenience in Fig. 6.4. The problem requires

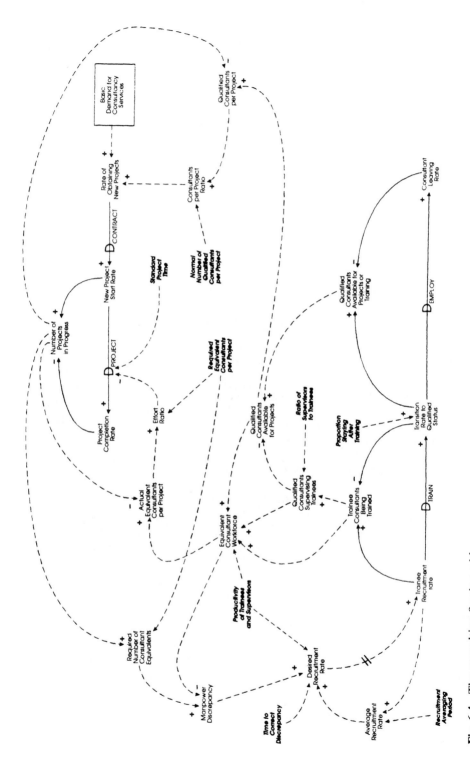

Fig. 6.4 The consulting firm's problem.

some new modelling techniques and has some very interesting behaviour. Again, the equations are shown in groups, each of which is followed by an explanation. The reader should study the group before reading the explanation.

The management problem

What, however, is the problem? The narrative on p. 49 mentioned that the firm felt that it was falling behind its market potential and that it never seemed to have the right numbers of trainees and consultants. This is rather vague, but it is typical of practical problems in that the problem proprietors often find it hard to articulate their difficulties. It sometimes turns out that they are worrying unnecessarily. We shall see what happens when we run the model.

The model equations

The reader should first study Fig. 6.4 and decide on variable names, work out the dimensions of each variable and assign types to the variables. The variable names used below are mnemonics based on the initial letters of the names used in the diagram. There are four levels and six rates, but it is *vital* that you try to work these out for yourself *before* looking at the solution. Nothing is learned by studying someone else's solutions; you have to make your own attempt and work out where, if at all, you were wrong.

The equations are discussed below, with the definitions given at the end of the explanation. The model name is FIG6.6.COS because Fig. 6.5 will be introduced before the model's graphs are shown.

```
* Figure 6.6 The Management Consultant Problem
note
note    file named fig6-6.cos
note
note
note    demand and business generation sector
note    ===================
note
note    basic demand
note
a bdc.k=base+swdp*(ramp(slp,slptim))+amp*
x sin(6.283*time.k/bcylen))
c base=50
c amp=10
c swdp=0
c slp=0.1
```

```
c  bcylen=48
c  slptim=10
```

After the usual opening lines note the way in which the model has comments for its major sectors and subsectors to improve clarity of explanation. BDC is modelled to offer a choice between constant demand and a rising oscillatory trend. The former is artificial, but is useful for debugging and testing, as the behaviour is not clouded by the sine wave effects.

It is always important to be aware of any hidden assumptions in a set of equations. In this case it is that BDC represents the demand for *this firm's* services; its share of the total market. As we shall see below, the firm may get more or less than its normal share, depending on the service it offers to clients, so the actual Rate of Obtaining New Projects may be greater or less than BDC.

Most consulting firms operate on a basis of charging for the time spent on a project at a rate which covers all costs and a profit margin. The fact that BDC exists suggests that clients are content with this arrangement, so money is, if not irrelevant, at least peripheral to this problem.

```
note
note    effect of qualified consultant availability
note    on rate of obtaining new projects
note
a  qcpp.k=qcafp.k/nproj.k
a  cpprat.k=qcpp.k/nnqcpp
c  nnqcpp=1.82
```

These variables and equations are simply those in the upper right-hand corner of Fig. 6.4. The sector is broken into several auxiliaries to make writing and debugging easier. Because the names state pretty clearly what the variables mean, the equation writing can flow easily. For example, Qualified Consultants per Project can only be the number of qualified consultants not supervising trainees divided by the number of projects. Careful attention to variable names will considerably assist modelling, partly for the reasons just given but also because clear names make it easier to write the correct dimensions and to get the confidence which comes from the model being known to be dimensionally valid, because the dimensional validity can more easily be checked by hand or by software.

```
a  cpmult.k=tabhl (ctab, cpprat.k, 0, 2, 0.5)
t  ctab=0/0.3/1/1.5/1.7
```

These equations represent the functioning of the link from the Consultants per Project Ratio to Rate of Obtaining New Projects in the top right-hand corner of Fig. 6.4. The rise and fall of CPPRAT represents the success of the firm in providing the right number of qualified staff to supervise the projects on hand, and this, as the problem description makes clear, affects the firm's ability to generate new business. As will be seen below, the multiplier, CPMULT, operates on the Basic Demand for Consultancy Services to help the firm to win or lose market share. This new variable, CPPMULT is not shown on the influence diagram and this is a typical case of drawing a diagram at something less than Level 4 so as to convey the right information to the client without confusing the issue with extraneous detail. For practice, the reader should alter Fig. 6.4 to show this new variable, the solution being shown in Fig. 6.5.

The TABHL function is used in preference to TABLE as there is no reason why CPPRAT should not exceed 2. The numerical values in the non-linear table CTAB are found by studying previous sales data and relating them to data on numbers of staff available, by asking experienced managers their views and by common sense. It is, for example, clear that no work will be obtained if there are no consultants, accounting for the 0 in the first element of CTAB. By definition, when CPPRAT=1 the situation is normal, so the third entry in CTAB is 1.

The important point is that the influence diagram's portrayal of the mechanisms of obtaining business has been agreed by the company as reasonable and it now forms a theory on which data analysis can be based. It is not a matter of collecting some data and seeing what theories are suggested; it is a question of having a theory and analysing data with its assistance.

Note also that NNQCPP is correctly defined in the D statements below as [PEOPLE/PROJECTS]. It is *very* common to see numbers carelessly called 'ratio' and given dimensions of [1] and that is usually a sign that the model's dimensions have not been properly checked. The ratio CPPRAT is, however, correctly dimensioned as [1].

It would also be easy to make the typing error

a cpprat.k=qcpp.k*nnqcpp

which is clearly wrong, because CPPRAT should be a ratio but [PEOPLE/ PROJECTS]*[PEOPLE/PROJECTS] is not [1]. Such errors can be found by dimensional analysis, but only if dimensional checking is actually done.

```
note
note    projects sector
note    ========
note
note    projects being obtained and started
note
r ronp.kl=bdc.k*cpmult.k
r npsr.kl=delay3(ronp.jk,cdel)
c cdel=2
```

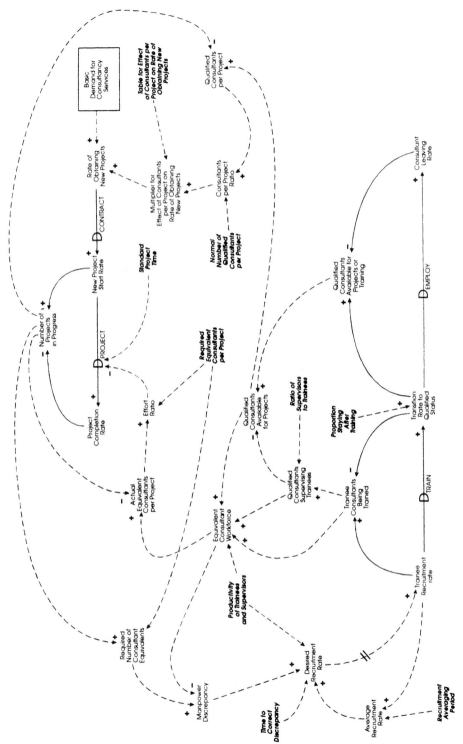

Fig. 6.5 The consulting firm's problem (modified).

The projects sector is a straightforward flow of projects starting after a delay for contracts.

    l netdef.k=netdef.j+dt*(bdc.j−ronp.jk)
    n netdef=0

A new variable is introduced to measure the cumulative difference between projects available, BDC, and those obtained, RONP, as a measure of the extent to which the firm is, or is not, falling behind its market potential. A positive value of NETDEF means that the company is not meeting all its demand, while a negative value would mean that market share is being created. NETDEF is not shown on the influence diagram as it is not part of the system. It is, in fact, an additional variable introduced to help the analyst to gauge system performance.

Note that this involves feeding an auxiliary variable, BDC, into a level, whereas so far we have seen only rates feeding levels and, indeed, Fig. 4.1 implies that auxiliaries *cannot* feed levels. This is an example of how the flexibility of *most* of the system dynamics packages allows one to bend the strict rules to simplify modelling provided *one understands the rules and knows when and why they can be broken.* Using the flexibility of the software to fiddle one's way out of a difficulty is very bad practice. In this case, the alternative would seem to be to define BDC as a rate, but that involves the difficulty that rates *must* be reserved for things which actually flow into or out of physical levels in the real system. NETDEF is not such a level; it is an artefact created by the modeller to help measure system performance. In technical terms it is a simple example of the **objective functions** which were mentioned in Figs. 1.3 and 1.4 and which we shall encounter in more detail in Chapters 8 and 9.

In the equation for NETDEF a typing error such as

    l netdef.k=netdef.j+dt*(bdc.j+ronp.jk)

could not be detected by dimensional analysis, but only by running the model and carefully examining why NETDEF always turned out to be very strongly positive. Dimensional analysis is a necessary but not sufficient condition that a model is correct, and there is no substitute for carefully studying the output and asking oneself whether this is the same thing that the real system would have done, *and for the same reasons.*

    note
    note   projects in progress
    note
    l nproj.k=nproj.j+dt*(npsr.jk−pcr.jk)
    n nproj=242

A simple level for projects. The initial value of 242 is derived from company data, as are the other numerical values.

Case studies in modelling

note
note   project completions
note
r pcr.kl=nproj.k/dproj.k
a dproj.k=spt*effrat.k
a effrat.k=recpp/acepp.k
a acepp.k=ecw.k/nproj.k
c spt=5
c potas=0.5
c recpp=2

A deceptively simple sector in which $D_{PROJECT}$ is modelled as dependent on the variables identified in the influence diagram, though the numerical value for RECPP was not mentioned in the narrative. The reader should work through the equations from DPROJ to ACEPP, checking against the influence diagram and understanding how they are derived. The definitions and dimensions appear below. After working through them they will look like very simple common sense and we have already remarked that disciplined common sense is very helpful in doing system dynamics. Most of the technique *is* very simple, because unless we can work in simple steps we shall have little practical chance of modelling the complexities of managed systems.

The key feature of this sector is that PCR is modelled as a *first-order* delay of PSR, as can be seen by comparing the equations for NPROJ and PCR with Fig. 4.8(a). It is important to remember that the D on a delayed link in an influence diagram does *not* state what type the delay is, it merely states that there *is* a delay. Why, though, did we use DELAY3 for the connection between RONP and PSR while here we have not even used DELAY1 but have explicitly written the equations for a first-order delay? There are two reasons.

The first is that DELAY3 is standard in system dynamics and requires only one equation. The second is that we do not need the contents of the delay between RONP and PSR, but we do need the contents between PSR and PCR, because they drive several other factors in the model. Modelling the contents, NPROJ, requires an initial value, and that value would have to be exactly consistent with the three internal levels of DELAY3, had we used that. Getting initial conditions to balance in such equations is very difficult, so it is easier to do it in an explicit first-order delay.

The general rule is that, where the contents of a delay affect its input or its output, the delay equations have to be written explicitly in order to get the initial conditions to balance correctly.

Some packages will allow the contents of delay functions to be available, but it is better to make things explicit. In any case, as we saw in Fig. 5.2, there is not very much difference between the behaviours of the different delay types when there is a continuous input to the delay, which is why we did not add complication to the model by writing the explicit equations for a third-order delay. System dynamics concerns itself with overall behaviour under the influence of policies, not with an illusory attempt to get very precise values when the data are usually fairly imprecise in any case.

```
note
note    trainee consultants sector
note    ===============
note
note    desired trainee recruitment rate
note
a ecw.k=qcafp.k+potas*(tcbt.k+qcst.k)
a dtcrr.k=(mandis.k/potas)/rat+arr.k
a mandis.k=rnce.k-ecw.k
a rnce.k=nproj.k*recpp
c rat=10
```

A set of auxiliaries corresponding closely to the influence diagram. Again, the reader should work through them.

```
note
note    non-negativity constraint on recruitment
note
r trr.kl=max(dtcrr.k,0)
note
note    average recruitment rate
note
l arr.k=arr.j+(dt/rat)(trr.jk-arr.j)
n arr=4.0
```

The non-negativity constraint is modelled as was explained on p. 123. The reader should review that section to recall the difference between the two types of non-negativity constraint. There is a simple smoothing equation for ARR.

```
note
note    trainee consultants being trained
note
l tcbt.k=tcbt.j+dt*(trr.jk-tcr.jk)
n tcbt=96
```

A simple level for TCBT.

```
note
note    trainees becoming qualified
```

```
note
r tcr.kl=tcbt.k/dtrain
r trqs.kl=tcr.kl*psat
c dtrain=24
c psat=0.9
```

TCR is modelled as a first-order delay, for the reasons explained above. The equation for TRQS is another example of bending the rules of Fig. 4.1, in this case to allow a rate to affect a rate. TCR is a physical flow from TCBT and TRQS is also a physical flow into QCAPT. Multiplying by PSAT simply recognizes that one is not the same as the other.

```
note
note   qualified consultants sector
note   ================
note
note   number of people
note
l qcapt.k=qcapt.j+dt*(trqs.jk−clr.jk)
n qcapt=441
note
note   consultant leaving rate
note
r clr.kl=qcapt.k/demploy
c demploy=120
```

This group of equations is easily followed by analogy with the trainees. In effect, the trainees' sector and the consultants' are identical modules, and recognizing such cases is a useful skill as it speeds up modelling, provided one is very careful not to jump to conclusions and too readily assume that modules are identical without thinking about it. Some packages support vectors and arrays, though these features should be used only after careful thought about the problem.

```
note
note   consultant division of labour
note
a qcst.k=tcbt.k*ratstt
c ratstt=0.1
a qcafp.k=qcapt.k−qcst.k
```

This is simple arithmetic to subtract from QCAPT the small number of consultants who are supervisors. With 96 trainees, about 10 consultants will be required for supervision, which is so few out of 440 consultants that this aspect of the narrative is probably irrelevant and might, on the face of it, be dropped from the model. The reason why it cannot be dropped is that the partner in charge of training is a very powerful and senior figure in the firm.

It is highly unlikely that he would take the slightest notice of the model if his interests were not represented in it. Such political aspects often play a role in practical modelling. Indeed, it is sometimes a fine judgement to decide between including detail for political reasons and making the model so complicated that it ceases to be well suited to its purpose of policy evaluation and design.

```
note
note   control and output
note   ===========
note
c dt=0.125
c prtper=5
c pltper=1
c length=120
print 1)bdc,ronp,netdef,npsr,pcr
print 2)nproj,qcpp,cpprat,cpmult,dproj
print 3)qcapt,qcafp,qcst,clr,acepp
print 4)dtcrr,trr,arr,tcbt,trqs
print 5)effrat,ecw,rnce,mandis
plot bdc=a(0,120)/nproj=b(0,400)/qcapt=c,
x tcbt=d(0,600)/netdef=e(0,300)
```

DT is governed by the shortest delay in the model, CDEL, which is 2 months long and third-order. That suggests that DT cannot be greater than 1/6, to which 0.125 is the next lower acceptable value. The value of LENGTH is set to the employment period of consultants, which is the longest delay in the model, and should allow all the interesting dynamics to be seen. (Experiment with LENGTH=240.) *All* the variables are printed, for debugging, and a selection are plotted. The commands used in the PLOT command are special to COSMIC, and your package may be different. The scale for NETDEF is suitable for the two cases tested here, but, in general, we have no idea what its magnitude will be, or even whether it will be positive or negative.

```
note
note   simulation experiments
note
run A Constant Demand
c swdp=1
run B Demand Growth and Business Cycle
```

Two experiments to create Fig. 6.6.

```
note
note   definitions of variables
note
```

d acepp=(people/projects) actual consultant equivalents per project
d amp=(projects/month) amplitude in the business cycle
d arr=(people/month) average recruitment rate of trainees
d base=(projects/month) base level of demand for company's services
d bcylen=(month) length of the business cycle
d bdc=(projects/month) basic demand for consultancy
d cdel=(month) delay in signing consultancy contract
d clr=(people/month) consultant leaving rate
d cpmult=(1) multiplier reflecting the effect of consultant availability
x   on the rate at which new projects are obtained
d cpprat=(1) ratio of ability of consultants to do projects to number
x   of projects on hand
d ctab=(1) table of values for cpmult
d demploy=(month) consultants' average employment time before
x   leaving the company
d dproj=(month) actual project completion time
d dt=(month) solution interval
d dtcrr=(people/month) desired trainee consultant recruitment rate
d dtrain=(month) training delay
d ecw=(people) equivalent consultant workforce
d effrat=(1) ratio of required and actual equivalent consultants per
x   project
d length=(month) simulated duration
d mandis=(people) manpower discrepancy
d netdef=(projects) net deficit between BDC and RONP
d nnqcpp=(people/projects) normal number of qualified consultants
x   per project
d nproj=(projects) current number of projects
d npsr=(projects/month) new project start rate
d potas=(1) productivity of trainees and supervisors
d pcr=(projects/month) project completion rate
d pltper=(month) plotting interval
d prtper=(month) printing interval
d psat=(1) proportion of trainees who stay with the company on
x   completion of training
d qcafp=(people) qualified consultants available for work on projects
d qcapt=(people) number of qualified consultants available for
x   projects or training
d qcpp=(people/projects) qualified consultants per project
d qcst=(people) qualified consultants supervising trainees
d rat=(month) recruitment adjustment time
d ratstt=(1) ratio of consultants to trainees
d recpp=(people/projects) required equivalent consultants per
x   project for normal project duration

d rnce=(people) required number of equivalent consultants
d ronp=(projects/month) rate of obtaining new projects
d slp=(projects/month) slope for growth in market
d slptim=(month) time at which slope is allowed to start
d spt=(month) standard project time
d swdp=(1) switch to turn demand pattern on and off
d tcbt=(people) trainee consultants being trained
d tcr=(people/month) rate of completion of training
d trqs=(people/month) transition rate of trainees to qualified status
d time=(month) simulated time
d trr=(people/month) trainee recruitment rate

As with the previous model, the definitions are in alphabetical order at the end of the model.

Model validation and testing

The behaviour of the model with constant demand and with a rising sine wave is shown in Fig. 6.6. With constant demand (Fig. 6.6A), the company declines very slowly in terms of numbers of consultants and trainees. The number of projects not obtained, NETDEF, reaches about 200 over 10 years, though that is a small proportion of the 6000 projects which might have been obtained. On the whole, the model is broadly consistent with the reference mode of the behaviour of the firm as described in the narrative.

When the sine wave is implemented (Fig. 6.6B), to simulate a slowly rising market with an oscillation due to the business cycle, the behaviour is more complicated. The number of consultants, QCAPT, oscillates gently, though the peaks and troughs are almost exactly out of phase with those of BDC. Since there are relatively large numbers of consultants during the troughs of the cycle, there is a surge of new projects at the beginning of the upswing of the cycle, for example, from month 40 to month 50. As the cycle reaches a peak around TIME=60, the Rate of Obtaining New Projects ceases to rise sharply and there is a slow decline in the rate of obtaining work until the next upswing of the cycle. The numbers of trainees rises sharply just as the cycle passes its peak. NETDEF, the projects lost, oscillates quite severely, with numbers of projects lost during the peak and downswing of the cycle (from TIME=0 to TIME=25, for example). In the early stages of the upswing, market share is built and NETDEF falls to a minimum at about TIME=48. At the end of 10 years, NETDEF is 289, though that can only be found from the printout.

Overall, the simulation corresponds to the rather vague worries that the company does not have the right numbers of people at the right times. What can be done?

It is quite obvious that the company needs to find some way of increasing

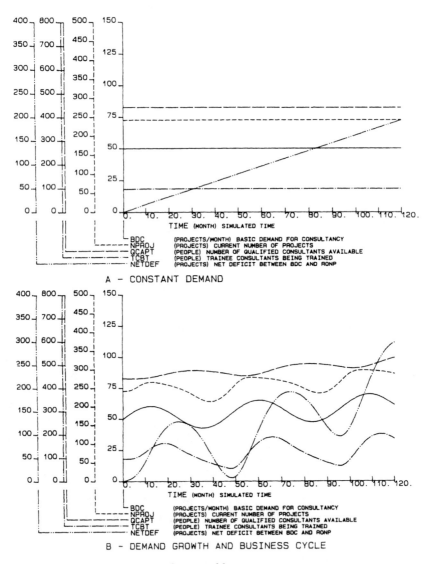

Fig. 6.6   The management consultant problem.

the productivity of trainees and consultants, and the effects of doing so are
shown in Fig. 6.7A in which POTAS=0.7. The striking thing is that
NETDEF *increases*, the value at the end of 10 years actually being 649! The
company is doing worse with more productive trainees. One of the standard
system dynamics tests is to subject the model to extreme, even ridiculous,
values to see if its behaviour is correct for those conditions. Figure 6.7B
shows the effects of lower productivity (POTAS=0.3) of consultants and
trainees, the result being *better* in that NETDEF becomes negative, reaching −379 by the end of the run. A negative deficit means that the company

is building market share. It is contrary to intuition that improving productivity makes matters worse and that reducing productivity makes things better, hence the caption 'Counter-intuitive Consultants' on Fig. 6.7.

A further insight into this strange behaviour is given by the second part of Fig. 6.7[3] on p. 171 which shows comparative plots for the total number of qualified consultants, QCAPT, and for the net deficit, NETDEF, against the three cases of productivity. The salient feature is that lower productivity leads to better market share and to more consultants, whereas higher productivity does the opposite. A little thought shows the reasons. With lower productivity, more trainees are recruited who, after about two years, become qualified. Since it is qualified consultants not supervising trainees who generate business, it is not surprising that more business is generated.

The reason for the effect is the non-linearity in CTAB, and we have no really solid data on CTAB. One might worry, therefore, that the results are not 'correct', since CTAB is not known accurately. While it is certainly true that changing the values of CTAB would change the detailed results, it would not change the *type* of result as long as the *shape* of CTAB is correct.

No-one in their right mind is going to set out to *reduce* productivity; the implications of these experiments are that one needs to be quite careful about increasing it unless the recruitment policy is also reconsidered. The real implication is that it is consultants who matter and that ways of increasing the effect of their numbers need to be found.

We referred to 'pet theories' in Fig. 1.3 and someone in the firm argues that 'we need to get rid of some of the old stagers and bring in some new blood', so the reader should experiment by reducing D_{EMPLOY}. Another theory is that there are better ways of managing projects and of looking after clients, both of which would have the effect of reducing NNQCPP from its current value of about 1.8.

The reader should revisit this problem later in the book; for the present we shall turn to the combat model, which also has some strange behaviour and illustrates some more of the features of system dynamics modelling.

THE SIMPLE COMBAT MODEL

In Chapter 3 we studied a simple model of two forces in combat. Although military modelling with system dynamics has been a highly successful application area, the reader may not feel particularly interested in the topic. Nevertheless, this model has many interesting features of general applicability, so the effort of studying it will be worthwhile. The influence diagram is reproduced in Fig. 6.8.

---

[3] The second part of Fig. 6.7 is produced by model FIG6-7M.COS on the disk, which has been set up to produce the comparative plots on the same page, purely to make the COSMIC plotter match the layout of the book.

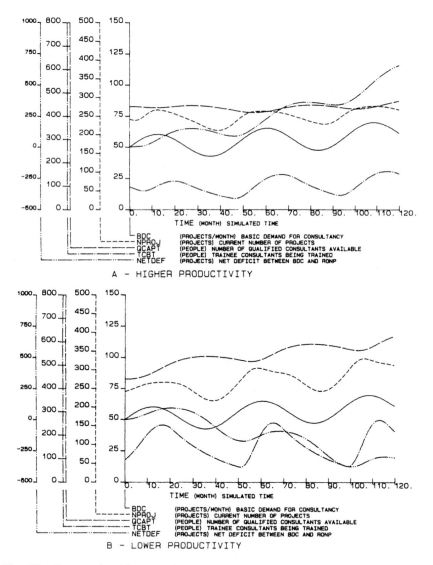

Fig. 6.7 Counter-intuitive consultants.

The defence planning problem

The purpose of the model is to help Blue identify the 'pressure points'
where investments in assets or changes in the commander's intentions for
fighting the battle might affect Blue's chances of winning and thereby, one
hopes, deter Red from attacking. In practice, Blue's government will have
a limited amount of money to spend on defence, so it will be interesting
to see where money can be spent to have the greatest effect on Blue's

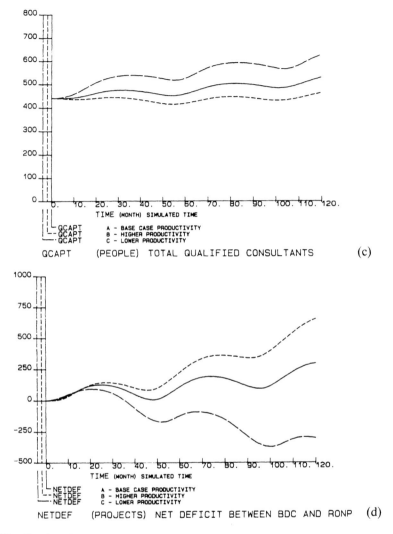

Fig. 6.7 *Continued*

defence capabilities. We shall test this below and again when we study optimization.

The model equations

As usual, the equations are shown in groups, each followed by a brief explanation. However, only some of the groups are explained, and the reader is invited to write the rest. The definitions appear at the end of the equations and the reader should note the distinction made between, say,

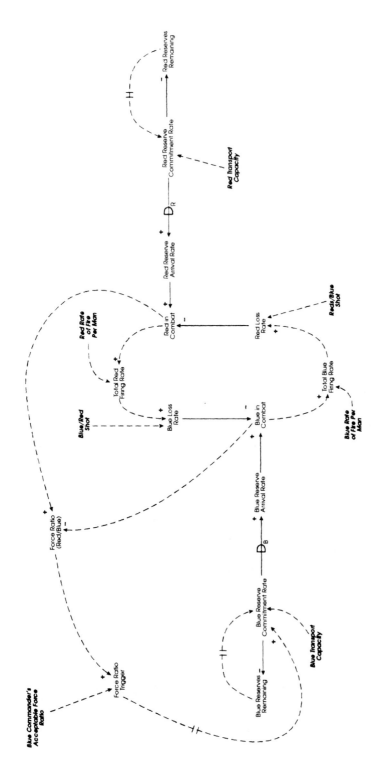

Fig. 6.8 The simple combat model.

Red men and Blue men. This is done to make dimensional checking as rigorous as possible. The full model appears on the disk as FIG6-9.COS.

```
* Figure 6.9 The Simple Combat Model
note
note    file named fig6-9.cos
note

note    start of war
note    = = = = = =
note
a wswitch.k=step(1,5)
```

It is often good practice in system dynamics models to prevent anything happening in the first few time periods. This helps to ensure that the dynamics subsequently produced are genuinely those of the model's relationships and not due to faulty initial conditions. In this case, WSWITCH allows the battle to start when TIME=5.

```
note
note    blue force sector
note    = = = = = = = = =
note
note    blue forces in combat
note
l blue.k=blue.j+dt*(brar.jk−blr.jk)
n blue=iblue
c iblue=10000
```

Blue forces in combat are a simple level. The initial value is set to a constant to allow different balances of force distribution between front line and reserves to be tested easily.

```
note
note    blue losses
note
r blr.kl=trfr.k*bprs
```

This is what is shown in the influence diagram, Fig. 6.8. Note that Total Red Firing Rate, TRFR, is modelled as an auxiliary, not the rate which one would expect. This is another case of the flexibility of the software allowing one to have two options. Since there is, in the simple model, no stock of Red ammunition to be depleted, there is no need for TRFR to be a flow rate. When the stock is modelled, TRFR could, and should, be a rate, and the equation would be

```
r blr.kl=trfr.kl*bprs
```

which again breaks the strict rules of the connections between levels, rates and auxiliaries which were shown in Fig. 4.1. It is important to grasp that, though the rules *can* be broken, one should only do so after careful thought!

```
note
note   blue reinforcement arrivals
note
r brar.kl=delay3(brcr.jk, bmdel)
c bmdel=6
note
```

A simple flow from Reserve Commitments.

```
note   blue reinforcement commitments
note
r brcr.kl=min(blres.k/dt,bmcap)*frtrig.k
a frtrig.k=clip(1,0,frat.k,bcafr)*wswitch.k
c bcafr=2.0
c bmcap=4000
```

Blue Reserve Commitment Rate depends on the Blue commander wanting to commit reserves, which only happens when Force Ratio, FRAT, exceeds a desired level, thereby making the Reserve Commitment Trigger change to 1. Note the use of the non-negativity constraint, BLRES.K/DT. Note also the use of auxiliaries to break the equations into simple components which are easier to check than one complicated equation for BRCR. The use of WSWITCH prevents Blue from initiating the provocative act of committing reserves before the war has started, even if his front line troops are outnumbered.

```
note
note   blue reserves
note
l blres.k=blres.j+dt*(−brcr.jk)
n blres=iblres
c iblres=20000
```

A simple drain of Blue reserves as they are committed.

```
note
note   blue firing rate
note
a tbfr.k=blue.k*brfpm*wswitch.k
c brfpm=150
c rpbs=0.001
```

The total rate at which the Blue force fires depends on how many troops there are and how fast each man fires, once the war has started.

```
note
note   force ratio
```

```
note   ======
note
a frat.k=red.k/(blue.k+beps)
c beps=0.001
```

The equation for Force Ratio illustrates the need for protection against division by 0 if BLUE.K becomes 0. The use of a small number, such as BEPS, is a useful trick.

```
note
note    red force sector
note   ========
note
```

The reader should now write the equations for the Red force, recalling that Red's 'policy' is simply to commit reserves as fast he can, once the war has started. Use the following numerical values.

```
c ired=15000
c rmdel=6
c rmcap=2500
c irres=50000
c rrfpm=150
c bprs=.00033
```

```
note
note    output and control
note   ==========
note
c pltper=0.25
c prtper=10
c length=60
c dt=0.25
print 1)blue,blres
print 2)brcr,brar
print 3)blr,rlr
print 4)red,rres
print 5)rrcr,rrar
print 6)frtrig
plot blue=a,red=b(0,60000)
plot blres=a,rres=b(0,60000)/frtrig=c(0,2)
```

Although the *format* is used in some of the system dynamics packages, but not all, the *principle* of printing *all* the variables so that the model can be checked is universal. The plotting interval, PLTPER, is set to be the same as DT to give the smoothest possible curves in Fig. 6.9. The scale for FRTRIG is set to force it to be in the middle of the page, where it will be easier to see.

```
note
note    simulation experiments
note    = = = = = = = = = = = =
note
run Base Case
```

A simple test to see how well the model works.

```
note
note    definitions of variables
note    = = = = = = = = = = = =
note
d bcafr=(red/blue) blue commander's acceptable force ratio – the
x                       force ratio at which he commits reserves
d beps=(blue) small number of blue to prevent division by zero
d blr=(blue/hour) blue loss rate
d blres=(blue) blue reserves remaining
d blue=(blue) blue forces in contact with red
d bmcap=(blue/hour) blue movement capacity
d bmdel=(hour) delay in before blue reserves committed to combat
x                   become effective
d bprs=(blue/rshot) blue killed per red shot – the red standard of
x                       marksmanship
d brar=(blue/hour) rate at which blue reserves enter battle
d brcr=(blue/hour) rate at which blue reserves are committed to battle
d brfpm=(bshot/hour/blue) blue rate of fire per man
d dt=(hour) simulation time interval
d frat=(red/blue) force ratio of red: blue
d frtrig=(1) switch to trigger commitment of blue reserves
d iblres=(blue) initial blue reserves
d iblue=(blue) initial blue force in contact
d ired=(red) initial red force in contact
d irres=(red) initial red reserves
d length=(hour) simulated period
d pltper=(hour) plotting interval
d prtper=(hour) printing interval
d red=(red) red forces in contact with blue
d rlr=(red/hour) red loss rate
d rmcap=(red/hour) red movement capacity
d rmdel=(hour) delay before red reserves committed to combat
x                   become effective
d rpbs=(red/bshot) red killed per blue shot – the blue standard of
x                       marksmanship
d rrar=(red/hour) red reserve arrival rate
d rrcr=(red/hour) rate at which red reserves are committed to combat
```

d rrcsw=(1) switch to control commitment of red reserves
d rres=(red) red reserves remaining
d rrfpm=(rshot/hour/red) red rate of fire per man
d tbfr=(bshot/hour) total blue firing rate
d time=(hour) simulated time
d trfr=(rshot/hour) total red firing rate
d wswitch=(1) switch to denote onset of combat

Testing the model

The behaviour in the base case is shown in Fig. 6.9, in which the scale values for RED, BLUE, BLRES and RRES are to be multiplied by 1000 to get the actual values. The plots show that something is very seriously wrong with this model[4]. In the upper graph, the error is that BLUE reaches about −100000 and RED goes to about 160000, despite having started with only 65000 in total.

The behaviour in the lower graph is a little more plausible. The war starts at TIME=5, at which point Red starts to commit reserves as fast as he can. Since he has 50000 reserves and a transport capacity of 2500 men/hour, it should take 20 hours to commit all reserves, and it does. Similarly, it is also correct that Blue commits all his reserves in 5 hours. Both sets of reserves stop at 0, so the non-negativity constraint is working. The dynamics of FRTRIG are strange. Blue starts to commit reserves at TIME=12 and continues to want to do so after TIME=16, even though there are no more to commit. At about TIME=34, Blue suddenly stops wanting to commit reserves, which is odd.

The reader *must* pause and try to work out what is wrong before going any further. To do so, run the model, look at the output, and think.

The error is that we allow Red and Blue to continue inflicting casualties, even when all the opponents have been wiped out. In other words, there is no non-negativity constraint on Blue Loss Rate, or on Red's, since there is a possibility that Blue might be the victor.

The corrected model is FIG6-10.COS, the behaviour of which is shown in Fig. 6.10. Alter the model, and the diagrams in Figs. 6.8, 3.8 and 3.14, to get this behaviour.

Figure 6.10's dynamics are much more plausible. Blue starts to suffer casualties when the war starts and, by TIME=12 is outnumbered by 2:1. At that point, the Blue commander starts to commit reserves and the curve for BLUE starts to rise a few hours later. Unfortunately, there are insufficient reserves to turn the tide, and Blue is wiped out by about TIME=33. Red's remaining reserves continue to arrive on the field, and RED ends

---

[4] Congratulations to the reader who spotted the deliberate error in the model and didn't get the behaviour shown in Fig. 6.9!

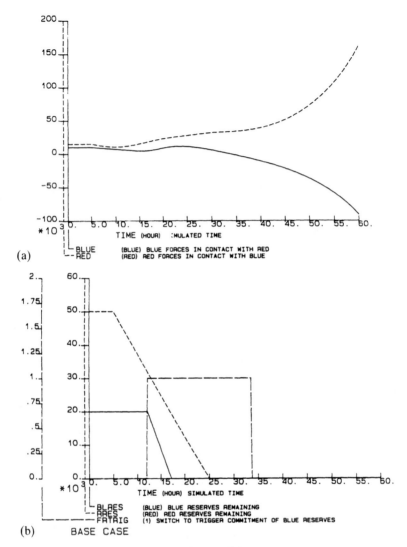

Fig. 6.9 Output from the simple combat model.

the battle with a total strength of about 33 000 men. The lower graph shows the previous behaviour for RRES and BLRES, but FRTRIG, having switched on at TIME=12, stays on, showing the Blue commander's wish to commit reserves continues even after the battle is lost, which is plausible.

Unfortunately, *even though the graphs look right, the model still contains errors.* Run the model, looking at the tables and, perhaps, plotting a few more variables to see if you can detect the error. Plotting all the variables is usually just confusing: the tables of output are more informative.

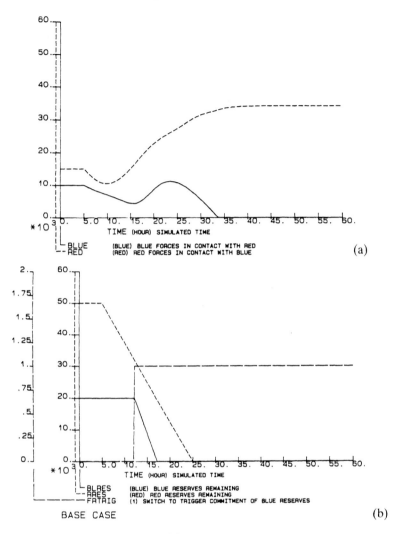

Fig. 6.10 The corrected combat model.

The error is that Red still continues to fire even after Blue has been completely wiped out. In other words, TRFR does not go to zero when BLUE does. In a sense, this does not matter, as neither Red's nor Blue's ammunition stocks feature in the model. It is, however, a logical error and one can have little confidence in a model which contains errors, let alone looking rather foolish and destroying the client's confidence if he or she happens to question that aspect of the model. FIG6-10M.COS on the disk contains the final corrected model.

We shall study this model in more detail in Chapters 7 and 9, but a few final comments are called for.

In the first place, this model differs from the two previous examples in that it has no external driving force. The first two models are *driven*, but this one is *closed* and its dynamics arise entirely from its initial conditions and structure.

Secondly, although non-negativity constraints need to be added to the Blue and Red Loss Rates in Fig. 6.8 they are not needed for Blue Ammunition Reserve Remaining, Blue Ammunition Stocks, Engineer Supplies Remaining and Long Range Ammunition Stocks in Fig. 3.14[5]. For the first of those, the problem statement implied that they were so large that they could never be used up. In fact, they are probably irrelevant to the problem and could be omitted. If they are retained, it would be as well to add a constraint, partly for consistency but also to be on the safe side. The other three all work by affecting the rate at which they are used; for instance Blue Ammunition Stocks influence the Blue Rate of Fire per Man and, when stocks reach zero, consumption automatically ceases. It would not be wrong to add the constraint, but it would be an unnecessary complication, which is bad practice. A model should be kept as simple as possible.

THE PREDATOR/PREY MODEL

Predator/prey systems, such as that shown in Fig. 3.5, and reproduced in Fig. 6.11, present some interesting modelling problems. They are usually 'closed' or undriven models in which the dynamics arise entirely from the structure, initial conditions, and parameter values (usually including sharp non-linearities) and there is no external driving force, as was the case for the washing machine company and the management consultants. They also differ from the closed combat model examined earlier in that the dynamics cannot be prevented from taking place in the early stages of the simulation, as the war switch allowed us to do. That means that it is difficult to check for false dynamics due to faulty initial conditions or errors in the equations. To add to all that there are usually no 'managers' to tell us what the system is supposed to achieve.

The purpose of the model

The purpose of the model is to represent the dynamics of predator/prey behaviour and to study how those dynamics change with uncertainties in the data. When the system has been well understood, the purpose might develop into understanding how the system might be managed by human intervention. For instance, foxes might have to be culled and wild rabbits are a delicious food.

---

[5] The point is not only to do with military modelling; it could arise in any other type of problem.

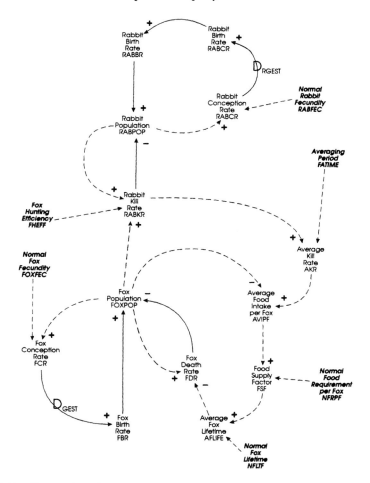

Fig. 6.11 The predator/prey system.

Developing the equations

You will probably be able to write most of the equations for yourself if you have assiduously studied the previous examples. It is suggested that you try to do so, concentrating on the equations and leaving the numerical values blank for the moment.

The following suggested solution contains roughly plausible numerical values which produce some interesting dynamics. After studying the solution and its behaviour the interested reader may care to try to find some more 'accurate' data to see what effect they have.

```
* Figure 6.12 The Predator/Prey Model
note
note file named fig6-12.cos
note
note    rabbit population sector
note    ==============
note
note    rabbit population
note
l rabpop.k=rabpop.j+dt*(rabbr.jk−rabkr.jk)
n rabpop=inrab
c inrab=300
note
note    rabbit reproduction
note
r rabcr.kl=rabfec*rabpop.k
c rabfec=0.1
r rabbr.kl=delay3(rabcr.kl,drgest)
c drgest=4
note
note    fox population sector
note    ===========
note
note    fox population
note
l foxpop.k=foxpop.j+dt*(fbr.jk−fdr.jk)
n foxpop=ifox
c ifox=20
note
note    fox reproduction
note
r fcr.kl=foxfec*foxpop.k
c foxfec=0.05
r fbr.kl=delay3(fcr.jk,dfgest)
c dfgest=7
```

The equations so far are probably self-evident, though the reader will wish
to check them.

```
note
note    fox predation of rabbits
note
r rabkr.kl=fheff*rabpop.k*foxpop.k
c fheff=0.005
```

The Rabbit Kill Rate is modelled by equations which are also used to model
the spread of an infectious disease from the infected to the uninfected parts

of the population. The parameter for Fox Hunting Efficiency, FHEFF, has the dimensions [[1/week]/fox] which is the fraction of the Rabbit Population which a fox can kill in a week.

The underlying idea is that R rabbits occupy an area of X square miles. A fox can search Y square miles per week so the number of rabbits it encounters per week will be $R \times Y/X$. If the conditional probability of a rabbit being killed, given that it has been encountered, is P, then the weekly rate of rabbit kills per fox will be $P \times (Y/X) \times R$. The Fox Hunting Efficiency, FHEFF, is $P \times (Y/X)$. In other words, FHEFF is the fraction of the rabbit population which a fox will encounter in a week of hunting, multiplied by the fraction of those encounters which end in a fox killing a rabbit.

```
note
note    fox food supply
note
l akr.k=akr.j+(dt/fatime)*(rabkr.jk−akr.j)
n akr=rabkr
c fatime=2
a avipf.k=akr.k/foxpop.k
```

The Rabbit Kill Rate is smoothed over a time constant which represents the period for which a fox can go hungry before hunger starts to have an effect on its lifespan.

```
a rlife.k=rabpop.k/(akr.k+reps)
c reps=0.00001
```

It will be useful to have an indicator of the average lifespan of a rabbit. One could devise a second smoothing time to calculate that, but we use FATIME for simplicity. REPS is introduced to prevent division by zero if, for any reason, AKR reaches zero.

Although most of the software packages would accept it, it would be logically incorrect to write the equation as

```
a rlife.k=rabpop.k/(rabkr.KL+reps)
```

RABKR.**KL** is the kill rate which is about to occur in the impending DT, so it is wrong to calculate RLIFE.**K**, something which exists now, by using something which has not yet happened, which is what the .KL postscript represents.

RABKR.**JK** cannot be used, because the time-shifted window has moved to TIME.K, so the .JK values no longer exist. In any case, RABKR.KL may vary pretty sharply and it would be better to get a smoother view of RLIFE.

```
note
note    food supply effect on fox lifetime
note
a fratio.k=avipf.k/nfrpf
c nfrpf=5
```

```
a fsrf.k=tabhl(tfsrf,fratio.k,0,1.0,0.2)
t tfsrf=0.0/0.024/0.12/0.8/0.98/1
```

This is the formulation discussed in Chapter 5.

```
note
note    fox death rate
note
r fdr.kl=foxpop.k/(dlife.k+minlt)
c minlt=1
a dlife.k=nfltf*fsrf.k
c nfltf=120
```

Finally, Fox Death Rate is modelled using a first-order delay with fox lifetime dependent on the food supply. It would have been easy, though somewhat more laborious, to have written explicitly the equations for a third-order delay (recall Fig. 4.8), which would have allowed us to distinguish between immature, mature and old foxes, probably with different tables to reflect the possibility of old and immature foxes dying of hunger more easily than mature ones. Whether that would add much to the credibility of the model is a matter of judgement. It would certainly impose greater demands for numerical data.

```
note
note    control and output sector
note    ===============
note
c dt=0.125
c prtper=5
c pltper=1
c length=300
print 1)rabpop,rabbr
print 2)foxpop,fcr,fbr,fdr,
print 3)rabkr,akr,rlife
print 4)fratio,fsrf,dlife
plot foxpop=a(0,40)/rabpop=b(0,800)/fdr=c(0,4)
plot dlife=a(0,120)/rlife=b(0,40)
```

Choosing a value for DT requires us to evaluate two candidates. The third-order delay for Rabbit Gestation is 4 weeks long, which would require DT<4/12. Fox Lifetime is a first-order delay with a minimum value of 1 if there is no food at all for foxes. That would require DT<1/4. To be on the safe side, we put DT=0.125, half the value required by MINLT.

```
note
note    simulation experiments
note    =============
note
```

run A – Trial of Model
c inrab=600
run B – Larger Initial Number of Rabbits

Two experiments, to test the effects of uncertainty about the numbers of rabbits in the area.

note
note   variable definitions
note   = = = = = = = = =
note
d akr=(rabbits/week) average kill rate of rabbits
d avipf=((rabbits/week)/fox) average food intake per fox
d dfgest=(week) average length of fox pregnancy
d drgest=(week) average length of rabbit pregnancy
d dlife=(week) average fox lifetime
d dt=(week) solution interval
d fatime=(week) averaging time for fox food intake
d fbr=(fox/week) fox birth rate
d fcr=(fox/week) fox conception rate
d fdr=(fox/week) fox death rate
d fheff=((1/week)/fox) fox hunting efficiency – the proportion of the
x   rabbit population that a fox will kill during
x   a week
d foxfec=(1/week) fox conceived per fox per week
d foxpop=(fox) fox population
d fratio=(1) food intake of foxes relative to normal
d fsrf=(1) reduction factor on fox life span due to relative availability
x   of food
d ifox=(fox) initial number of foxes
d inrab=(rabbits) initial rabbit population
d length=(week) simulated duration
d minlt=(week) minimum lifespan of foxes
d nfltf=(week) normal lifespan of a fox when food supply is normal
d nfrpf=((rabbits/week)/fox) food intake per fox for normal life span
d pltper=(week) plotting interval
d prtper=(week) printing interval
d rabbr=(rabbits/week) rabbit birth rate
d rabcr=(rabbits/week) rabbit conception rate
d rabfec=(1/week) rabbits conceived per rabbit per week
d rabkr=(rabbits/week) rabbit kill rate
d rabpop=(rabbits) rabbit population
d reps=(rabbits/week) small number to prevent division by zero
d rlife=(week) average lifetime for a rabbit
d tfsrf=(1) table for food supply reduction factor
d time=(week) time within simulation

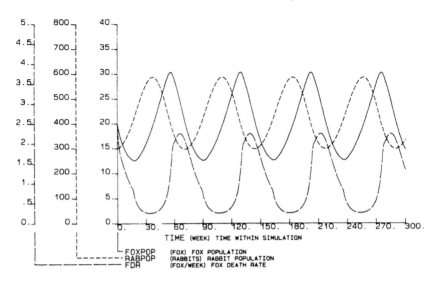

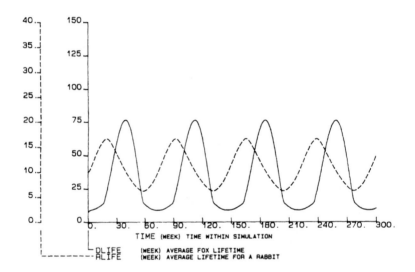

A - TRIAL OF MODEL

Fig. 6.12 The predator/prey model.

Testing the model

The model is on the disk as FIG6-12.COS, and the behaviour for two different initial values of Rabbit Population is shown in Fig. 6.12. The behaviour is characteristic of predator/prey models in that the two

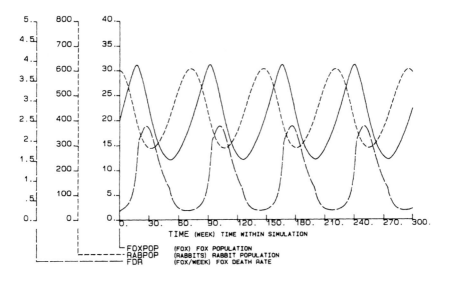

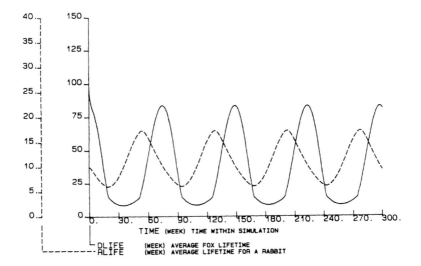

B - LARGER INITIAL NUMBER OF RABBITS

Fig. 6.12 *Continued*

populations show sustained oscillations. It must be emphasized that the
oscillations arise entirely from within the model and are not due to outside
influences.

In the first case, the upper half of the graph suggests that the 300 rabbits initially present are not sufficient to support a population of 20 foxes, so the Fox Death Rate is relatively high. The fall in Fox Population allows rabbits to live long enough to breed prolifically and Rabbit Population rises. Naturally, the increased availability of food causes a rise in Fox Population, and so on. The period of the oscillation is about 70 weeks and the slight irregularities in the curve for Fox Death Rate are due to the changes in slope in the piece-wise linear approximation in the TABLE function.

The lower half shows that the average lifespan of both species fluctuates considerably, but at the times which one would expect from the rise and fall of the respective populations. The lifespans at the troughs are rather short, probably less than the time required to become capable of breeding. At these points, rates of conception are low, so the error is not large. It would be easy to complicate the model to prevent breeding from happening when the average lifespan is less than a certain value.

With 600 rabbits, the pattern is similar, but in reverse. The model starts with ample fox food, so Fox Population rises, after which the same pattern of oscillation sets in, with the same period. The contrast between Figs. 6.12A and 6.12B suggests that there might be some initial number of rabbits which would stabilize the system. We shall examine that topic in a moment, but first we must test the model in another way.

The value of DT was set to 0.125, but, if we run the model with 300 rabbits and successively set DT to 0.25, 0.5 and 1, we obtain the comparative results shown in Fig. 6.13 (FIG6-13.COS on the disk). The behaviour patterns with DT=0.125 and DT=0.25 are not appreciably different, though there is a slight indication that the oscillations become larger with DT=0.25. With the two larger values of DT, the oscillations certainly become greater, to the point where RABPOP goes off the graph scale. The period of the oscillation also lengthens.

Clearly, the value of DT has an effect on the results, and this arises from the assumption, discussed in Chapter 4, that DT is so small that the rates can be assumed to be constant during DT. This is called **Eulerian integration**. Some of the system dynamics packages use other integration methods, but, in all cases and for all models, it is essential to double and halve the value of DT initially selected[6] to see if it makes any noticeable difference to the results from the model. The reason for this is that DT is chosen, by the modeller, to get the equations which purport to represent the system to run on a computer. It has nothing to do with the real system, so its value should not significantly affect the behaviour of the real system as simulated by the model. If DT has to be reduced to an absurdly low value in order to get invariant behaviour there is something seriously wrong with the model.

---

[6] Recall that the acceptable values of DT are 0.5, 0.25, 0.125 etc. There are the decimal equivalents of exact binary fractions and are used in order to avoid rounding error when functions such as PULSE and STEP are activated by comparison with the imbuilt TIME variable.

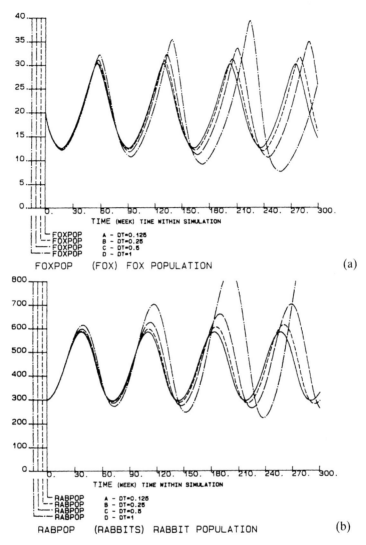

Fig. 6.13   DT in the predator/prey model.

The pattern of oscillation in Fig. 6.12 implies that there might be population levels which make the system stable. A little experimentation shows
that, with 411 rabbits and 20 foxes initially, the populations do stabilize,
though with rather short lifespans. This is shown in Fig. 6.14, the model
being FIG6.14.COS on the disk. With 300 rabbits, there is no value of fox
population which stabilizes the system. One might conclude that fox culling
will not stabilize the system and that the correct policy is to introduce
enough *rabbits* to bring the population up to 411 at a time when there are

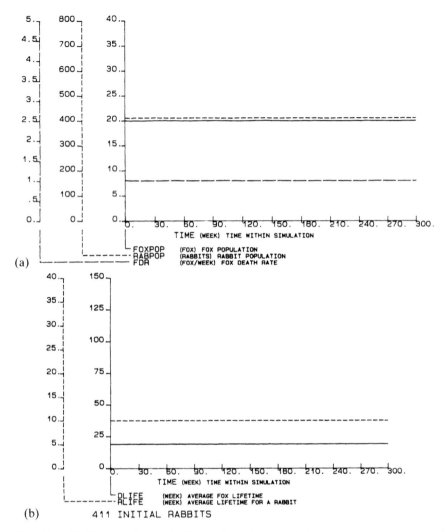

Fig. 6.14 Stable populations in the predator/prey model.

20 foxes present. This is an example of the way in which system dynamics models generate novel insights into problems.

The dynamics of a non-linear system are heavily dependent on its initial conditions. We saw that the behaviour when INRAB=300 is rather different from that with INRAB=600. When INRAB is set to 30000, the difference in behaviour becomes very dramatic, as seen in Fig. 6.15 (FIG6-15.COS will produce that illustration). The cycles are still present, but the period has extended to about 150 weeks and the peaks of the cycles

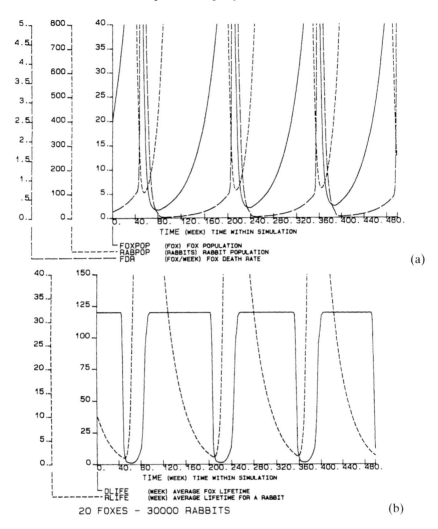

Fig. 6.15   Larger initial numbers in the predator/prey model.

steadily decay. If the model is run for 5000 days the cycles do reach approximate stability.

The behaviour is so strange that a little further investigation is called for. It turns out that RABKR reaches a maximum value of 3616 rabbits per week, which means that the average fox is killing 180 rabbits per week, or 26 times the food intake it requires for normal life. It is possible that the model should be modified to prevent a fox killing more than, say, 10 rabbits per week. What would the effect of that be with this large initial population of rabbits?

Other applications of this model

The techniques used here, particularly the modelling of Rabbit Kill Rate as dependent on the interaction of both populations, are quite widely applicable. A model of this general type might, for instance, be used to study the practicalities of reindeer harvesting when there is also a population of wolves to manage. The same general approach can be used in modelling the spread of a disease. Similar ideas are used in some types of combat modelling, such as air defence and submarine warfare.

THE RULES FOR TIME POSTSCRIPTS FOR VARIABLES

In Chapter 4, the discussion of the fundamental equations of system dynamics stated some rules for the postscripts which are permissible on the right-hand side of an equation for a given variable type on the left-hand side. In essence, these are:

L ALEV.K=ALEV.J+DT* (RATEs.**JK**)

L SLEV.K=SLEV.J+(DT/TCON) (RATE.**JK**−SLEV.J)

R RATE.KL=f (LEVs.**K**,AUXs.**K**,PARAMETERS)

R RATE1.KL=DELAYn (RATE2.**JK**,DEL)

A AUX1.K=f (LEVs.**K**,AUX2s.**K**,PARAMETERS)

There are two possibilities for a level: ALEV is a pure accumulation *of* a physical flow, while SLEV is the smoothing of information *about* a physical flow. Similarly, there are two possibilities for a RATE, depending on whether or not there is a delay in the equation.

The lower case 's' shows that there may be more than one of the indicated variable, and numbers, such as AUX1 and AUX2s, suggest that AUX1 can depend on a number of other auxiliaries, though not, of course, on itself. The postscripts in **bold** are those which Fig. 4.1 and the fundamental equations of system dynamics permit on the right-hand side. Strictly, neither levels nor auxiliaries can appear on the right-hand side of a level equation.

In practice, however, and depending on the software package, these strict rules can usually be broken where to do so makes it easier to write equations. We have seen some examples of this in the models studied in this chapter. Figure 6.16 shows the postscripts which will be accepted by the software, those in **bold** being the standard cases, the others involving bending the strict rules.

The rules for a level must be carefully thought about. If the level is a pure accumulation, then it is very inadvisable to use anything other than rate variables within the parentheses on the right-hand side. The level should be an accumulation of some physical flow, and rates are the variables of physical flow. The software might well accept a level or an auxiliary, but to write anything other than a rate may mean that the modeller is seriously confused about how the physics of the system works and is incorrectly creating a memory of the physical consequences in it. That becomes even clearer when we recall the rules for the dimensions of levels and rates given on p. 88.

If, on the other hand, the level is the smoothing type which creates a memory of information it is perfectly permissible to have a level or an auxiliary as the variable being smoothed.

In the manufacturing problem, we used Average Order Rate to represent the smoothing of the physical inflow of orders. It would, however, have been perfectly logical to model the smoothed (averaged) backlog or discrepancy using equations such as:

L SBLOG.K=SBLOG.J+(DT/TCON) (BLOG.J−SBLOG.J)

and

L SEPS1.K=SEPS1.J+(DT/TCON) (EPS1.J−SEPS1.J)

Equations such as these could be written, for example, as

L SEPS1.K=SEPS1.J+DT*RCSEPS1.JK
R RCSEPS1.KL=(EPS1.K−SEPS1.K)/TCON

with RCSEPS1 representing the rate of change of SEPS1. This would make a level, SEPS1, depend on a rate, RCSEPS1, which, in turn, depends on a level, SEPS1, and an auxiliary, EPS1. That is formally consistent with the

	Variable on the Right is		
Variable on the Left is	Level	Rate	Auxiliary
Level (accumulation)	J	**JK**	J
Level (smoothing)	**J**	**JK**	**J**
Rate (no delay)	**K**	KL	**K**
Rate (delay)	J	**JK**	J
Auxiliary	**K**	KL	**K**

For explanation see text.

Fig. 6.16 The rules for time postscripts.

fundamental equations, but is clearly an unnecessary fuss. The strict rule is too restrictive for practical use.

For flow rates, we used a rate-dependent rate to model the transition of trainees to the state of qualified consultant by writing:

R TRQS.KL=TCR.**KL***PSAT

Again, a rate-dependent rate is not, strictly, allowed by the fundamental equations. In practice, TCR is a rate dependent on levels and it would be carrying adherence to strict rules to silly extremes to waste effort by writing the equation for TCR on the right-hand side of the equation for TRQS and then multiplying it by PSAT. The equation above is perfectly clear and only cosmetically violates the strict rules. The main thing in writing equations in which a rate depends on a rate is to be sure that one is genuinely taking a legitimate short cut and not using the flexibility of the software to fudge a real difficulty.

Where a delay is involved, it would, however, be *very* undesirable to have LEV.J or AUX.J within the parentheses on the right-hand side. Again, the reason is that a delay is a physical process, while levels represent accumulations and auxiliaries are usually parts of the fine structure of decision processes within the system. It is hard, therefore, to see these as logically feeding directly into a physical flow. Again, correct application of dimensional analysis will usually be a great help in avoiding such errors.

In short, most of the system dynamics packages permit flexibility in writing equations, but the privilege should not be abused without careful thought.

SUMMARY

This chapter has shown, in careful steps, the building of four models. In each case, they were tested in various ways, and we shall use some of them for policy analysis and optimization in later chapters. You should study these examples carefully and further extend that knowledge by building for yourself some of the problem models given in Appendix B.

CHAPTER 7

Policy experiments with system dynamics models

INTRODUCTION

In Chapter 1, we discussed the idea that the main aim of system dynamics is to develop policies which improve the dynamic behaviour of a system. This is obviously the intention when the clients are the management of a business firm, but it is also the objective if the 'client' is the academic audience for a book, as was the case for the decline of the Maya, studied in Chapter 3. At first sight, it might seem that the purpose of such a model would be simply to understand why the system had behaved in a certain way, but, inevitably, the researcher becomes interested in how the collapse might have been avoided had the Maya rulers acted differently, and that is, by definition, policy analysis.

In Chapter 6, we developed models of four problems and showed that, when the models were tested, their modes of behaviour could be changed quite noticeably. The behaviour was not necessarily *better* than the 'base case', but it was *different*, and knowing that behaviour can be changed lends encouragement to the idea that we might be able to *design* the best possible robust behaviour into the system in question.

This chapter will apply that idea to some of the models from Chapter 6. The aim is *not* to imply that the solutions to be developed are in any sense 'ideal' or representative of good practice. The intention is simply to show how models can be experimented with to test policy effects and to indicate the wide range of behaviour which is usually found. At the end of the chapter, you should have developed confidence in working with models and have had your imagination stimulated with ideas about what can be done with them. In Chapter 9, we shall take this idea of design much further when we study the use of optimization for policy design in system dynamics models.

POLICY IN SYSTEM DYNAMICS MODELS

Before we can say much about policy analysis, we should be clear about what 'policy' means in system dynamics modelling and how policies are represented. Since we shall use the Domestic Manufacturing Company, DMC, for our first policy studies later in this chapter, let us also use it as the basis for discussing this point.

The Raw Materials Manager orders raw materials on the basis of:

r rmor.kl=eps2.k/tarms+apr.k

The equation says that the Raw Materials Manager has a policy of ordering raw materials so as, firstly, to keep actual raw material stock up to a target level, correcting any discrepancies over a period of TARMS weeks, and, secondly, of replacing the consumption of raw materials in the production process.

There is no law which states that this *must* be his policy: it is simply that which he chooses to use, probably having inherited it from his predecessor.

An alternative policy would be to order raw materials so as to ensure that Backlog could be brought down to its desired level within a reasonable time and to match the current level of Orders. That policy would require the different equation:

r rmor.kl=**eps1.k**/tarms+**aor.k**

These two equations would require two different influence diagrams. Links would, as it were, have to be switched on and off to depict the two possibilities. When we study policy in the DMC problem, we shall introduce a notation for doing so.

There are, however, *very many* other possibilities, and there is no way of knowing in advance which will give the best performance for the business as a whole. The only way forward is to experiment with different policies with the intention of designing that which is most satisfactory.

It is, however, critical to understand that *policy has two components*. The first is its **structure**; the form of the equation and the corresponding set of links on the influence diagram. The second is the **parameters** within a given structure; the numerical value of TARMS might have to be changed considerably in order to get the best results from each of the two options suggested above.

EXPERIMENTATION VERSUS DESIGN

In the last section we referred to 'experimenting' with models and to 'policy design', and it is necessary to understand the important distinction between the two. We shall use 'analysis' to encompass both 'experimentation' and 'design'.

System dynamics models often have many parameters. As we shall see, the DMC model has seven in its basic form, and it will have many more by the time we have finished with it. When we introduce structural variations into the model, the number of parameters increases yet again, and the number of combinations of those parameters is very large, even though this is a fairly small model. A large model will have so many parameters and structural possibilities that the number of combinations becomes enormous.

With such large numbers of options, coherent analysis becomes very difficult, so policy analysis usually proceeds in two phases, as was shown in Fig. 1.3. In that diagram, stage 5A of the system dynamics approach involves experimenting with the model to see what happens when at least some of the parameters and structures are changed. Parameter changes are usually quite large, such as halving or doubling existing values. The purpose is to deepen understanding about why the model behaves as it does, to discover the range of behaviour of which it is capable and to stimulate imagination about the structural changes which might be possible. With that deeper understanding, analysis can move on, in stage 5B, to use optimization methods, which, guided by measures of what the system is supposed to be achieving, search the parameter and structural possibilities much more thoroughly than one could hope to do by 'suck it and see' experimentation. That is the real design phase of the system dynamics method[1]. This chapter concentrates on policy experimentation; Chapter 9 will deal with optimization.

BUSINESS PROCESS RE-ENGINEERING

The process of policy analysis in system dynamics models involves, as we shall see in the DMC model, analysing flows of information in the model to see how the actions thus produced generate consequences in the behaviour of the system. The information/action/consequences paradigm discussed in Chapter 1 is brought into play as an analytical concept as well as the basis on which models can be constructed and tested. This process of experimenting with and designing the flows to produce good dynamic behaviour is not new, but is the essence of what has recently become popular as business process re-engineering.

THE DOMESTIC MANUFACTURING COMPANY ⚡

The problems of the domestic manufacturing company (DMC) were encountered earlier and this will be the first case to be studied. Although the model is fairly simple, it contains the seeds of so many solutions to DMC's

---

[1] The author realizes that some of his system dynamics colleagues, whose views he respects, will differ from this interpretation. They will argue that the feedback loops give all the guidance one needs to design good behaviour into a system. The difficulty is that a model of even moderate size can contain many thousands of feedback loops.

problems that it is a good training ground in letting the imagination loose in a disciplined way to explore a system.

For convenience, the influence diagram from Fig. 2.7 is redrawn in Fig. 7.1. The layout of the diagram has been changed somewhat to allow space for additional influence links which will be introduced as we study experimental possibilities. Figure 7.2 reproduces, for ease of reference, the behaviour for the base case and with smoother responses, which was shown in Figs. 6.3B and C. The extremely unstable behaviour mode in the base case is credible enough for our purposes, and we have already decided that it would be a waste of effort to complicate the model to represent catastrophic behaviour more realistically. The model is on the disk as FIG7-2.COS.

The Raw Materials Manager's pet theory

The Raw Materials Manager is heavily criticized within the firm for the extreme volatility in Raw Material Stocks, and by DMC's suppliers for the great variations in Raw Material Order Rate. His defence is that the large number of different raw materials used in making washing machines, and the inadequate clerical and computer support which he has, make it impossible to keep even reasonable track of what has been ordered but not yet received. It is, therefore, impossible for him to avoid the over-ordering and excessive cutbacks which underlie the extreme swings in the raw material variables. In short, his view is that the company's troubles are due to its failings, not to his incompetence. His solution, or pet theory, is that he should be allowed to invest a considerable sum of money in acquiring, and subsequently operating, a Management Information System (MIS) to allow him to control the quantity of raw materials in the delivery delay pipeline.

Our first task is to find out whether or not it is true that the Raw Materials Manager's 'pet theory' will bring the system under control. Testing the effects of putative information systems in the inexpensive world of the model is a most fruitful area of system dynamics application. Of course, we have already seen in Fig. 7.2B that there is a solution to the problem of achieving stability which involves no more than changing the policies which control the system, and that has the attraction that it would cost nothing to implement. If the MIS also achieves stability, the question will be whether it is 'better' than the policy solution and, if so, whether the improvement is worth the cost. There might, of course, be other free solutions which are better still, and we shall search for them when we have tested the MIS.

The first stage of testing this pet theory is to change the influence diagram to represent the new system. This requires us to model the contents of the pipeline between RMOR and RMAR. This is the actual Raw Material on

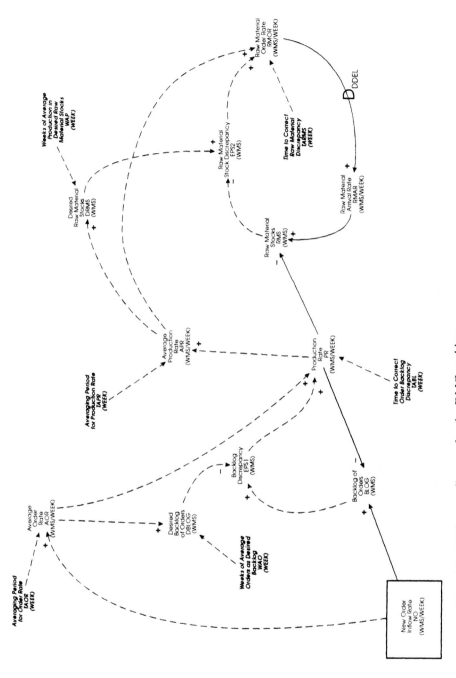

Fig. 7.1 The original influence diagram for the DMC problem.

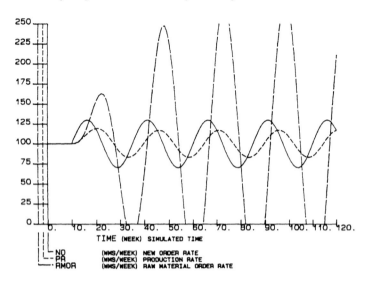

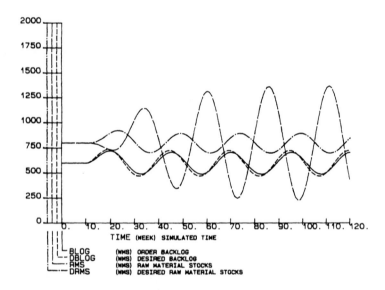

A - RESPONSE TO SINE WAVE

Fig. 7.2   The domestic manufacturing company.

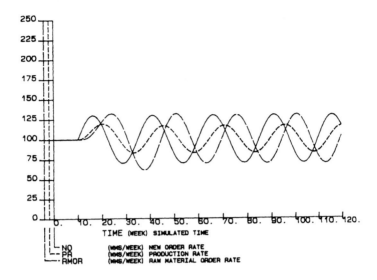

TIME (WEEK) SIMULATED TIME

NO	(WMS/WEEK)	NEW ORDER RATE
PR	(WMS/WEEK)	PRODUCTION RATE
RMOR	(WMS/WEEK)	RAW MATERIAL ORDER RATE

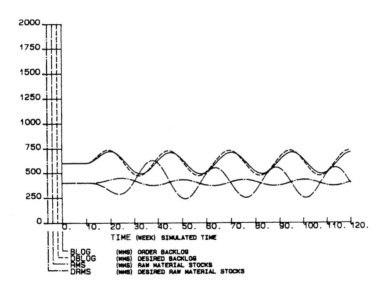

TIME (WEEK) SIMULATED TIME

BLOG	(WMS)	ORDER BACKLOG
DBLOG	(WMS)	DESIRED BACKLOG
RMS	(WMS)	RAW MATERIAL STOCKS
DRMS	(WMS)	DESIRED RAW MATERIAL STOCKS

B - LOWER STOCKS AND MORE SMOOTHING

Fig. 7.2 *Continued*

Order, RMOO, and, if it is required to *control* that variable, we must introduce a negative feedback mechanism to do so.

As was pointed out in Chapter 1, negative feedback requires an actual state, a desired state, a comparison between the two and a policy for reacting to discrepancies. To meet those requirements, we must introduce a new variable for the Desired Raw Material on Order, another for the discrepancy between desired and actual raw material on order and a time constant for correcting that discrepancy. This will give the simplest form of control: proportional control. It is important to grasp that we *must* introduce these factors *because they are required by the laws of negative feedback*, which must be satisfied if control is to be achieved.

The laws of negative feedback lay down what must be achieved, but not *how* it is to be done. Thus, the designer of managed systems has complete liberty to model factors such as the Desired Raw Material on Order, DRMOO, in whatever way gives the best performance for the system. It is this degree of freedom which makes designing control policies and structures for managed systems a practical possibility and an interesting task. By analogy with Desired Raw Material Stock, DRMS, which is also a target condition for a negative feedback loop, we shall initially base DRMOO on the Average Production Rate, which is the rate at which raw materials are used up, multiplied by the number of weeks of raw material usage which it seems plausible to have in the raw materials pipeline.

These influences are shown on the right-hand side of Fig. 7.3, the new links being emphasized by thicker lines. The reader should carefully study that diagram before proceeding. For instance, have solid and dotted lines been correctly used? Are the signs correct? Notice the new notation for Switch 1, with an arrow pointing to a solid circle on the link from EPS3 to RMOR. This is to indicate that we may wish to run the model with and without the MIS, the latter being the current system. If the MIS is turned off by having Switch 1 set to zero, the new variables will still be calculated within the model, but they will have no effect. Switch 1 is, therefore, a **structural parameter**.

The equations required to model these changes are fairly straightforward.

```
r rmor.kl=eps2.k/tarms+apr.k+switch1*eps3.k/tcrmp
c tcrmp=4
```

The order rate equation acquires a new term, activated or suppressed by Switch 1.

```
note
note   new sector for management information system
note   ===========================
note
note   switch to test effects of MIS
```

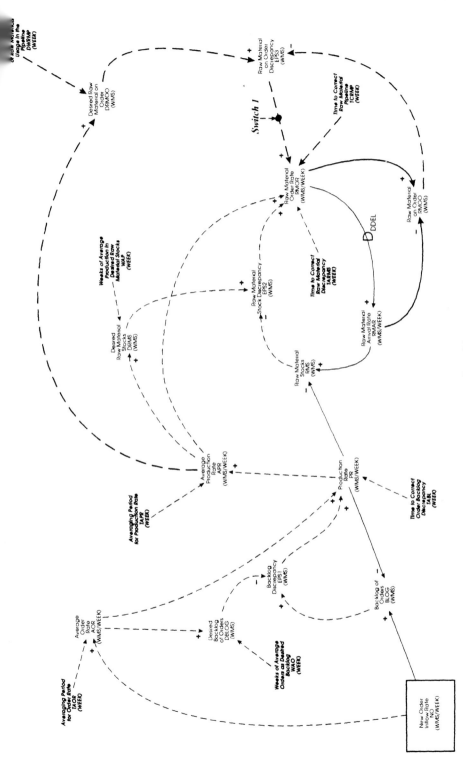

Fig. 7.3 The Raw Materials Manager's information system in the DMC problem.

```
note
c switch1=1
note    raw material on order
note
l rmoo.k=rmoo.j+dt*(rmor.jk−rmar.jk)
n rmoo=drmoo
note
note    desired raw material on order
note
a drmoo.k=apr.k*dwrmp
n dwrmp=ddel
note
note    pipeline discrepancy
note
a eps3.k=drmoo.k−rmoo.k
```

The equations for the raw materials pipeline and its control are exactly like those for, say, BLOG. Note the use of

```
n dwrmp=ddel
```

It seems sensible for DWRMP to be the same as the delivery delay, and this use of the N label sets up a 'computed constant' in which one constant is determined from another. This prevents errors such as changing DDEL, to represent a change to suppliers who can deliver more quickly, but forgetting to change the numerical value of DWRMP.

```
d drmoo=(wms) desired raw material on order
d dwrmp=(week) desired number of weeks of raw material usage in
x                the raw material pipeline
d eps3=(wms) difference between desired raw material on order and
x                actual raw material on order
d switch1=(1) switch to test effects of introducing the raw materials
x                manager's management information system
d tcrmp=(week) adjustment time on raw material in the pipeline
```

The amended model is on the disk as FIG7-4.COS.

The effects of implementing this fairly costly MIS are shown in Fig. 7.4. The upper graph shows that the MIS does, indeed, bring the system under control, in the sense that the explosive behaviour of Fig. 7.2A has been replaced by sustained oscillations, but the behaviour of RMOR is worse than that in Fig. 7.2B. The control of RMS is poorer and stock levels are higher because WAP has not been changed from the base case.

When TCRMS is increased (Fig. 7.4B), the behaviour of the system with the MIS becomes worse. The oscillations in RMOR are of greater amplitude, as are those of RMS. This difference of behaviour between the two

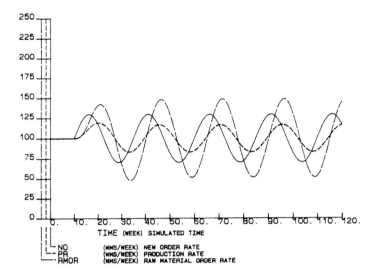

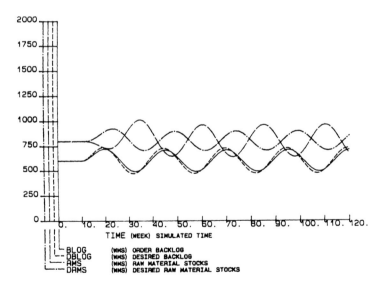

A – RESPONSE TO SINE WAVE

Fig. 7.4   The management information system.

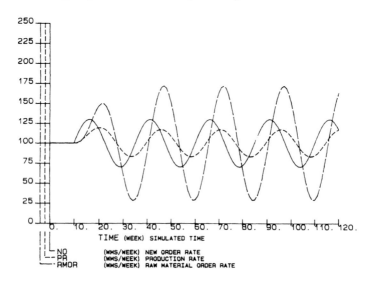

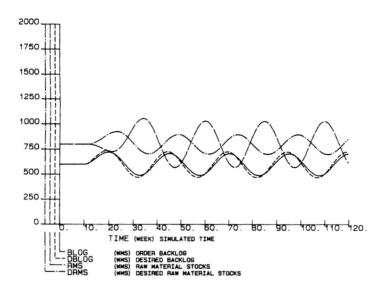

B - MIS WITH SLOWER REACTIONS

Fig. 7.4 *Continued*

halves of Fig. 7.4 drives home the essential insight that simply implementing an MIS is unlikely to be effective in any system, not just DMC, unless one carefully studies the system's policies. Introducing the MIS has changed the system, so it is intuitive that the old policies might not be good for the changed system.

We conclude from this that implementing the MIS, *with the existing policies*, is by no means as good as changing the policies. Of course, it is possible that implementing the MIS *and* changing the policies might produce even better behaviour.

Parameter analysis

The range of behaviour already found suggests that some more experiments are called for, but it might be as well to be a little more systematic than simply trying a few at random, as we did in Fig. 7.2B. For these experiments we shall switch off the MIS and not consider its parameters.

Apart from the parameters which specify the behaviour of NO, the DMC model has seven constants:

- WAO: the number of weeks in advance which the customers place their orders. This would be difficult to change without upsetting them.
- DDEL: the delivery delay on raw materials. Again, hard to change without renegotiating the whole system of raw material contracts.
- WAP: the number of weeks of average production to be held in Raw Material Stock.
- TAOR, TAPR: the two averaging times for New Orders and Production Rate, respectively.
- TABL, TARMS: the two adjustment times for Backlog and Raw Material discrepancies, respectively.

The last five parameters are entirely at the choice of DMC's managers and can be changed freely. They are, however, of two distinct types.

- WAP acts as what is called a **gain** in control engineering. That is, it acts to multiply the variations in APR when deriving DRMS. WAO is also a gain.
- TAOR, TAPR, are delays involving the smoothing of information, as was discussed in Chapter 4, p. 105.
- TARMS and TABL are parameters in control equations and, as explained on p. 102, they also act as delays.

It is a rule of thumb in control engineering that reducing gains and increasing delays is likely to reduce instability. When that rule is applied to the gains and delays in the model, using FIG7-4.COS on the disk, Fig. 7.5 is the result.

In Fig. 7.5A, the stability of RMOR is very slightly better than Fig. 7.2B, showing that the extra parameters, TABL and TAOR, have had a practi-

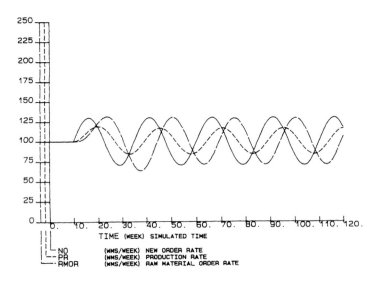

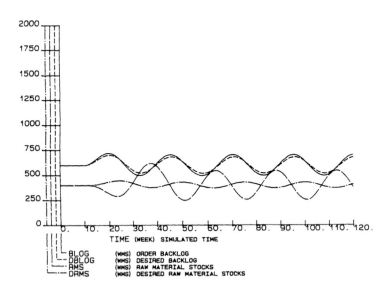

A — DOUBLE AND HALVE GAINS AND DELAYS

Fig. 7.5 Parameter sensitivity for DMC.

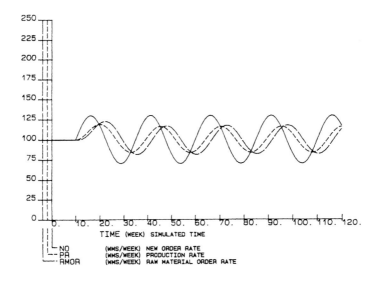

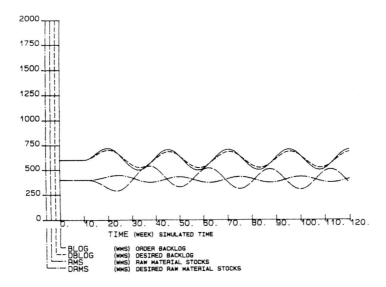

B - MIS AND GAIN/DELAY CHANGES

Fig. 7.5 *Continued*

cally negligible effect. That is not surprising, as those parameters relate to the production system and there is no feedback relationship between the production and raw materials departments.

The conclusion we draw from Fig. 7.5A is that the policy solution to the problem of bringing the system under control is spectacularly better than the MIS solution, shown in Fig. 7.4A. Since it is also a free solution, there seems to be no contest. It is, of course, possible that the costly MIS combined with the free policy solution will produce even better behaviour, and the test of that is shown in Fig. 7.5B.

The addition of the MIS, in Fig. 7.5B, improves the stability of RMOR quite significantly, and produces a slightly better match between RMS and DRMS. In both Figs. 7.5A and 7.5B, the match between BLOG and DBLOG is virtually perfect. Stability of RMOR is nice to have from the suppliers' point of view and probably eases DMC's cash flow a little, but the truly significant variables for DMC are the ability to match BLOG to DBLOG and to keep RMS close to DRMS. The performance in Fig. 7.5B is so little improved in these respects that the MIS seems to be hardly worthwhile.

Structural analysis

The preceding experiments with parameters only show that quite considerable improvements can be made to the quality of the system's performance, and this encourages us to become more wide-ranging in our thinking.

Consider, for instance, the equation for Raw Material Order Rate, RMOR:

r rmor.kl = eps2.k/tarms + **apr.k**

The APR term exists in this model of the Raw Materials Manager's decision-making because he wishes to ensure that he continues to match the current level of usage, as well as correcting stock discrepancies. We know, however, that the delivery delay on raw materials is 6 weeks, which happens to be exactly how far ahead the customers order their washing machines. It might, therefore, make sense for him to use AOR in his equation so that he orders raw materials now for delivery 6 weeks hence, which is just about when the orders now being received will be in production. He does not do so, because the poor personal relations between him and the Production Manager mean that order rate information is not passed on. This may seem crazy, but is exactly the kind of thing which happens in real systems and which can considerably complicate the task of the analyst.

If, however, some more senior manager directed that the Raw Materials

[2] At the risk of repetition, this case study is adapted, and somewhat simplified, from a consultancy project. The company really did have the policies and problems described.

Manager should have access to order rate information he could use the equation:

r rmor.kl=eps2.k/tarms+**aor.k**

Intuitively, one would expect this to give an improved performance, but it might be that a mixed policy would be even more satisfactory. Replacing the existing equation for RMOR with:

r rmor.kl=eps2.k/tarms+(alpha*apr.k+(1−alpha)*aor.k)

will allow us to experiment conveniently with different options.

ALPHA is a weighting factor between two streams of information, and is shown in Fig. 7.6. It is a **structural parameter** because, if the best performance is found when ALPHA=0, the dotted link from APR to RMOR could be deleted and the thicker solid link from AOR to RMOR would be retained, thus changing the structure. That would also imply that the firm's management information structure had been fundamentally wrong; in the real world, the Raw Materials Manager would have to switch to basing order rate decisions on AOR and abandon APR, except for determining DRMS.

If, on the other hand, the best performance is achieved when ALPHA= 0.5, then both links are required and, in the real world, the Raw Materials Manager should use both information streams, giving them equal weight in his decisions.

Of course, Switch 1 is also a structural parameter, but it can only have the values of 1 and 0; the management information system which is the Raw Materials Manager's 'pet theory' either exists or it does not. Switch 1 is said to be a **binary structural parameter**. ALPHA, on the other hand, is a **continuous structural parameter**, which can have any value from 0 to 1, inclusive. Naturally, the combinations of ALPHA and SWITCH1 are all valid: ALPHA not being 1 does not rule out SWITCH1 being set to 1 to represent the combination of the changed information flows and the MIS[3].

Testing the model with the changed equation for RMOR (see FIG7-7.COS on the disk) produces the behaviour shown in Fig. 7.7. Because using AOR in the RMOR equation is intuitively appealing, ALPHA is set to 0 and the MIS is turned off in Fig. 7.7A and on in Fig. 7.7B.

Comparing Fig. 7.7A with 7.5A, the structural change involving ALPHA has had very little effect. The oscillations in RMOR are about as large, but they are now slightly closer to matching those in PR. There has been no detectable effect on the behaviour of RMS and BLOG compared with their respective targets.

---

[3] As we remarked earlier, the system dynamics approach to evaluating the structure of a management information system before one considers hardware and software solutions has proved to be very fruitful.

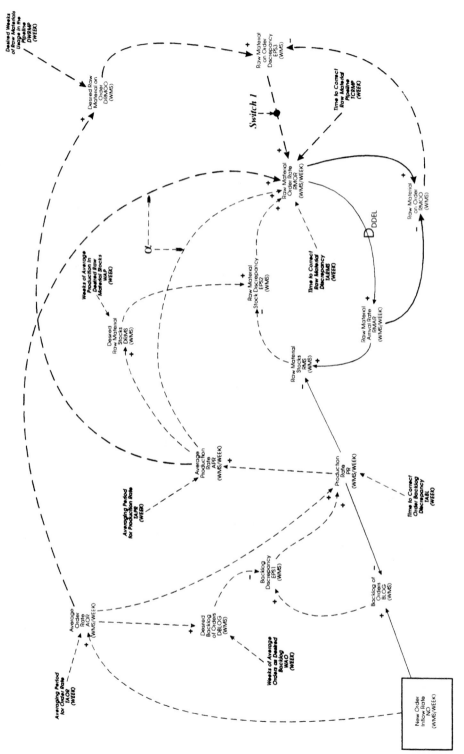

Fig. 7.6 A structural option for the DMC problem

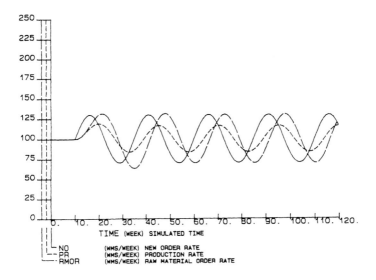

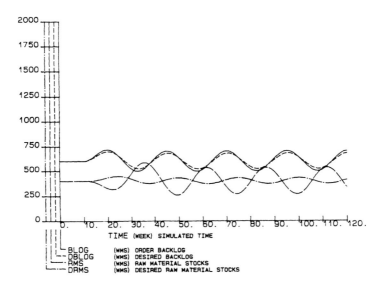

A – ALPHA=O AND NO MIS

Fig. 7.7 Output from structural option for the DMC problem.

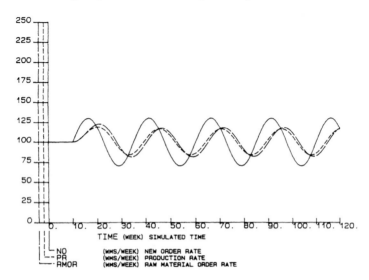

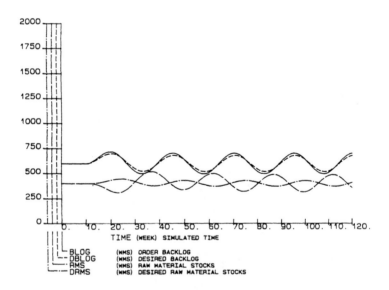

B - ALPHA=O WITH MIS

Fig. 7.7 *Continued*

When Fig. 7.7B is compared with 7.5B, the addition of the MIS has brought RMOR into practically perfect agreement with PR, though there is, again, no visually detectable difference between the behaviour of RMS versus DRMS and BLOG versus DBLOG between the two graphs. Once more, the MIS seems to be hardly worth the cost.

Just because ALPHA seems to have had little effect on the system does not invalidate the idea of changing structures, and it is worth looking imaginatively at Fig. 7.6 to see what further changes suggest themselves. A collection of possible changes is shown in Fig. 7.8, although that does not exhaust all the options. For instance, the EPS1 and AOR policy for RMOR, mentioned on p. 196, is not shown.

The changes involving the continuous structural parameters GAMMA and DELTA are very similar to that involving ALPHA. In control engineering terms, all three of these parameters involve **feedforward**. The effect of the external driving force, NO, as measured by AOR, is transmitted directly forward into the system. The corresponding effects on the equations for DRMS and DRMOO are:

a drms.k=wap*(gamma*apr.k+(1−gamma)*aor.k)

and

a drmoo.k=dwrmp*(delta*apr.k+(1−delta)*aor.k)

Obviously, if SWITCH1=0, the value of DELTA has no effect on behaviour, as DRMOO is effectively neutralized.

Note that in FIG7-9.COS (and in FIG7-7.COS), the equations involve, for example,

a drms.k=wap*(gamma*apr.k+(one−gamma)*aor.k)
c one=1
d one=(1) the number 1

This is good practice for any system dynamics software package which supports dimensional analysis (some do not), as it defines the number 1 on the same basis as GAMMA. This allows the dimensional analysis software to make a valid comparison between the two to ensure a thorough check.

Assuming that Switch 1 and Switch 2, which will be described in a moment, are both zero, the parameter BETA will introduce a third feedback loop into the system. It has the effect of linking PR to its own average, thereby introducing an element of **damping** into the system. The effect of this should be to reduce the variations in PR, on the grounds that changing production rates is usually a very expensive business.

Switch 2 has the effect of modifying PR according to the availability of raw materials. If EPS2 is positive, raw materials are in fairly short supply, so it might be a good idea to ease back on production to prevent the problem getting any worse. If EPS2 is negative, raw materials are plentiful,

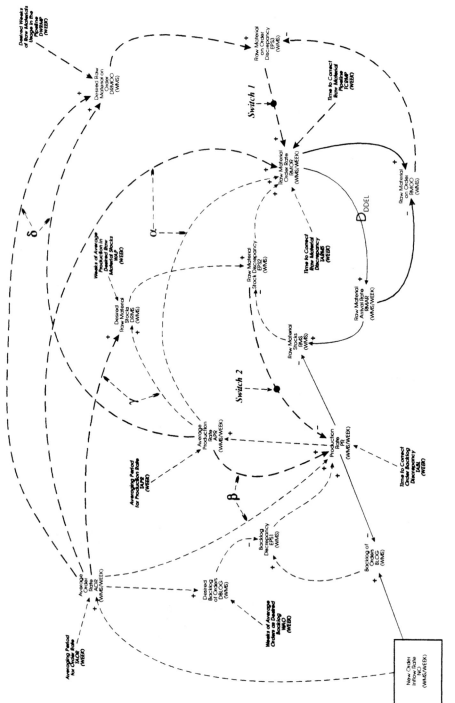

Fig. 7.8 A collection of structural options for the DMC problem.

so it might be worth taking advantage of the surplus to reduce backlog so as to be in a better position to cope with any surge in orders which might happen.

BETA and SWITCH2 alter the production rate equation to:

r pr.kl=eps1.k/tabl+(beta*aor.k+(1−beta)*apr.k)
x −switch2*eps2.k/tarms

Note the minus sign before SWITCH2. If EPS2 is positive, we need to *reduce* production.

Strictly speaking, we should not use TARMS in the above equation. TARMS represents the time constant in adjusting raw material stocks by ordering new supplies. Here we are using raw material stock to control PR, and there is no reason why the same time constant should be used. The reader should alter FIG7-9.COS to use a new parameter TARMS2 in the equation for PR. This has the advantage of creating a fresh policy analysis option, as it may be that different values for TARMS and TARMS2 produce the most effective behaviour for the system.

Switch 2 is potentially a very important feature because of the effect it has on the system's feedback loop structure. The original loop 1, shown in Fig. 2.7, controls production, but loop 2, and the new loops which are created if SWITCH1 is 1, operate only on the raw materials sector. Loop 1 drives loop 2 through PR but, without Switch 2, there is no feedback from raw materials to production. Switch 2 therefore has the potential to create an overall linkage tying the two parts of the system together. Since the new loop is negative, it ought, in principle, to improve the ability to control the system.

What values seem plausible for these new parameters? Let us keep ALPHA to 0 on the grounds that raw material ordering should be geared to the level of business. We set BETA to 0.5, because it seems a good idea to strike an even balance between PR being flexible to meet changes in AOR and damped to avoid unnecessary changes. GAMMA=0.5 also seems a plausible value, though the reader should work out why. We shall leave SWITCH1 and SWITCH2 at 0, so as to allow you the chance to make your own experiments.

The dynamics with these values appear in Fig. 7.9, and we shall need to compare that with Fig. 7.5, which was, on balance, the best option found so far.

In case A, without the MIS, the structural changes shown in Fig. 7.9 are a general improvement over Fig. 7.5. The oscillations in RMOR are a little smaller and come practically into phase with those of PR. However, as we have already mentioned, this improvement is nice to have, but is not the crux of the matter, which is to control BLOG and RMS to their respective targets. In those respects, Fig. 7.9 shows a slight improvement in the RMS/DRMS comparison at the expense of a slight deterioration in BLOG vs. DBLOG.

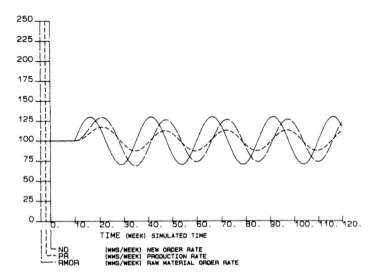

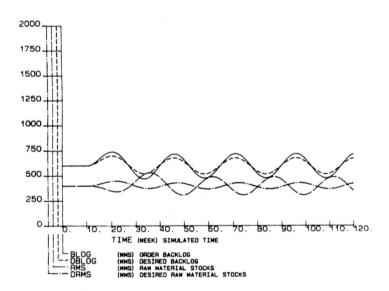

A - BETA, GAMMA = 0.5, ALPHA = 0 WITHOUT MIS

Fig. 7.9 Output from other structural options for the DMC problem.

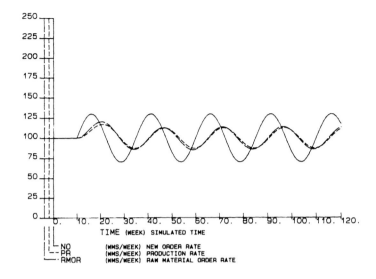

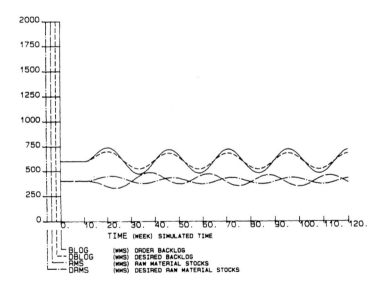

B - BETA, GAMMA = 0.5, ALPHA = 0 WITH MIS

Fig. 7.9 *Continued*

With the MIS, case B, the structural changes in Fig. 7.9 leave RMOR practically indistinguishable from PR, and the control of RMS becomes much improved, though there is a slightly greater deterioration in the control of BLOG to its target of DBLOG. Whether the improved control of RMS is worth the extra cost of the MIS is a question we cannot resolve. To some extent, careful costing studies might show the answer, though it would be hard to find convincing cost values for the slight degradation in BLOG control.

Why, though, is there such a perfect agreement between PR and RMOR? The reason is not far to seek. Basing RMOR on AOR instead of PR, and ensuring that RMS does not go wildly out of control by keeping track of the quantity within the delivery pipeline ought to give good control, and does.

It should now be evident that there are a large number of solutions to the problem of keeping DMC under control. Some are only fairly good, such as Fig. 7.4, which showed the effects of simply buying an MIS and plugging it in. All the others are much better than that. Reducing gains and increasing delays (Fig. 7.5) gives a very good result, to which some of the structural options add further improvements. The MIS added to the parameter and structural options seems to add relatively little.

There are, however, other structural options in Fig. 7.8 which have not yet been tried, and Fig. 7.10 shows yet one more, which is to damp RMOR by taking into account its own average, ARMOR. If that is to be considered in combination with the possible use of APR and/or AOR, an additional parameter, KAPPA, is needed. One way of writing the equation for DRMOO would be:

a drmoo.k=dwrmp*(kappa*armor.k+(1−kappa)*
x (delta*apr.k+(1−delta)*aor.k)))

Summary of DMC's problems

The reader should now see that a system dynamics model can contain very rich policy analysis options. We have used DMC to illustrate this phenomenon, but that model is by no means unique. One of the skills of building system dynamics models is to recognize the policy options they contain and to be able to adapt the model to test the options which have been detected. This is a skill which comes with practice, so the reader is urged to experiment further with this model, trying some of the variations which have been suggested, and others which can be discovered for oneself, such as adding noise to NO.

It will also be evident that simply trying parameter and structural options like this is rather a hit-and-miss process. One is using visual judgement of the output to assess whether or not matters have improved, and one can get trapped into an almost endless series of experiments in the hope that something will be found. Paradoxically, this is an essential step on the way

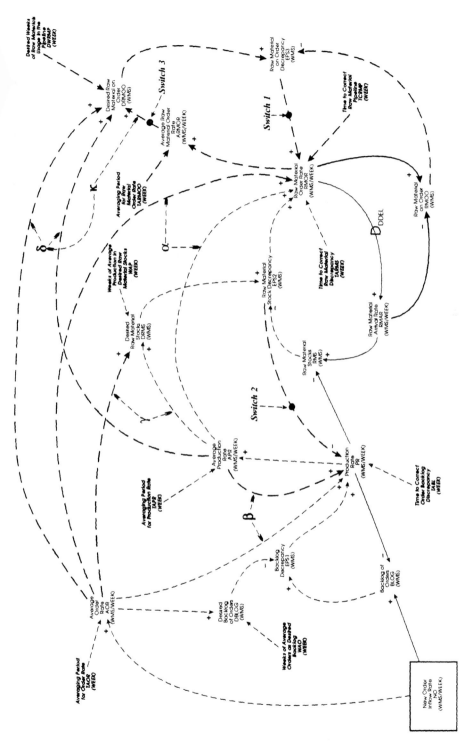

Fig. 7.10 Yet another structural option for the DMC problem.

to deep understanding of a model, but it is also clear that there ought to be a more comprehensive approach. Fortunately, the use of optimization, which we shall study in the next two chapters, is such an approach. Before we use it, however, we must experiment with another model.

PRESSURE POINTS IN THE COMBAT MODEL

The DMC model is a case in which one experiments with the parameters and the structure in order to obtain good performance. The system is, however, so heavily dominated by the external driving force, in that case the incoming new orders, NO, that the overall mode of behaviour tends to change in relatively fine detail, once the designer has cured the explosive behaviour mode of the base case. There is, however, another class of model in which the behaviour mode is determined entirely by the structure and initial values, not by a continuous external driving force; the combat model studied in Chapter 3 is such a case. In these **undriven** models, the behaviour mode can change quite remarkably depending on variations in parameters and initial values. These parameter changes usually correspond to invest-ments which its managers might make in aspects of the system; we referred to those aspects as '**pressure points**' in Chapter 3.

Pressure points and policies

The concept of a pressure point is rather different from that of policy which was discussed earlier in this chapter. We saw there that 'policy' meant a guiding rule which would be applied continually as time passed and circum-stances changed. It specified how actions should be taken in the informa-tion/action/consequences paradigm, and there was an emphasis on the repetitiveness of the actions. Pressure points, on the other hand, carry an implication of an investment or a choice which is made only once, after which the system will run under the influence of that choice. Naturally, one seeks to find the pressure points which will make it run in the most effective manner possible, so pressure point analysis is still policy analysis, albeit of a slightly different type. Equally naturally, there is no hard-and-fast divid-ing line between the two modes of analysis and, in practice, a given study may involve both types.

The combat model is a good example because it is relatively simple and contains numerous pressure points. This multiplicity of possible invest-ments leads very naturally to the idea of the optimal investment when funds are limited, which we shall explore in detail in Chapter 9. Military modelling has been a very successful application area for system dynamics, so the model is interesting in its own right though it is very simple and does not begin to show the true subtlety of system dynamics models of military problems. Of course, since it involves an interplay between two non-cooperating sides, it has implications for the modelling of competition between businesses and nation states.

Experiments on the pressure points

For convenience of reference, the influence diagram is reproduced in Fig. 7.11. Three of the pressure points are emphasized by thicker ellipses, as they are the ones we shall test. The reader has several other options to investigate, though the Information Collection point is for the advanced student. FIG7-12.COS on the disk sets up experiments to produce the graphs shown in the successive pages of Fig. 7.12 for our chosen pressure points. It is also set up to print values of the variables at the beginning and end of each run. This use of printed *and* plotted output is far better than relying simply on graphs of a few variables. As a basis for comparison, Fig. 7.12 A shows the base case in which Blue is defeated by TIME=33, Red having 34140 troops left out of his original total initial strength of 65000.

Figure 7.12B shows the effects of increasing Blue's transport capacity from 4000 men/hour to 6000, perhaps by buying more vehicles. This is a fairly significant change, but pressure point testing is best done in large steps initially to get directions of improvement. Blue's defeat is staved off by about 2 hours, Red being left with 29800 survivors; a better performance by Blue. The interesting feature of the lower half of Fig. 7.12B is that Blue starts to commit reserves at TIME=12, as before, the 20000 men available being just sufficient to restore the force ratio to the acceptable value. Unfortunately, by TIME=24, Blue is again too heavily outnumbered, so the Force Ratio Trigger goes back to 1, but there are no reserves left to commit.

Figure 7.12B showed the same phenomenon as DMC: differences in performance relative to Fig. 7.12A, but within the same overall behaviour mode. In Fig. 7.12C, Red Transport Capacity is reduced from 2500 men/ hour to 1500, perhaps by Blue sabotage teams. It is not assumed that this change would cost the same as the previous one; it is just another experiment. This change produces a change in behaviour mode. Blue now wins, with nearly 5700 men surviving. The reason is that Red's reserves trickle slowly onto the battlefield over a period of 33 hours and Blue's reserve commitment at TIME=12 is enough to give him an overwhelming advantage from which Red can never recover.

Figure 7.12D is the result of the experiment of reducing Blue's movement delay from 6 hours to 3, perhaps by buying surveillance equipment to locate Red forces so that Blue's reserves can be positioned more effectively. Blue now survives until TIME=37, and Red is left with 22800 at the end. The behaviour of Blue's reserve commitments is interesting. Some are sent forward at TIME=12, as before, but commitment stops a few hours later, when there are still a few reserves left. Blue's commander needs to commit reserves again at TIME=26, but there are too few to save the day. Nonethe-

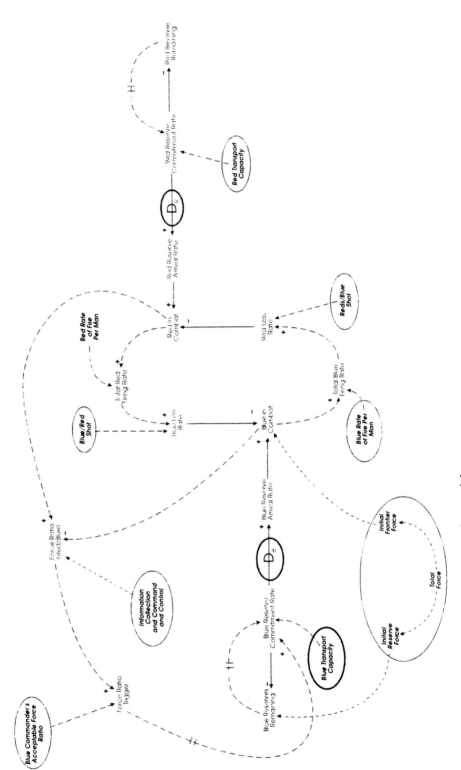

Fig. 7.11 Pressure points in the simple combat model.

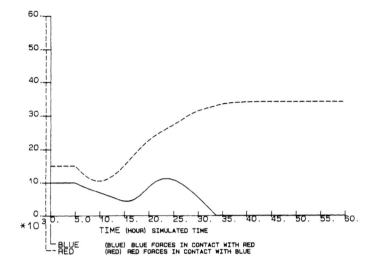

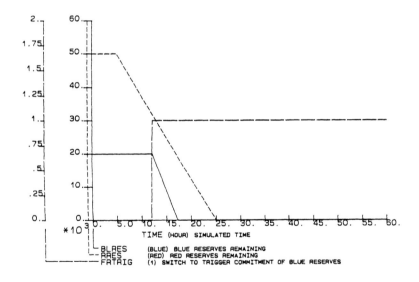

A - BASE CASE

Fig. 7.12 The combat model's pressure points (output).

less, case D is an improvement over the base case, even though the behaviour mode has gone back to being the same as the base case's.

Since Case C is very favourable to Blue, and Case D is at least to his comparative advantage, it is intuitively obvious that the two combined will be even better than either in isolation. Figure 7.12E shows the phenomenon

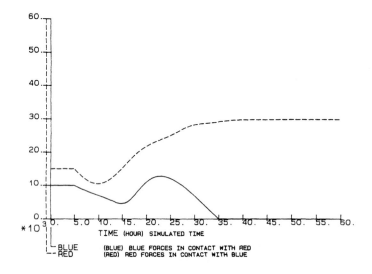

TIME (HOUR) SIMULATED TIME

```
— BLUE        (BLUE) BLUE FORCES IN CONTACT WITH RED
-- RED         (RED) RED FORCES IN CONTACT WITH BLUE
```

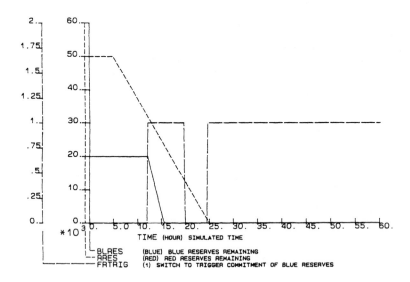

TIME (HOUR) SIMULATED TIME

```
— BLRES        (BLUE) BLUE RESERVES REMAINING
-- RRES         (RED) RED RESERVES REMAINING
— FRTRIG       (1) SWITCH TO TRIGGER COMMITMENT OF BLUE RESERVES
```

B — INCREASED BLUE TRANSPORT CAPACITY

Fig. 7.12 *Continued*

of counter-intuitive behaviour which we saw in the consultancy problem in
Chapter 6. Blue is now winning at TIME=60, but only just, with 3100
survivors whereas Red has 2400. This is worse than case C, though better
than D, and the behaviour mode is different from the two previous ones of

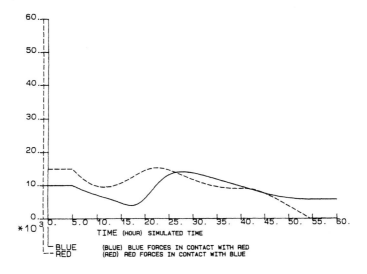

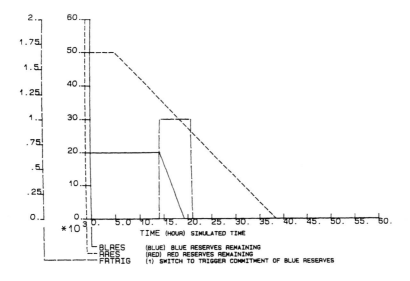

C - REDUCED RED TRANSPORT CAPACITY

Fig. 7.12 *Continued*

a decisive Red or Blue victory, respectively. Again, we see the strange behaviour of the Blue commander in committing some reserves slightly later at TIME=14, but having enough left to turn the tide of combat again at TIME=40.

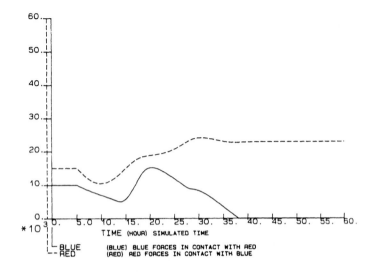

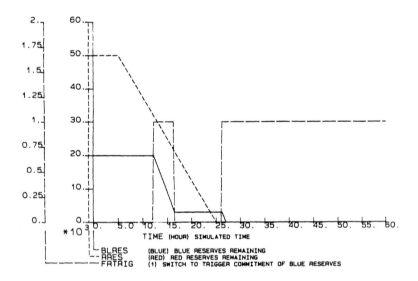

☐ - REDUCED BLUE MOVEMENT DELAY

Fig. 7.12   *Continued*

Policy as a pressure point

The strange behaviour of the Force Ratio Trigger in Fig. 7.12E suggests that the Blue commander's 'policy' of committing reserves only when he needs to do so is influencing the behaviour in unfavourable ways. Let us, there-

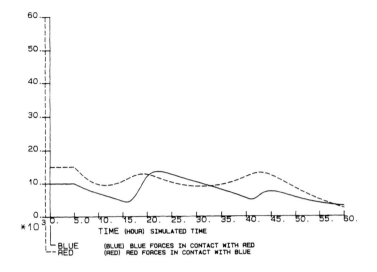

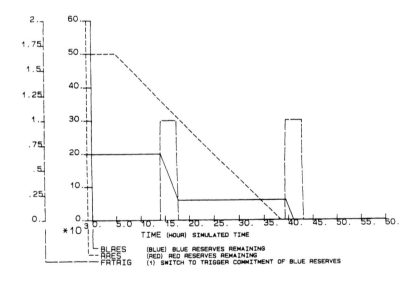

E – CASES C AND D COMBINED

Fig. 7.12 *Continued*

fore, alter the model so that, once he has started to commit reserves, he commits them all. To do so we need to transform FIG7-12.COS into FIG7-13.COS.

The change lies in one sector, shown below, with explanations following the equations.

```
note
note    amendments to Blue commander's decision on
note    reserve commitments
note    ===========================
note
note    the original equation
note
note a frtrig.k=clip(1,0,frat.k,bcafr)*wswitch.k
```

As stated, this is the original equation, for ease of comparison. It correctly does what the Blue commander wanted to do, that is to commit reserves when FRAT.K equals or exceeds BCAFR and the war has started. Its side effect is that, if FRAT.K goes back below BCAFR, then reserve commitment will stop, but will start again if FRAT.K goes back above BCAFR. This pattern will be repeated whenever FRAT.K changes relative to BCAFR.

What we require is to keep FRTRIG.K at the value of 1 once it has changed from 0. Since this book aims to be as independent as possible of any particular software package we shall solve the problem using only the common system dynamics formats.

```
note
note    revised equation using only standard System
note    Dynamics equations
note
l frtrig.k=frtrig.j+dt*rcpul.jk
n frtrig=0
```

The key is to realize that FRTRIG must 'remember' that it has been activated on the first occasion that reserves are committed. This 'memory' requirement means that we must use a level equation in which the dummy rate, RCPUL, will act as a PULSE to move FRTRIG.K from 0 to 1 within 1 DT.

```
r rcpul.kl=clip(1/dt,0,frat.k,bcafr)
x *clip(1,0,0,frtrig.k)*wswitch.k
```

The first CLIP makes RCPUL equal to 1/DT as soon as FRAT.K equals or exceeds BCAFR. Of course, if FRAT.K stayed above BCAFR, then RCPUL would be 1/DT all the time, so FRTRIG would move from 0 to 1 to 2 to 3, etc. with each DT for which FRAT.K equalled or exceeded BCAFR. What effect would that have elsewhere in the model?

To prevent that effect, and to keep FRTRIG to its logically correct value, the second CLIP permits only one PULSE, since it only produces the value 1 when 0, the third argument, is greater than or equal to FRTRIG. In other words, it will only allow a change to FRTRIG when there has been no previous change. The use of WSWITCH prevents any change until the war has started.

d rcpul=(1/week) pulse to turn on blue commander's switch for
x   commitment of reserves

As always, new variables must be defined.

The reader should run a small model, essentially consisting of only these few equations, and with FRAT.K made to STEP from 0 to 3, say, and study what happens to FRTRIG with the two different policies.

note
note   revised equation using COSMIC FLIP function
note
note a frtrig.k=flip(frat.k,0,bcafr,0)

Readers who have access to COSMIC can achieve the same effect more simply by using the FLIP function, described in the *COSMIC User Manual*.

When this change is implemented in the model, and the model is run to reproduce those cases in Fig. 7.12 in which FRTRIG was turned on and then off by the old policy, the results shown in Fig. 7.13 are produced. In case B, the overall outcome is not changed, the only difference is that FRTRIG, once turned on, stays on. In case D, Blue lasts about 6 hours longer and Red has 16 700 survivors, rather than 22 800. Case E now ends in a complete Blue victory, with 17 300 troops left. Notice, in the upper half of case E, Red's reserves are still obligingly marching up and being promptly destroyed, from TIME=31 until the last reserves arrive at about TIME=45. This demonstrates the working of the non-negativity constraint on Red losses.

Even in the original model, Blue's policy can have strange effects. As with the Raw Materials Manager in the DMC problem, the Blue commander takes no account of the reserves moving forward, or of the Red reserves, whom we shall assume he can see in the distance. Define two new variables, BRER and RRER, for Blue and Red reserves *en route*, respectively, write suitable level equations, and alter the equation for FRAT.K to:

a frat.k=(red.k+rrer.k)/(blue.k+brer.k+beps)

and see what the effect is.

SUMMARY

This chapter has shown the immense range of design possibilities which are usually to be found in system dynamics models. We examined two models, one in which the behaviour arose from the system's response to an external driving force and the other in which it derived from the interactions within a completely closed system. In the first case, we experimented by changing parameters and structures, in the second, only parameters were changed. We used the term 'pressure points' in that model to suggest that parameter

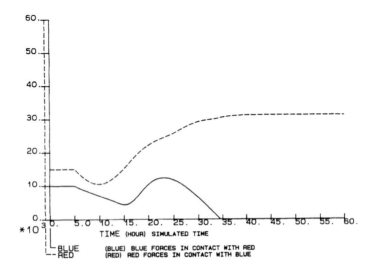

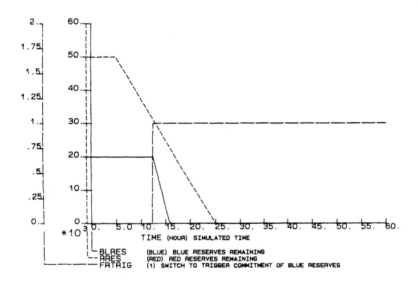

A — INCREASED BLUE TRANSPORT CAPACITY

Fig. 7.13   Combat model – once committed.

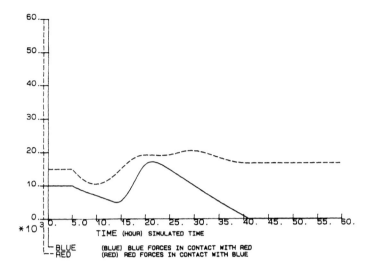

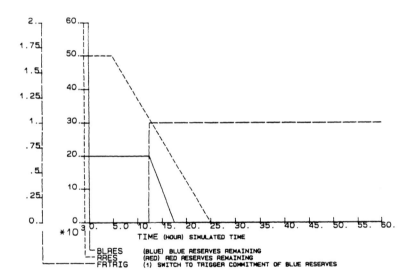

B - REDUCED BLUE MOVEMENT DELAY

Fig. 7.13 *Continued*

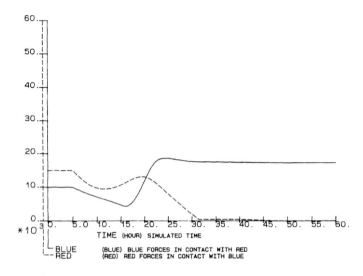

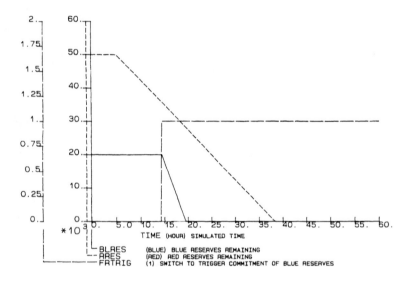

C - CASES C AND D COMBINED

Fig. 7.13   *Continued*

changes might require investment in different aspects of the system. To emphasize the difference, a parameter such as WAP in the DMC model can, in the real world, be changed simply at the behest of the managers, if they deem it to be in their interest to do so. On the other hand, a factor such as BMDEL, which is a parameter in the model, can only be changed, in the real world, by investing limited resources to do so. It seems useful to distinguish these costly parameters by a special term.

Naturally, the distinction between driven and undriven models does not mean that only the latter have pressure points and only the former have free parameters. In the combat model, FRTRIG is a free parameter and, in DMC, to change DDEL would probably require considerable time and expense to set up a new supplier network.

The whole point of this chapter is, we reiterate, to stimulate thinking about the wide range of changes which are possible in models and the dramatic effect that changes to parameters and, more especially, structures can have on system performance. The reader should now experiment further with these models, others developed in the previous chapter and the new problems in Appendix B.

However, it should also be clear that there is a limit to the amount of useful information about how to improve a system's performance than can be gained by simply experimenting with a few parameters at a time. Experimentation is essential to deepen insight into what can be done, but we require a more powerful method of searching the parameter and structural possibilities. That power is provided by dynamic optimization, and it is to that topic that we turn next.

CHAPTER 8

Optimization in system dynamics

INTRODUCTION

In the last two chapters, we built and experimented with several system dynamics models. The problems in Appendix B call for more model construction and exploration of model behaviour. The conclusion is that the behaviour of a model can be changed quite remarkably, for better or worse, by changes to parameters or pressure points. Usually, even more significant variations in behaviour stem from changes to the model's structure. The drawback is, however, that there is always a nagging doubt that, had one tried only one more experiment, something even better would have been found. Unfortunately, there is *always* yet one more experiment, so the process never stops.

It would, therefore, be highly desirable to have some automated way of performing parameter variations,[1] up to a certain number, and reporting to the analyst the best result found in that set of experiments. On the face of it, that could be provided quite easily by writing software which would test combinations of parameters and report the best result. In fact, that is not what is needed, because the number of possible combinations and conceivable numerical values of the parameters is usually colossal; in theory, it is infinite. One needs, therefore, some sort of *guided search* of the parameters to be considered and the numerical value each might have so as to seek out the result which is most rewarding in terms of enhancing the system's performance and not pursue blind alleys. Unfortunately, there is no perfect way of achieving that, but the principle of dynamic optimization comes very close to providing this subtle searching of the design possibilities of the system. This chapter will discuss the theory of dynamic optimization, illustrated by a simple example; the next will apply it to some of the models already developed.

It will be noticed that the first paragraph referred to 'experimentation' and the second has mentioned 'design'. This is the distinction which was

---

[1] Throughout this chapter, the term 'parameter' includes the structural parameters which were discussed in Chapter 7.

mentioned in Chapter 7 between *finding* the range of behaviour of which a system is capable, and deepening one's insight into it, by experiment, and *designing* policies and structures to give the best performance which can be attained. An important aspect of dynamic optimization is the development of measures of system performance: so-called **objective functions**. We shall have more to say about that later in the chapter.

Unfortunately, most of the system dynamics software products summarized in Appendix A do not support optimization. The two which do differ very considerably in the details of how they work, though they adhere to the general principles to be described here. In this book, we shall use the COSMOS optimization software to illustrate the underlying principles and the practical power of optimization.

THE SYSTEM RESPONSE AND THE PARAMETER PLANE

Figure 8.1 shows the first of the ideas we shall need to understand. The diagram shows a two-dimensional picture of a three-dimensional object. The two dimensions in the horizontal plane are labelled for two of the parameters in a model. It is essential to realize that these may be ordinary policy parameters, such as WAP in the washing machine model; they may be structural parameters, such as ALPHA and its cousins in the same model; or they may be pressure points, such as BMDEL in the combat model. Each parameter has a range within which it may lie, shown as, for example, $P1_{UPPER}$ and $P1_{LOWER}$. The initial, or base case, values are labelled $P1_1$ and $P2_1$ on the parameter axes. When these two values are projected into the 'parameter plane', following the dotted lines, they meet at Point A. When the model is run with those values, the response is, we shall suppose, rather poor, so a short line is drawn in the vertical direction to indicate that poor response. Naturally, this idea of the model's response at a point in the parameter plane is valid whether there are 2 parameters, 22, or any other number.[2] However, with 22 parameters one would need to draw a 23-dimensional object on a two-dimensional sheet of paper, which is rather difficult.

The third, vertical, axis is a measure of the quality of dynamic behaviour which the model produces for any given combination of parameters. For the moment, we shall simply label that scale as running from 'Bad' to 'Good'. Later in this chapter we shall explain how 'bad' and 'good' can be measured.

If the two parameters are now changed to $P1_2$ and $P2_2$, a new point B is defined which, we shall imagine, produces better performance and hence qualifies for a rather longer arrow in the vertical direction.

---

[2] If there are more than two parameters, the 'plane' becomes a hyper-plane in *n*-dimensional space. We shall stick to the simpler term of 'parameter plane'.

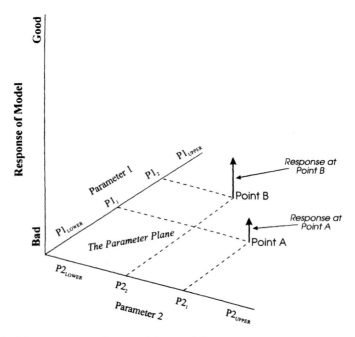

Fig. 8.1 The parameter plane and the model's responses.

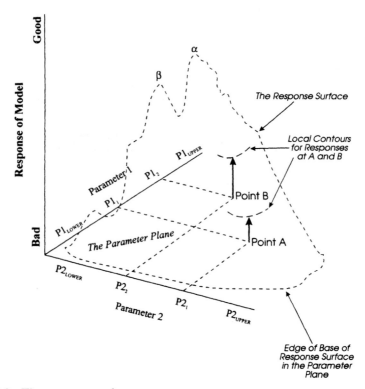

Fig. 8.2 The response surface.

Figure 8.2 now extends the argument to suggest a **response surface** in the three-dimensional space. The response surface is the surface traced out by the responses at all possible combination of P1 and P2. The heads of the arrows at Points A and B just touch this surface and can be imagined to lie on local contours of the 'hill'. The diagram tries to show that the response surface is a very rugged mountain, with several peaks. Two of the peaks, α and β, are of rather different heights and α is clearly 'better' than β. Figure 8.2 also suggests that there are all sorts of irregularities and gullies and that the mountain slope is much steeper in some places than in others. It helps to understand this (and Fig. 8.4, which we shall encounter later), if one looks at a photograph of a mountain range. The reason for the extreme irregularities lies in the complexity of system dynamics models and the non-linearities which they often contain.

We have just argued that we do not wish to test all possible combinations of parameters but, unless we do, the shape of the response surface will be unknown. The problem of dynamic optimization is: how does one find one's way to the top of a mountain of unknown shape?

HILL-CLIMBING OPTIMIZATION

The answer lies in the analogy of the blind man who is marooned on a mountain and wishes to find his way to the top.[3] His strategy is to feel the shape of the ground around the point where he is sitting (Point A). Having detected the direction in which the ground slopes up most steeply in that vicinity, he takes a cautious step or two in that direction and then feels the ground again. In this way, he hopes to find the top of the mountain, as shown by the sequence of arrows moving up the surface of the hill from Point A in Fig. 8.3. Unfortunately, the blind man's strength eventually fails him and he can go no further than μ. This is not even as high as β, so there is no guarantee that the blind man will reach his goal, α, or even another, lower, peak.

The idea of feeling one's way and taking steps in search of the top of the hill corresponds in optimization software to **iterations** of the model. Each iteration is a run of the model with a particular set of values of the collection of parameters being searched. The first iteration uses the base case parameter values. After the model has run, the value of the objective function is calculated as a measure of how good or bad the performance was. The trick comes in 'remembering' which sets of parameter values gave good results and using that information to predict how the set of values should be changed to carry out the guided search for good values, and not wasting effort examining parameter values which lead nowhere.

---

[3] A blind woman would be far more sensible and would try to find her way to the bottom, and safety. She would, however, use the same principle.

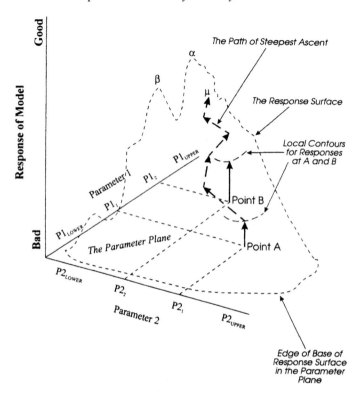

Fig. 8.3   The concept of hill-climbing optimisation.

The trick derives from the use of hill-climbing algorithms developed in
the mathematics of numerical analysis and adapted for use in system dynamics software packages. A number of good, efficient, algorithms are
known. The reader interested in the mathematical technicalities should
refer to the literature on numerical analysis; the pragmatist can rely on the
evidence that the approach works, which will be shown for a simple example later in this chapter and developed more fully in the next.

The equivalent of the blind man's strength failing him before he reaches
the top of the hill is not commanding sufficient iterations to be performed.
Since, however, the COSMOS software is very efficient, there is no harm
in demanding 500 iterations, if one wishes. Experience indicates that, for
many models, about 30 iterations are sufficient to find dramatic improvements in performance. Even with a large model, a hundred or more iterations take no more than a couple of minutes on a reasonably powerful PC.

It is important to understand that hill-climbing by the method of steepest
ascent is a **heuristic** method; that is to say, a method which is based on a
common-sense rule of thumb which is not guaranteed to produce the

optimal result. This is in distinction to the well-behaved mathematical functions, for which a guaranteed optimal result can be proved to exist and can be found by such standard methods as calculus or linear programming.

OVERCOMING THE LIMITATIONS OF HEURISTIC ALGORITHMS

Computer software is not limited by the blind man's strength, so that any number of iterations can be commanded. Even so, there is no guarantee of finding the maximum of the response surface. However, numerical hill-climbing algorithms are sophisticated and are capable of searching with something close to the intelligence that the blind man would use.

Consider Fig. 8.4. The vertical axis is still the model's response, but the horizontal axis should be imagined to be a 'cross-section' of the parameter plane, representing varying combinations of the two parameters. The hill-climbing search has started at Point A and followed a steep ridge until it reached B. From there, a valley leads forward into the hills, and there is a ridge to the left, but the locally steepest direction is towards C. On reaching C, it becomes clear that it is a false peak, from which all directions lead downwards. Since the man is not yet exhausted, he would remember C as the best point so far discovered and, reasoning that he can always go back to C, he accepts the loss of height and searches towards D. At that point, he discovers a slowly rising path which emerges from behind peak C, as shown by the faint line, and moves off to E and hence, we hope, to F. As soon as any point is reached which is better than C, it will be remembered as the point to which he can return if no better solution is found.

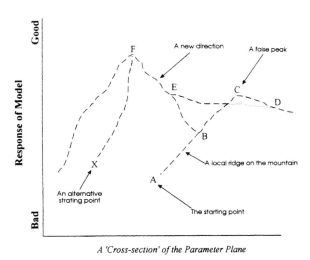

A 'Cross-section' of the Parameter Plane

Fig. 8.4 Continuing to search.

There is still no guarantee that this course of events will unfold, so another strategy for avoiding failing to find the real peak would be to recommence the search at a new starting point, X, in the hope that there is a ridge going directly to F. In practice, the improvement in behaviour found through optimization is usually so large that it is hardly worth the effort of repeating searches from new starting points. The main benefit of doing so would be to increase the experimenter's confidence that a good solution has been found.

THE PERFORMANCE OF A SIMPLE MODEL

Problem 9 in Appendix B is intended to give practice in developing a model, but it will also serve as an example of optimization. Figure 8.5 (FIG8-5.COS on the disk) shows the base case and is practically the same as Fig. S.9 (we hope that you are doing the problems as you work through the book!). The addition is that Desired Inventory, DINV, has also been plotted and the plotting scales have been changed, as we no longer need to show negative behaviour. Although the model now avoids the error due to the omission of the non-negativity constraint, its behaviour is clearly fairly disastrous. DINV and INV do not match until TIME=140, which is 2 years after the shocks in CONSR. Furthermore, PMBLOG rises to a peak of about 1100 at TIME=30, but then falls to zero at TIME=55 and stays there for 45 weeks. This corresponds to the factory having to be closed for nearly a year. All in all, a less than sparkling performance, but by no means untypical of the behaviour of real systems.

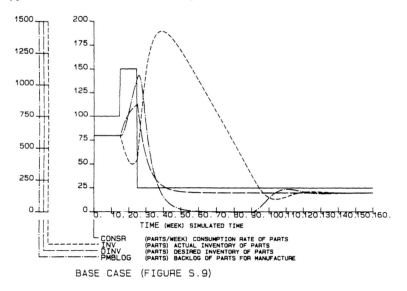

BASE CASE (FIGURE S.9)

Fig. 8.5 Simple model from problem 9.

How, though, can we *measure* that poor performance so as to have an indicator of improvement?

FORMULATING AN OBJECTIVE FUNCTION

As we remarked earlier, an objective function is a measure of system performance which is used to guide the optimization search. It should take account of what the system is trying to achieve and calculate the extent of its success. It has to be admitted that developing objective functions for system dynamics models is something of a black art, and that this is an area which is ripe for much more research.

Simple approaches based on common sense do, however, work well and, in this case, it seems reasonable to assume that the management objective is to make INV match closely to DINV, otherwise there seems little point in having a desired inventory. If, however, we wrote:

A OBJFUN.K=DINV.K−INV.K

we would mislead ourselves. The COSMOS package works by taking account of the value of the objective function only at the end of the run, when TIME=LENGTH. (Other packages may operate differently, though the underlying theme is the same.) It would be no good to optimize only for that one value of TIME, and we must take account of the behaviour over the whole of the run. This requires a level variable which will act as a memory of what happened during the run, and we might define an Inventory Penalty, INVPEN, as:

L INVPEN.K=INVPEN.J+DT*(DINV.J−INV.J)
N INVPEN=0

This will not do, however, as there is clear evidence from Fig. 8.5 that the difference between DINV.K and INV.K is sometimes positive and sometimes negative. To avoid the positive and negative discrepancies simply cancelling each other we square them and use:

L INVPEN.K=INVPEN.J+DT*(DINV.J−INV.J)**2
N INVPEN=0

INVPEN is, of course, the area which would lie under a graph of $(DINV-INV)^2$.

The numerical value of INVPEN will be some strange number, so, to make performance comparisons easier, we introduce a new variable, OBJFUN.K, defined by

A OBJFUN.K=INVPEN.K/SCALE

where SCALE is chosen to make OBJFUN, the objective function we shall actually use, equal to 100 at the end of the base case run. In this case,

SCALE is equal to 46.7933×10^4 which is found by running the model with SCALE=1 and the base case parameters, observing the value of INVPEN at the end of the run, and calculating SCALE accordingly. In this instance, it is evident that we wish to minimize OBJFUN, which is tantamount to the blind man finding his way down the crater of a volcano (hopefully extinct) by the method of steepest *descent*.

These changes are added to the model to give FIG8-5M.COS on the disk.

THE SIGNIFICANCE OF THE OBJECTIVE FUNCTION

It is very important to be quite clear about the significance of objective functions.

In the first place, they are extra equations added to the model *for the analyst's benefit*. They are not part of the real system and have no physical meaning. They are only there to help the analyst, and the software which serves him, to keep track of how improvements to behaviour can be found.

Secondly, the dimensions of the objective function *do not have to have a real-world meaning*. The dimensions of INVPEN are [WEEK*PART^2], which does not correspond to anything in the real system. For this reason, the dimensions of INVPEN and OBJFUN in FIG8-5M.COS are given as [ARBUNIT] to denote arbitrary units with no physical meaning, though one could equally well use [WEEK*PART**2]. If ARBUNIT is used, dimensional analysis, whether by hand or by software, will suggest that the equation is dimensionally invalid, but that can be ignored.

Obviously, there is nothing wrong in having an objective function which *is* dimensionally valid, such as minimizing average inventory over the course of the run, but the objective function and its components do not *have* to obey the strict requirement for dimensional consistency which applies to all of the rest of the model.

Thirdly, one has to be very careful about choosing objective functions. On the face of it, minimizing average inventory is an attractive thought, but a moment's consideration shows that the true minimum inventory is zero, and a firm which has zero inventory is usually not going to sell very much and could rapidly head for bankruptcy. In other words, choosing an objective function is not a trivial matter. Simply getting smooth dynamic behaviour is not always ideal if the smoothness has nothing to do with what the firm is really trying to achieve.

OPTIMIZATION EXPERIMENTS

When the optimization software is activated, one has to specify the parameters to be searched and to state the upper and lower values of each one. In this problem, there are four parameters, TAC, TCI, TTCAS and

TEBLOG. TTCAS is a gain, the rest are delays. To start with, we allow all four to be in the parameter plane to be searched. All the base case values are 6, and, after careful discussion with the management of the company, to see what kind of changes they might be prepared to tolerate, we allow them all to lie in the range from 2 to 10. The software is instructed to do 30 iterations. Beyond that, the precise details of how COSMOS works need not concern us; the results are what we are interested in.

The optimal values of the parameters are reported to be TAC=10, TCI=2, TTCAS=2 and TEBLOG=2. The value of OBJFUN falls from 100 to 4.03. This reduction of practically 96% is by no means unusual. All the parameters have been driven to their extreme points.

It is interesting to note that the gain, TTCAS, has been reduced, but that only one of the delays, TAC, has been increased, while the other two delays have been reduced. In Chapter 7 we stated that it is a rule of thumb in control engineering that reducing gains and increasing delays is likely to increase stability. It is indeed true that that is the rule of thumb, but it has clearly not worked in this case. Blind adherence to such rules can be seriously misleading.

It is usually not a good idea simply to throw all the parameters into the optimizer and take the results on trust. To do so is to abandon thought and rely on computation. In this case, we might feel that to increase TAC too much might make the system too insensitive to the unpredictable changes in CONSR, so let us try another optimization in which only TCI, TTCAS and TEBLOG are allowed to enter the parameter plane, with the same ranges as before. The results are that all three are driven to their lower value of 2 and OBJFUN is reduced to 5.268. On the face of it, this is nothing like as good as the previous case, because OBJFUN is about 30% larger. In fact, the visible differences are very slight, as shown in Fig. 8.6 (FIG8-6.COS). The only difference between the two plots is that INV reaches a maximum of 562 with four parameters and 581 with three.

The optimization with three parameters has driven the value of TTCAS to 2, which means that the firm is trying to operate with only 2 weeks of stock cover. This is fine when demand is constant, but might be quite insufficient for any noise in the demand pattern and provides little protection against another upsurge in CONSR. When it sees these results, management therefore feels that it has been a little rash in allowing TTCAS to have a range from 2 to 10, and wishes the optimization to be done with TCI and TEBLOG ranging from 2 to 10, but TTCAS in the range from 4 to 10. The optimization produces all three parameters at the lower limit of their ranges and OBJFUN reduced to 12.84. In management terms, the parameter values correspond to holding a prudent level of stock, but reacting quickly to changes. The behaviour is shown in Fig. 8.6C, and is clearly not as good as either cases A or B in Fig. 8.6. For instance, the maximum value of INV is now 857.

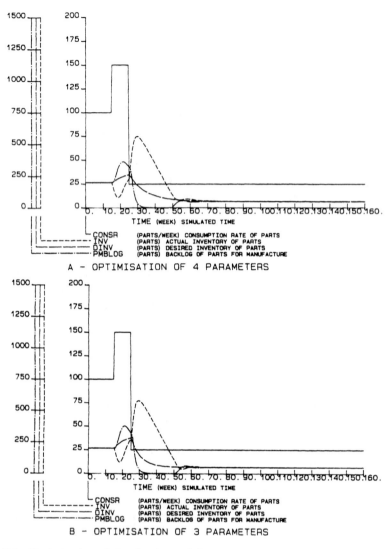

Fig. 8.6 Optimization of the simple model.

At this stage, and for this *very* simple model, optimization has now run its course and management must choose, on criteria which it would probably find hard to articulate, which of the results it prefers. If, of course, management *can* articulate its objectives more precisely, a fresh, and more subtle, objective function could be developed and the process repeated. We have, however, covered enough of the theory and a demonstration of the practice to serve the purposes of this chapter; more detail will be seen in the next. For the moment, we must turn to some final observations on the concept of optimization.

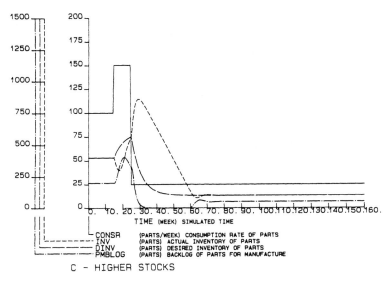

Fig. 8.6 *Continued*

SIMULATION BY REPEATED OPTIMIZATION

The concept that the top of the hill is sought by repeatedly running the model makes it clear that the technique is optimization by repeated simulation. However, the two other cases, with reduced numbers of parameters and with different ranges should make it clear that the real underlying theme is *simulation by repeated optimization*. The model is optimized and something is learned. That leads to further optimization and more learning, and so on. The power of the optimization calculation is to provide a much more powerful guided search of the parameter space than could possibly be achieved by ordinary experimentation, but the principal aim is still to experiment and to understand so that a better experiment can be designed.

The idea of simple experimentation to see what the model *might* be capable of is a fundamental precursor to optimization. One cannot devise an intelligent objective function without first having 'played' with the model, and an unintelligent objective function is worse than useless.

SUMMARY

This chapter has covered the underlying theory of optimization as applied to system dynamics models and showed something of the power of the approach by a simple example. The main point to grasp is that optimization, for all its superficial attractiveness, is not a panacea and it does not guarantee good analysis. There are, indeed, some limitations to the approach.

The first limitation is that the hill-climbing algorithm does not guarantee an optimal solution. In practice, that matters less than might be thought because most managed systems perform so badly that any improvement is welcome and the differences between different optima are usually much less than the objective function values might imply, as we saw in Figs. 8.6A and B.

Secondly, the optimization technique does not, of itself, give any guidance on the development of a good, subtle, objective function. A poor objective function, such as minimizing inventory, might be truly disastrous.

The final weakness is that the thought of optimizing something is so seductive that the naïve analyst might stop thinking.

In practice, it is only the second and third limitations which are serious and, provided one thinks, optimization is probably the most powerful development in system dynamics since the field was first developed.

CHAPTER 9

Optimization in practice

INTRODUCTION

In the previous chapter we examined the theory of optimizing system dynamics models, drew attention to some of the limitations of heuristic algorithms and suggested how they might be overcome, and illustrated the argument by optimizing the behaviour of a very simple model. We placed great emphasis on the idea that optimization is not a panacea which absolves the analyst from the need for thought about the problem, and we suggested that the correct point of view is to think in terms of *simulation by repeated optimization* rather than *optimization by repeated simulation*. In this chapter we shall take that line of thought a long way forward by studying, in detail, the optimization of two of the models developed in earlier chapters.

In particular, we shall examine the problems of **unconstrained** and **constrained** optimization. The difference between these two forms is that in, for example, the DMC model, parameters representing management policies and the structure of the information flows can be changed without limit or cost if management chooses to do so. The question is, therefore, to identify the changes which will give the greatest benefit to the performance of the system. In the combat model, on the other hand, money must be spent to bring about changes in the various pressure points we detected. The question is still to identify the changes which will give the greatest benefit to the performance of the system, *but without exceeding given financial resources.*

THE IMPORTANCE OF EXPERIMENTATION
BEFORE OPTIMIZATION

We studied the DMC problem in some depth in Chapters 6 and 7. In particular, in Chapter 7 we explored the wide range of behaviour which that model can produce and we saw that there is a large number of possible changes one can make to this system. Some of the changes are simple parameter variations to test the effects of alternative forms of a given policy; for example, increasing TARMS, the time to eliminate raw material stock discrepancies, corresponds to management taking a more relaxed attitude about mismatches between desired and actual Raw Material Stocks. Other changes, such as making ALPHA equal to 0, correspond to switching management attention from one information stream to another. Yet other modifications, such as setting SWITCH1 to 1, create new sectors of the model, in that case reflecting the purchase of the management information system. We also saw that, if, say, SWITCH1 is set to 1, it might still be advantageous to change the values of ordinary parameters so as to get the best out of the MIS.

In short, this model, *like most system dynamics models* has an enormously rich seam of options to investigate *and one would not have detected that range of options without first having 'played' with the model*, as we did in Chapter 7. Playing is an important part of any learning, and that is as true in system dynamics as it is in childhood. Thus, optimization should not be used until one has fully understood the range of options which exist in the parameter plane. To do otherwise is to hope that the numerical power of the optimizer can substitute for the intellectual power of insight and imagination; that is a false expectation which will lead to far poorer results than can be had by combining thought with calculation.

To emphasize this point, Figs. 9.1 and 9.2 reproduce Figs. 7.1 and 7.8 showing, respectively, the original influence diagram for DMC and most of the options seen in Chapter 7, though we suggested in Fig. 7.10 that there may be yet more options to be imagined. Even so, the difference between the two diagrams is quite striking and the reader should review them before proceeding. The heavier lines used in Chapter 7 to indicate structural alternatives have been deleted from Fig. 9.2. One loop is shown in heavier lines for reasons which are explained later.

We do not know, however, whether the improvements we have found, such as in Fig. 7.9, are the best that can be achieved. It is the purpose of optimization to help us to be sure.

THE DMC PROBLEM

Having thought about the options which exist in the parameter plane, it is time to turn our attention to constructing an objective function for DMC. Let us start by using FIG9-3.COS on the disk. This contains all the options we have already studied but, for the moment, they are turned off in order to reproduce the base case behaviour. You may find it useful to review the discussion of the DMC model in Chapter 7 to remind yourself of the effects on the system of the various parameters and switches.

Criteria of performance

Figure 9.3 shows the base case performance, which we already know to be rather lamentable. How, though, can we assess whether

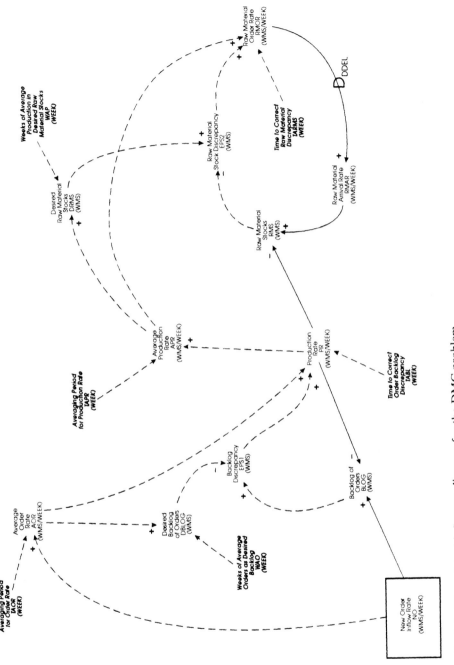

Fig. 9.1 The original influence diagram for the DMC problem.

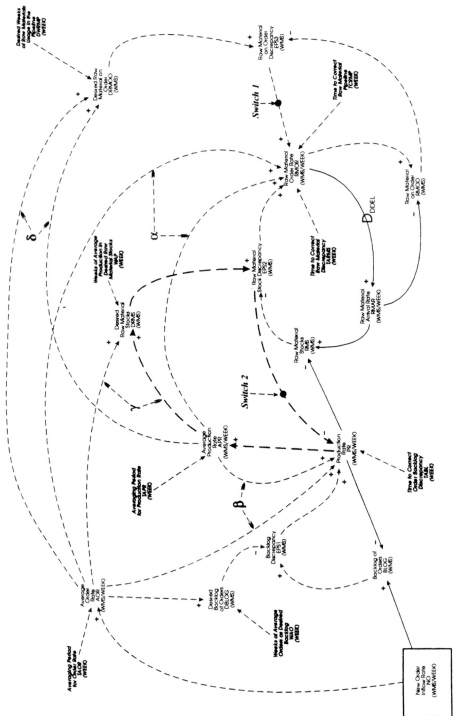

Fig. 9.2 Options for the DMC problem.

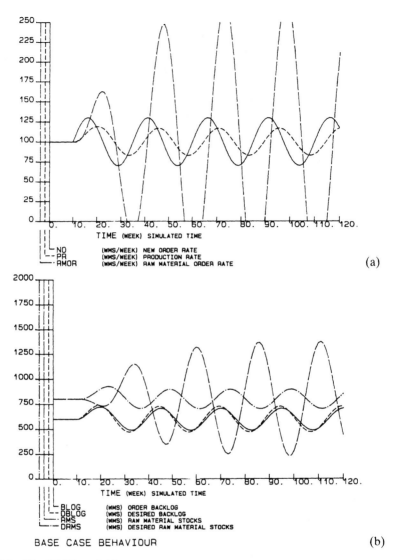

Fig. 9.3   The DMC model base case.

improvements have been achieved by any of the numerous changes we might make?

On the face of it, the original problem definition merely called for variables such as Raw Material Order Rate to be brought under control, but we can clearly see that there is more to it than that. For instance, Raw Material Stock matches very badly against Desired Raw Material Stock, although, on the other hand, Backlog tracks very closely with Desired Backlog. Production Rate oscillates quite nicely within the oscillations of New Order Rate, but it would be desirable to decrease its amplitude a little, as changing

Production Rate is, in the real world, rather an expensive undertaking. If you have been running the models and studying the output closely you will have seen that PR oscillates between 83 and 120 WMS/WEEK in the base case, but the amplitude is from 87 to 117 in Fig. 7.9 and from 93 to 115 if SWITCH2 is set to 1 (which has the effect of introducing the new policy of using Raw Material Stock to regulate Production Rate). We thus know that Production Rate can be reduced in amplitude, and it may be worthwhile to do so.

This close examination of the output from Fig. 9.3 suggests that there are four objectives to be achieved:

- to make Raw Material Stock come closer to Desired Raw Material Stock
- to ensure that Backlog does not move too far away from Desired Backlog
- to avoid unnecessary fluctuations in Production Rate
- to bring Raw Material Order Rate under control.

We have rather arbitrarily put the apparently obvious objective last to suggest that these four factors are not necessarily of equal importance. Indeed, common sense suggests that they are manifestly not of equal significance to the firm, a fact we shall have to allow for in the final objective function.

Developing an objective function

To develop an objective function we start by formulating equations to penalize failure to meet the target factors discussed above.

For the first, we wish to penalize differences between Desired Raw Material Stock and actual Raw Material Stock, DRMS and RMS, and, by analogy with the equation formulated in Chapter 8, we must use level variables to accumulate the discrepancies over the length of the run and we need to square the differences to avoid positive and negative discrepancies cancelling each other out. Thus:

L PEN1.K=PEN1.J+DT*(DRMS.J−RMS.J)**2
N PEN1=0

should fit the bill. It should be obvious that it does not matter whether one writes (DRMS.J−RMS.J) or (RMS.J−DRMS.J); we could equally well have used EPS2.J. PEN1 simply denotes the first of the four factors which are to be penalized in the final objective function.

As was mentioned in Chapter 8, the components of an objective function do not necessarily have to be tested for dimensional validity to the same extent as those parts of the model which represent the real system. In this case, PEN1 has dimensions of [WMS**2*WEEK], which is meaningless, so we shall simply call them Arbitrary Units, ARBUNIT. The same arbitrary unit notation will be used for the other penalty measures deduced below, though it will become clear that the arbitrary units for the last two are [[WMS/WEEK]**2*WEEK]. It does not matter if we lump all these different units together under the same title of ARBUNIT, as the components of the objective function are something which we are developing to help us measure the performance of the model, and they do not correspond to aspects of the real system.

Similarly,

L PEN2.K=PEN2.J+DT*(BLOG.J−DBLOG.J)**2
N PEN2=0

should suffice for the variations in backlog.

For these two factors there is a clear target, such as DRMS, against which the actual, RMS, can be compared (and similarly for the backlog), but there is no such target for Production Rate. We shall have to invent one, and it is one of the attractive characteristics of optimization that it frequently requires the modeller to use his imagination and creativity. It is, of course, one of the snags of optimization that, if one chooses silly objectives, one will get silly results.

We cannot use BLEV, the base level for the demand pattern, as a target for PR, because BLEV is not known within the firm; it is, as it were, known only to the deity whom these managers have to serve. In any case, that would gear our optimal policies and structures too heavily to a stable sinusoidal demand pattern, which would not be very robust. Instead, we can compare Production Rate to the Average Production Rate, on the grounds that, if Production Rate does not veer too far from its own average, relatively fewer production changes will have taken place and the costs of production changes will have been avoided. This will give:

L PEN3.K=PEN3.J+DT*(PR.JK−APR.J)**2
N PEN3=0

It would not be reasonable to use Average Order Rate, as we wish to avoid production variations, not chase order changes. By the same reasoning, we could penalize differences between Raw Material Order Rate and its own average, Average Raw Material Order Rate. That is a new variable which does not exist in the model, so we shall have to add it. The penalty equation will be:

L PEN4.K=PEN4.J+DT*(RMOR.JK−ARMOR.J)**2
N PEN4=0

These four penalties now require to be combined to give an overall objective function, OF. Since OF will involve the sum of four levels, it has to be an auxiliary. However, as we have already stated, the four factors to be penalized are not of equal significance to management, so we will introduce four weighting factors and get:

A OF.K=(W1*PEN1.K+W2*PEN2.K+W3*PEN3.K+W4*PEN4.K)

As in Chapter 8, it will be neater if OF has some nice round value, such as 100, in the base case, so we introduce a scaling factor to achieve that and have:

A OF.K=(W1*PEN1.K+W2*PEN2.K+W3*PEN3.K+W4*PEN4.K)
X /SCALE

where the numerical value of SCALE is found as described in Chapter 8.

When these changes are added to FIG9-3.COS the result is FIG9-6.COS which should be studied closely, not least to discover the numerical values for the four weights, W1 to W4. Note that *all* the new variables and constants have been defined and the new variables which comprise the objective function have been added to the output commands. Keeping up with documentation as a model develops is essential good practice. Printing out the values of new variables is also essential, as much experience shows that relying on plotted graphs does not give enough information to enable the modeller to be sure that the revised model is doing the right things for the right reasons.

Study of the printout for the base case run shows that the penalty components of the objective function have significantly different values. In fact, $PEN1 = 16.99 \times 10^6$, $PEN2 = 50.581 \times 10^4$, $PEN3 = 7.944 \times 10^3$ and $PEN4 = 62.111 \times 10^4$ at the end of the run, PEN1 actually being about 2000 times as large as PEN3. Does this matter?

The answer is that it does not. The penalties measure correctly, and in the correct units, what we have decided to be important in the behaviour of the system and, if the penalty for raw material stocks has turned out to be very large indeed, then that is how matters stand. To attempt to scale the penalties so that they were all of the same numerical magnitude might look more elegant, but would be to second-guess what we have carefully considered to be the requirements for the system.[1] In any case, even apparently crude objective functions tend to produce good results from a model.

Having developed an objective function, the next question is how to use it. A good way to start is to identify what can be controlled in the system. In this case, there are two control variables, PR and RMOR, which are, of course, the domains of the two managers. (There is rarely such a neat correspondence between control variables and management responsibilities.) For each of the control variables, it is useful to draw up a policy option diagram, POD. That for PR is shown in Fig. 9.4. This is intended to show the options which affect PR or which *could* affect it, which is why it is

---

[1] In the real case on which this textbook example is based a good deal of effort went into deciding what the system should be achieving, though the insight into what it *might be capable of achieving* had been generated by the preceding programme of experimenting, or playing, with the model.

called an *option* diagram. The solid lines show what affects PR in the base case of the present system. If we ignore BETA for the moment, TABL, WAO and TAOR are all parameters on the chain which leads to PR. WAO and TABL are shown at a so-called Policy Level because they regulate the daily operation of the fixed, currently chosen, control structure shown by the solid lines.

The firm could, however, adopt a different control strategy of allowing EPS2 to affect PR, and this is hinted at by the broken lines, to imply parameters which can only come into play when SWITCH2 is 1. In such a case, the dotted lines would have to become solid, and EPS2, DRMS and RMS would move up into the policy level to indicate that TARMS and

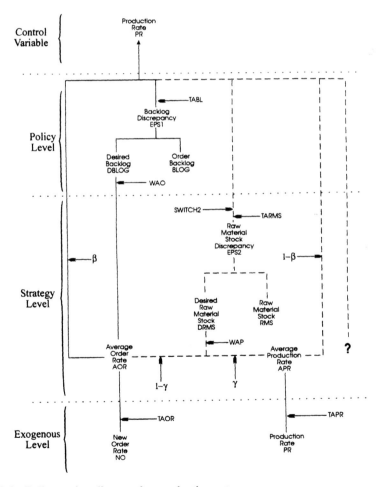

Fig. 9.4 Policy option diagram for production rate.

WAP are now policy parameters for this new strategy. In practice, the effort of redrawing the diagram would not be worth it. In fact, once one has understood the concept of a POD and had some practice, it may not be worth the effort of drawing the diagram at all, except, perhaps, for the benefit of a client; one simply does it in one's head. It is always a good idea not to let the conventions of diagrams get in the way of doing the analysis.

A further variation would be to allow GAMMA to be less than 1, which opens up the options of to some extent basing DRMS on AOR as well as the obvious, base case, usage of APR. Similarly, if BETA was allowed to be less than 1, the dotted line at the right-hand side with 1-BETA pointing to it would be brought into action.

The ? at the extreme right-hand side of Fig. 9.4 suggests that there may be other structural options not yet conceived of. We shall optimize the model with the options identified; the reader should experiment further.

The POD for RMOR is shown in Fig. 9.5, which is to be understood along the same lines as Fig. 9.4. Note that WAP and TARMS are strategic, or structural, parameters for PR because they only have an effect if SWITCH2 is 1. In Fig. 9.5, these same parameters are at the Policy Level, because they apply to the base case. The opposite arguments apply to TABL and WAO.

The external driving forces are shown as the Exogenous Level, with TAOR and TAPR as the parameters. For PR, the true exogenous force is NO. If, however, in Fig. 9.4 SWITCH2 is 1 and GAMMA is larger than 0, the passive influences from PR to DRMS and hence to EPS2 and PR at the top of the page would become active and dotted lines would become solid, as shown in Fig. 9.5. PR has become an exogenous force for PR; in other words, a new feedback loop has been brought into existence, and this is the loop shown with emphasized lines in Fig. 9.2. Policy option diagrams are often quite useful in suggesting feedback loops which one might not otherwise have thought of.

The practical difference between the Policy and Strategy Levels is that the policy parameters can usually be changed without too much fuss or upset in the organization. Any change to a strategy parameter, on the other hand, usually corresponds to changing someone's way of working, and that is not always politically desirable in the real world. Fortunately, in the world of the textbook we do not have to worry about such considerations.

The first optimization we shall test is the policy level of allowing WAP, TABL, TAOR, TAPR and TARMS to form the parameter plane. WAO is excluded for the moment as it represents the basis of the relations between DMC and its customers and is unlikely to be changed easily. DDEL is similarly excluded as changing DDEL would involve a complete upheaval in DMC's dealings with its suppliers; a distinctly non-trivial undertaking. Each of the selected parameters is allowed to lie within a range from 2 to 10.

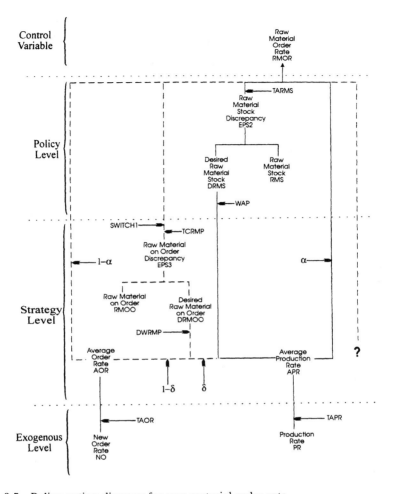

Fig. 9.5   Policy option diagram for raw material order rate.

The precise details of the COSMOS optimization procedure do not concern us, though they are fully explained in the COSMOS manual.[2]

The result of this policy optimization is that, after some 30 iterations, the values in Table 9.1 are produced.

This rather dramatic fall of more than 90% in the value of OF is mirrored in the dynamic behaviour with these values, shown in Fig. 9.6A. Note that the scale has been changed to allow the differences between successive optimizations to be seen more clearly.

[2] COSMOS supports two other optimization approaches, Base Vector optimization for iden-tifying especially sensitive parameters, and Simplification, which can be used to determine which parts of a model are really important to its behaviour. They are fully described in the manual but are not considered in this textbook as we aim to be as independent as possible of specific software.

Table 9.1 Results of first optimization

Parameter	Final value	Original value	Lower limit	Upper limit
WAP	4.00	8.00	2.00	10.00
TABL	10.00	4.00	2.00	10.00
TAOR	8.63	4.00	2.00	10.00
TAPR	10.00	4.00	2.00	10.00
TARMS	10.00	4.00	2.00	10.00
Initial value of objective function	100.00			
Final value of objective function	7.78			

Figure 9.6A shows that control can be achieved by simple policy changes. The fact that all the delay parameters, TABL, TAR, TAPR and TARMS, have been increased in the parameter values shown in Table 9.1 suggests that the change required is to react more slowly to discrepancies. The reduction in WAP corresponds to stock targets being set to lower values.

Four of the parameters have been driven to the rather arbitrary limits chosen for them, and it may be that using wider limits would produce even better results. The reader may wish to experiment, but, in practical studies, one has to give close attention to changes which the people concerned might be prepared to accept and not lose a good result by pushing them too far in search of an even better one.

TAOR ends up with the rather strange value of 8.63. There is no particular significance to that number. How much difference in behaviour would there be if it was rounded off to 10?

Having achieved a good result at the simple policy level, it is worth seeking out the effects of policy and structural changes. The results in Table 9.2 are achieved, in which the parameters have been grouped to separate policy parameters from their structural counterparts.

The objective function is reduced by about 30% and all the policy parameters are driven to their extrema, as are *some* of the structural ones. These results require some thought to interpret.

That ALPHA and DELTA are driven to 0 implies that the information structure in the base case is fundamentally misconceived, a point which the reader should confirm by studying the influence diagram. BETA, however, is only slightly reduced from the base case value, suggesting that the Production Manager should continue to give most weight to Average Order Rate, but that a little attention to Average Production Rate will also be helpful.

GAMMA remains at 1.0, implying that it is rather important to keep stock related to production.

There is, however, a deeper subtlety in these results, which is that the beneficial effects of the *structural* changes are dependent on the *policy*

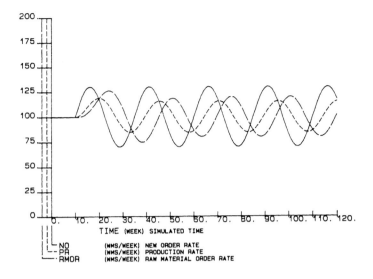

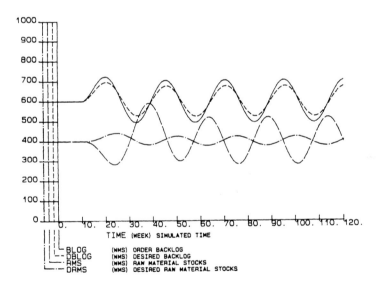

A - SIMPLE POLICY OPTIMISATION

Fig. 9.6 Optimal results for the DMC model.

Table 9.2   Results of second optimization

Parameter	Final value	Original value	Lower limit	Upper limit
WAP	4.00	8.00	2.00	10.00
TABL	10.00	4.00	2.00	10.00
TAOR	10.00	4.00	2.00	10.00
TAPR	10.00	4.00	2.00	10.00
TARMS	10.00	4.00	2.00	10.00
BETA	0.82	1.00	0.00	1.00
GAMMA	1.00	1.00	0.00	1.00
ALPHA	0.00	1.00	0.00	1.00
DELTA	0.00	1.00	0.00	1.00

Initial value of objective function   100.00
Final value of objective function   5.61

changes, and vice versa. Put another way, it will be necessary to change policies to get the best results from the reorganization of structure, and it will be necessary to reorganize to get the best results from new policies. This is a most fundamental insight into management theory, which helps to explain why corporate reorganizations often fail and why policy changes sometimes fail to have noticeable effects.

The dynamics of the second optimization are shown in Fig. 9.6B. Compared with Fig. 9.6A, the control of Raw Material Stock is rather better and that of Backlog is a little worse. Raw Material Order Rate is shifted in time so that it is closer to the cycles in Production Rate. Close study of the printed output shows that Production Rate oscillated between 84 and 119 in Fig. 9.6A and between 86 and 118 in the second case. That difference is hard to detect from the plotted graphs, so close study of printed output is always useful in detecting these second-order effects.

It sometimes happens in real analysis that powerful interests or personalities are simply unwilling to countenance changes, no matter how beneficial those changes might be to the wider organization. To illustrate that concept, the next optimization does not use TABL and DELTA and produces the results shown in Table 9.3.

Surprisingly, this is actually slightly *better* than the previous case. Again, the fundamental insight is that the policies and structures required are quite different when some options are ruled out. The parameter plane and the response surface have changed considerably, but optimization has found the best that can be done in these different circumstances.

The graph in Fig. 9.6C is also significantly changed from Fig. 9.6B. The control of Raw Material Stock is better, as is that of Backlog. The oscillations in the two backlog variables are much steeper and match more closely, and the amplitude of Production Rate is now from 81 to 121. Studying the dynamics is an essential aspect of interpreting optimization

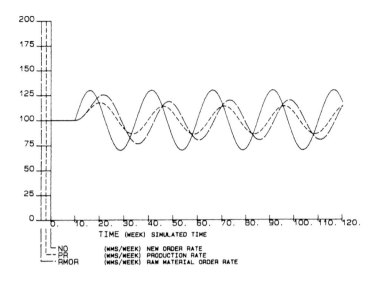

TIME (WEEK) SIMULATED TIME

NO   (WMS/WEEK) NEW ORDER RATE
PR   (WMS/WEEK) PRODUCTION RATE
RMOR (WMS/WEEK) RAW MATERIAL ORDER RATE

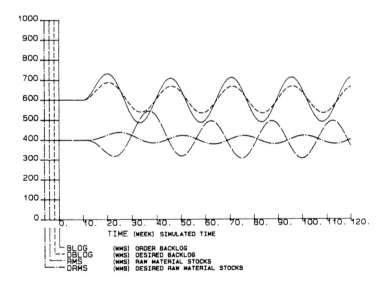

TIME (WEEK) SIMULATED TIME

BLOG   (WMS) ORDER BACKLOG
DBLOG (WMS) DESIRED BACKLOG
RMS   (WMS) RAW MATERIAL STOCKS
DRMS   (WMS) DESIRED RAW MATERIAL STOCKS

B - POLICY AND INFORMATION STRUCTURE OPTIMISED

Fig. 9.6 *Continued*

Table 9.3 Optimization without TABL and DELTA

Parameter	Final value	Original value	Lower limit	Upper limit
WAP	4.00	8.00	2.00	10.00
TAOR	2.87	4.00	2.00	10.00
TAPR	10.00	4.00	2.00	10.00
TARMS	10.00	4.00	2.00	10.00
BETA	0.77	1.00	0.00	1.00
GAMMA	1.00	1.00	0.00	1.00
ALPHA	0.00	1.00	0.00	1.00

Initial value of objective function 100.00
Final value of objective function 4.19

Table 9.4 Optimizing structures in isolation

Parameter	Final value	Original value	Lower limit	Upper limit
BETA	0.66	1.00	0.00	1.00
GAMMA	1.00	1.00	0.00	1.00
ALPHA	0.00	1.00	0.00	1.00
DELTA	0.00	1.00	0.00	1.00
SWITCH1	1.00	0.00	0.00	1.00
SWITCH2	1.00	0.00	0.00	1.00

Initial value of objective function 100.00
Final value of objective function 3.11

output. The objective function value for this third case is better than it was for the second, but whether the behaviour is really better might be a rather moot point. The objective function should be understood for what it is: a mathematical artefact to guide the optimization search, not an arbiter of what is good and what is not. The ultimate choice of whether or not to accept and implement results derived from optimization should be made by reviewing the output carefully. In practical cases, reviewing the output might make one decide that the components of the objective function and/ or their weighting factors did not adequately reflect the objectives of the problem proprietors. For that to happen is usually good, as the optimization has helped to clarify understanding, as was suggested in Fig. 1.4.

Having looked at policies on their own, it might be illuminating to examine the optimization of structures in isolation, the results being shown in Table 9.4.

The objective function is improved still further, and the pattern of earlier results tends to be confirmed in that ALPHA and DELTA are forced to 0,

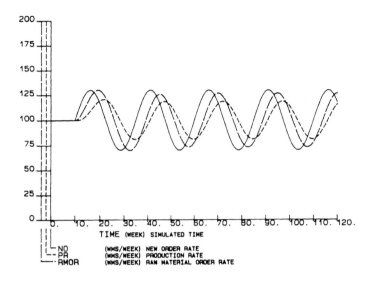

TIME (WEEK) SIMULATED TIME

NO   (WMS/WEEK) NEW ORDER RATE
PR   (WMS/WEEK) PRODUCTION RATE
RMOR (WMS/WEEK) RAW MATERIAL ORDER RATE

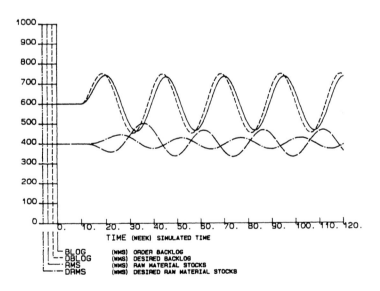

TIME (WEEK) SIMULATED TIME

BLOG   (WMS) ORDER BACKLOG
DBLOG (WMS) DESIRED BACKLOG
RMS   (WMS) RAW MATERIAL STOCKS
DRMS   (WMS) DESIRED RAW MATERIAL STOCKS

C - REFUSAL TO CHANGE TABL AND DELTA

Fig. 9.6 *Continued*

Table 9.5 Optimization using all policy and structural parameters

Parameter	Final value	Original value	Lower limit	Upper limit
WAP	4.00	8.00	2.00	10.00
TABL	10.00	4.00	2.00	10.00
TAOR	8.68	4.00	2.00	10.00
TAPR	10.00	4.00	2.00	10.00
TARMS	10.00	4.00	2.00	10.00
BETA	0.76	1.00	0.00	1.00
GAMMA	1.00	1.00	0.00	1.00
ALPHA	0.00	1.00	0.00	1.00
DELTA	0.01	1.00	0.00	1.00
SWITCH1	1.00	0.00	0.00	1.00
SWITCH2	1.00	0.00	0.00	1.00

Initial value of objective function   100.00
Final value of objective function   1.78

GAMMA remains at 1 and BETA is still weighted towards the use of AOR in the Production Rate decision. Interestingly, both SWITCH1 and SWITCH2 reach their limits of 1, suggesting that the MIS should be bought and that the raw material stock discrepancy should be used in the Production Rate decision. As the MIS is expensive, it would be interesting to run the model with the above parameters, apart from keeping SWITCH1 at 0, to see how much difference there is. The dynamics of this solution are to be seen in Fig. 9.6D, and it is now time for the reader to start assessing the quality of that performance against earlier solutions.

The next set of results (Table 9.5) are obtained when all the policy and structural parameters are used in the parameter plane.

There is a further improvement in the objective function and the pattern of results is maintained. The dynamics appear in Fig. 9.6E.

Finally, we optimize the model with the addition of TCRMP to the parameter plane. This is to drive home the idea that, if the MIS is implemented when SWITCH1 is 1 – a structural change – it is essential to examine the policy parameter(s) which might enable the best to be achieved with the changed structure. This gives results, shown in Table 9.6, which are consistent with the patterns achieved earlier, in the sense that most of the parameters are forced to their upper extrema. The exception is TCRMP, which is forced to its lower limit. In essence this means that the company should take a generally relaxed attitude to discrepancies, but, having decided to buy the MIS, the attitude should be to use it aggressively to manage the amounts in the pipeline. However, to enable those policies to have their full effect, changes to information structures will be needed. At the risk of repetition, this idea that policy and structure need to be harmoniously tuned is quite fundamental although, without the power of the

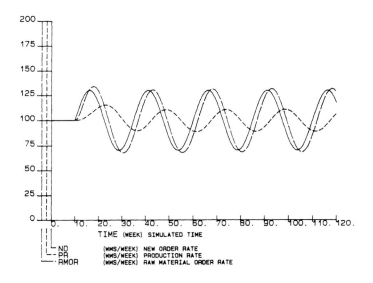

TIME (WEEK) SIMULATED TIME

NO   (WMS/WEEK) NEW ORDER RATE
PR   (WMS/WEEK) PRODUCTION RATE
RMOR   (WMS/WEEK) RAW MATERIAL ORDER RATE

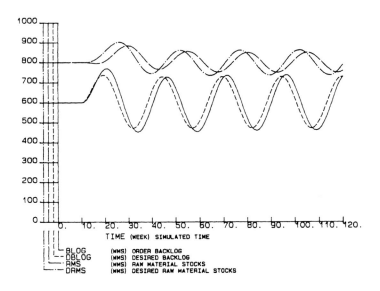

TIME (WEEK) SIMULATED TIME

BLOG   (WMS) ORDER BACKLOG
DBLOG   (WMS) DESIRED BACKLOG
RMS   (WMS) RAW MATERIAL STOCKS
DRMS   (WMS) DESIRED RAW MATERIAL STOCKS

D - STRUCTURAL CHANGES

Fig. 9.6 *Continued*

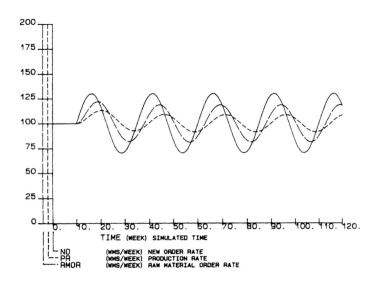

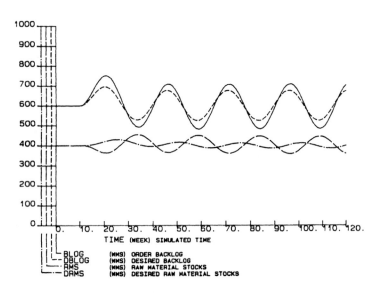

E - STRATEGY AND POLICY

Fig. 9.6 *Continued*

Table 9.6 Optimization using TCRMP

Parameter	Final value	Original value	Lower limit	Upper limit
WAP	4.00	8.00	2.00	10.00
TABL	9.99	4.00	2.00	10.00
TAOR	5.90	4.00	2.00	10.00
TAPR	10.00	4.00	2.00	10.00
TCRMP	2.00	4.00	2.00	10.00
TARMS	10.00	4.00	2.00	10.00
BETA	0.77	1.00	0.00	1.00
GAMMA	1.00	1.00	0.00	1.00
ALPHA	0.00	1.00	0.00	1.00
DELTA	0.00	1.00	0.00	1.00
SWITCH1	1.00	0.00	0.00	1.00
SWITCH2	1.00	0.00	0.00	1.00

Initial value of objective function 100.00
Final value of objective function   1.07

optimizer to search the options, it would run the risk of remaining little more than a pious intention.

The dynamics of this solution are shown in Fig. 9.6F. This is by far the best value so far found for the objective function, but you should form your own assessment of the dynamics shown in Fig. 9.6F relative to some of the previous cases. In fact, we have reached a stage at which you should pause and review all the optimizations so far done for DMC.

Summary of optimization of DMC

So far, we have discussed in some detail the approach to optimization, illustrating the method with the DMC problem. Before proceeding to assess the *value* of optimization, we must review the *process*.

At first glance, the idea of optimizing something is so attractive that a naïve analyst might jump to the conclusion that all one has to do is switch on the software and all will be revealed. However, we have tried to show that optimization is not quite so simple as that, and we have worked through several steps.

- After studying the base case performance of the model we decided on the factors to be aimed at; four in all. We were led to these four factors, rather than the obvious lone factor of achieving control of raw material ordering, by the experiments we had performed in Chapter 7 in which we explored the range of behaviour trade-offs which this model possesses. In a practical study, this would have involved much discussion with management. Interactive software and notebook computers are a great help in those consultations. Deciding what the system is supposed to achieve is,

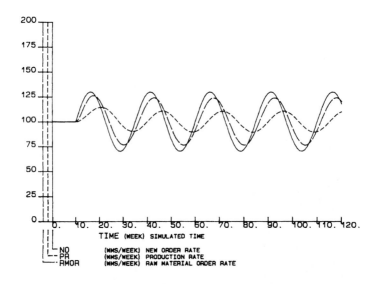

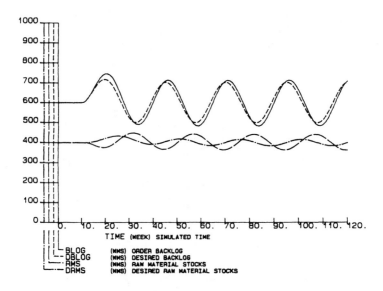

F - ADDITION OF TCRMP TO RUN E

Fig. 9.6 *Continued*

however, *the* crucial first step. We shall have much more to say about that at the end of the chapter, the point at which we shall explain why simply minimizing cost or maximizing profit are not usually suitable as objectives.

- Having decided on the objectives, the next stage was to formulate a measure of how well they are achieved: the objective function.
- Armed with the objective function, we next considered how to use it and suggested the policy option diagram as a pictorial support to thinking about the design options available. Again, we were led to those options by the previous work of experimenting, or playing, with the model. *It should be clear by now that sensible optimization cannot take place without prior experiments.*
- Even with the objective function and the options the optimization proceeded in several stages; this is simulation by repeated optimization at work! At each stage, the value of the objective function improved, but the dynamic behaviour did not necessarily improve to the same extent when it was studied by eye. Just as the eye enabled us to formulate the objective function, the eye should evaluate its results. In practical studies, visually evaluating optimization results often leads to an improved objective function. This is no great burden, as optimization is so computationally efficient in COSMOS that the extra cost is trivial. The real point is, however, that the objective function is not the arbiter of performance; that role belongs to human judgement after the objective function has fulfilled its role of controlling the hill-climbing computations.

This summary of the process does, however, beg the question of whether optimization is really of value. Not to put too fine a point on it, is optimization through several stages worth the effort and does optimization produce better results than the experimentation which must necessarily precede it? It is dangerous to generalize from a small sample, but we shall discuss the question, using DMC as an example, in the next section.

ASSESSING THE VALUE OF OPTIMIZATION

There are three bases on which we ought to weigh the value that optimization adds to system dynamics. The *first* is whether the last optimization, Fig. 9.6F,[3] is genuinely better than the first (Fig. 9.6A). In other words, has the study of policy, strategy and structural options taken us much further than simply allowing the software to search the obvious parameters? The *second* is whether the best optimization result is better than the 'best'

---

[3] Throughout this section, when we refer to Fig. such-and-such we mean the graph of the dynamic behaviour *and* the optimization results which accompany it.

experiment. The *third* is whether the simplest optimization is better than the 'best' experiment.

Notice that we have referred to the 'best' experiment in quotation marks. As we said at the beginning of this chapter, the snag with the experimental approach is that one never knows whether just one more experiment would have turned up something even better, and one can become rather old, and practical studies can turn out to be very expensive, if one continues to try just one more time. Optimization offers the prospect of much more powerful searches – the optimizations we have done in this chapter take literally a few seconds on a 486 DX2 PC running at 33 Megahertz – but how good are the results?

To answer the first question of assessing the value gained by repeated optimization, we need to contrast Fig. 9.6F's results with those of Fig. 9.6A, which gave a final objective function value of 7.78. On the face of it, Fig. 9.6F is more than 7 times better than Fig. 9.6A because its final objective function value of 1.07 is that many times as good as the 7.78 which was achieved in Fig. 9.6A. When the graphs are compared visually, however, it is very hard to say that 9.6F is *many* times better than 9.6A. It is, however, a very definite improvement, so the successive optimizations have served their purpose in guiding us to a better solution; we must simply be careful not to overstate matters and say that one solution is *x* times as good as another simply on the basis of the ratio of the two objective function values.

Why, however, did we bother about all these optimization stages? Why not simply allow the parameter plane to consist of all the options identified in the policy option diagram and let the software pick out the best solution? There are two answers.

The first answer is that the simplest change to a system is always the easiest to get accepted. People resist upheaval, so disturbance ought to be minimized, and a minimal alteration to established habits is much more likely to be accepted by the proprietors of a problem. This is true in fee-earning studies for organizational clients and it is also true for academic research aimed at policy recommendations. If, in this instance, the difference between Fig. 9.6A and Fig. 9.6F had been trivial, then the simple changes to base case policies would have been all that was necessary. There might, of course, be a mid-way point at which the improvement is good enough to be acceptable and the changes involved are small enough to be tolerable.

The second answer is that optimization is not a substitute for thought, and there is usually much to be learned about the system from simulation by repeated optimization just as there is from successive simulation experiments, as we saw in Chapter 7. In this instance, Fig. 9.6F includes the result SWITCH1=1; in other words, a recommendation to purchase the Raw Materials Manager's MIS. The experiments in Chapter 7 showed cases in which the MIS added little or nothing to performance, so the reader should re-optimize Fig. 9.6F without allowing SWITCH1 to lie in the parameter plane. The final value of the objective function may or may not be worse and the resulting parameter values may or may not be the same as Fig. 9.6F.

To see whether or not the final optimization is better than the 'best' experiment, the second question, consider Fig. 9.7 (FIG9-7.COS on the disk). That has been plotted to compare Fig. 7.9 with Fig. 9.6F, and the variables shown are the Production Rate and the two error terms, EPS1 and EPS2. Interestingly, the oscillations in PR are slightly *smaller* in Fig. 7.9, but the behaviour of EPS1 and EPS2 is significantly improved. It is very likely that someone who has a deep intuitive understanding of system behaviour, or who spends enough time experimenting, will come to results which are as good as optimization can achieve, but normally experimentation teaches enough about the properties of the system for optimization to be let loose in the profitable directions so as to find still better results. Even if the results are no better than the experiments, which is unlikely, one's confidence that they are the best that can be achieved is greatly increased.

Our third question was whether simple optimization, confined to the obvious parameters of the base case, is better that the 'best' experiment. To see that, consider Fig. 9.8 (FIG9-8.COS on the disk), which compares Fig. 7.9 with Fig. 9.6A. For all practical proposes there is no difference whatever between the two sets of graphs.

For this model, we can conclude that:

- A programme of successive optimizations does pay off and produces better understanding of the options available than simply choosing a set of parameters and letting the software loose on them.
- Optimization improves on, and adds confidence to, experimentation *when the results of the experiments are used to provide a basis of insight for optimization.*
- Optimization does not help very much when it is performed without thought, using obvious candidates for the parameter plane.

This all seems very much like common sense, which is comforting. It is typical of what has been found in numerous uses of optimization but, just as hill-climbing optimization is based on a heuristic rule of thumb, different patterns of result are sometimes found. A truly talented, or very lucky, experimenter may avoid getting stuck on a subordinate peak of the response surface, which can sometimes happen with optimization. Needless to say, a simple-minded objective function is likely to produce simple-minded conclusions, a point we shall review at the end of this chapter.

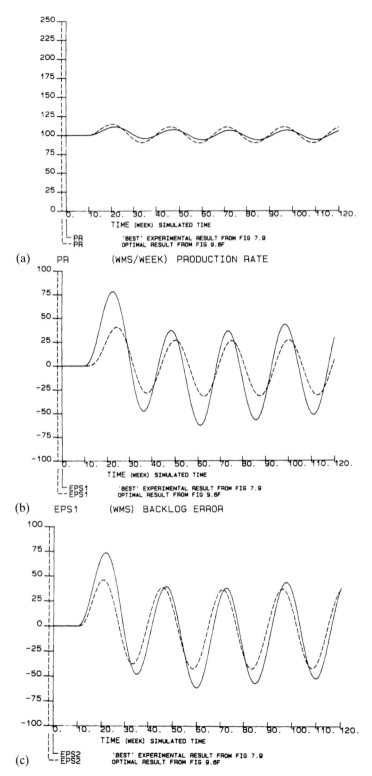

Fig. 9.7 Experimentation versus optimization.

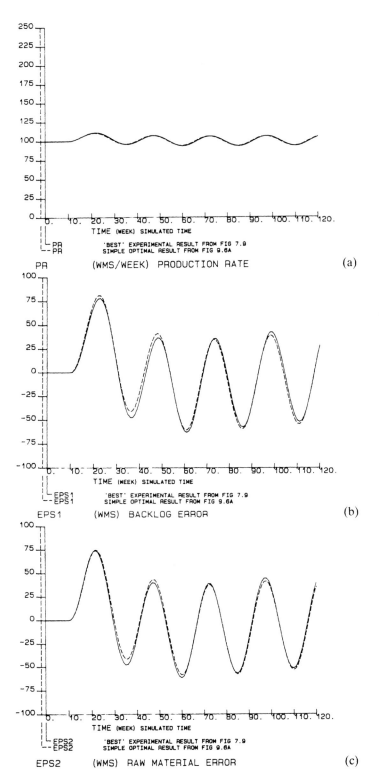

Fig. 9.8 Simplistic optimization.

CONSTRAINED OPTIMIZATION

So far, we have examined an optimization problem in which we were free to choose whatever parameters we wished for the parameter plane and the parameters could be changed without incurring any expense. We now need to consider the case of **constrained optimization**, in which money (or some other resource, such as people) has to be spent to make changes to the system. To do so, we shall use the simple combat model as an illustration (this is an *exceedingly* simple model and the reader should not think it is typical of system dynamics models developed for real defence studies!). System dynamics has been used very successfully in defence problems, and the methods for constrained optimization were first developed in that problem domain, though they have been equally successful in other problem areas.

The distinction between free and constrained optimization is, however, not absolute. Most constrained problems contain at least some free parameters, and the combat model is no exception. In particular, the parameter Blue Commander's Acceptable Force Ratio can be changed freely if he can be persuaded to change his mind about his intentions for fighting the battle. In a business strategy problem, the constraint is often on the amount of money which the firm can invest, but there are usually free parameters such as the ratio of long-term to short-term debt which management judges to be prudent. As we shall see below, the free and constrained parameters interact in ways which produce quite profound insights.

There is another distinction between the two forms of optimization, which is the shape of their respective response surfaces. In unconstrained optimization, the response surface is best envisaged as a mountain range, as we sought to indicate in Fig. 8.3. In constrained optimization, however, it is evident that, given enough resources, performance can be increased practically without limit. The snag is that there *are* limits, and a cost barrier will be encountered, which is why the optimization is constrained.

This is suggested in Fig. 9.9, which is a perspective sketch of an object viewed from behind and looking slightly to the right. The two parameters are in the horizontal plane; the performance index is vertical. From the origin, at which no money has been spent, the dotted lines indicate a slope of performance leading off into the three-dimensional distance, though it is not a smooth slope and contains many wrinkles and troughs. Instead of climbing a mountain, the blind man now feels his way up the most promising ridge. He has to pay to advance and, if he makes enough steps, he eventually encounters a cost barrier when his money runs out.

If, of course, increasing Parameter 1 paid off better than Parameter 2, then larger increases in the first would have to be paid for by smaller increases in the second, and the cost barrier should be imagined as being capable of rotating to favour the most beneficial parameter while

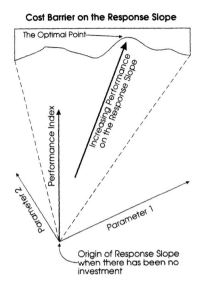

Fig. 9.9 Response slope for constrained optimization.

preventing total investment from exceeding the available funds, or whatever resource is being invested.

Having discussed these ideas, we shall now apply them to the simple combat model.

OPTIMIZATION OF THE COMBAT MODEL

For convenience of reference, the influence diagram for the combat model and its base case dynamics are shown in Figs. 9.10 and 9.11. Figure 9.10 also emphasizes the 'pressure points'; parameters which could be changed to alter the system performance, though usually only by spending money or investing in some way. Obviously, all the parameters in the DMC problem could be called pressure points, and one of them, SWITCH1, very definitely involves spending money, but it is useful to reserve the term 'pressure point' to emphasize in one's mind that the problem is one of constrained optimization.

As in the DMC case, the first step is to decide what the system, in this case the Blue government, is trying to achieve. The base case dynamics show a disastrous Blue defeat, with Blue's forces completely annihilated by TIME=33. A simple objective would be for Blue to survive as *long* as possible, but that implies an acceptance of eventual defeat. Further thought leads to the notion that Blue's forces should survive as *much* as possible, which is tantamount to moving the dynamics of Blue Forces in Contact

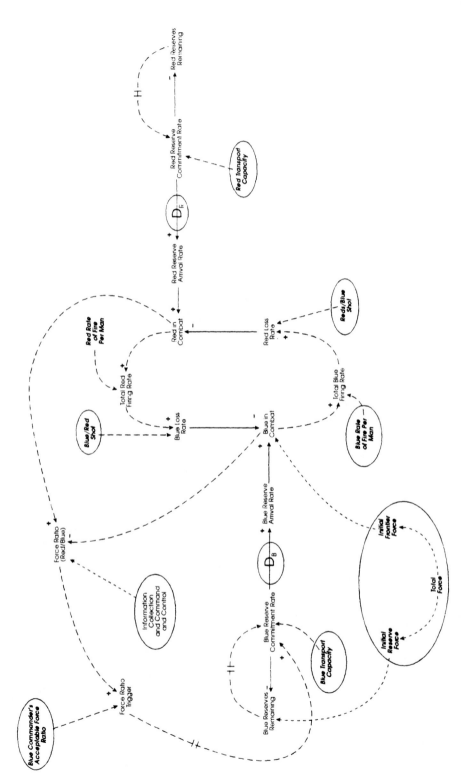

Fig. 9.10 Blue's pressure points in the simple combat model.

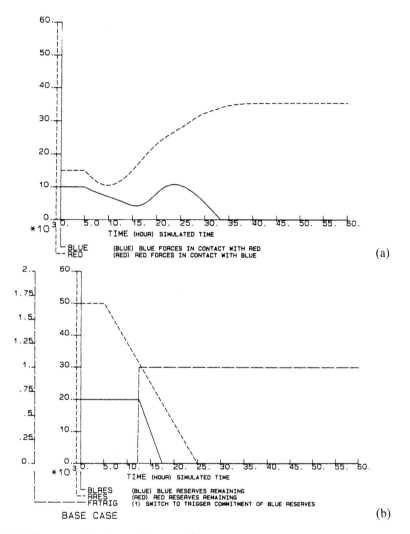

Fig. 9.11 Base case for the combat model.

With Red, BLUE,[4] as far along the time axis as possible and as far up its own axis as possible; maximizing the area under the curve, in fact. That suggests a Blue Performance Index, BPI, as shown below. It is obvious that BPI is something to be *maximized*, compared with INDEX in the DMC case, which was something to be *minimized*. COSMOS can minimize or maximize at the user's choice.

[4] In this section it will be useful to remember the distinction that 'Blue' means one of the sides in the conflict, while BLUE is the name of a variable in the model. Review FIG6-10.COS on the disk.

```
note
note    sector for objective function
note    = = = = = = = = = = = = = = = =
note
l bpi.k=bpi.j+dt*blue.j**2
n bpi=0
```

In this case, there is no actual need to square BLUE.J to prevent it going negative; the reason for doing so is partly to magnify the effect of larger Blue forces surviving, and partly to get into the habit of using squared values to avoid positive and negative effects cancelling each other. Acquiring the habits of good practice is an important part of successful modelling.

Alternatively, one might wish to base BPI simply on the numbers of Blue who survive at the end of the conflict, disregarding the dynamics of BLUE during the run. That would require:

```
a bpi.k=blue.k
```

In this case, there is no need for BPI to be a level, because we do not wish to accumulate a memory of performance during the run, only to store the value at the end of the iteration. The reader should experiment with this alternative approach.

BPI is only part of the story, because it is obvious that, if Blue spends enough money on increasing the size of its army, it is bound to reach a stage where it will win. Unfortunately, Blue does not have endless money, and the trick of constrained optimization is to subtract some very large number from the performance index whenever the optimization search tries to use a collection of parameters the cost of which would exceed the given budget. In this case, Blue has a budget of \$24 000 (mythical money!), so we subtract a huge penalty from BPI whenever that limit is exceeded:

```
a of.k=bpi.k/scale-10e06*(max(0,cost-budget))
c budget=24 000
c scale=20.5846e06
```

BPI is scaled to 100 for the base case.

Subtracting large penalties in this way is likely to make OF become large and negative in some iterations, which is tantamount to the blind man falling off a cliff. The software recovers from that misfortune, rather as cartoon characters can reverse in mid-air. It can be quite entertaining to watch the iterations!

How, though, are we to calculate COST?

The key idea is to understand that the software will search among the chosen set of parameters. Thus, we might allow the *initial value* of Blue in Contact With Red, IBLUE, to be in the parameter plane, in which case the hill-climbing will vary that parameter as it seeks to maximize BPI. If IBLUE is increased, money has to be spent. If IBLUE is decreased, money is made available to spend on other things. We need therefore to freeze the base case value of IBLUE, calling it, say, BIBLUE, so that the financial consequences of any changes to BLUE can be recorded. This is shown in italic type in the following equations:

```
note
note   blue forces in combat
note
l blue.k=blue.j+dt*(brar.jk-blr.jk)
n blue=iblue
c iblue=10000
c biblue=10000
```

This allows us to calculate one component of the cost effects as:

```
n cost=(iblue-biblue)*costreg
c costreg=1.2
```

A new value for IBLUE is chosen at the start of each iteration, so the cost will be fixed throughout that iteration and not vary during it. For that reason, we can use a computed constant to calculate COST from the values of other constants.

COSTREG [$/BMEN] is the cost of adding a soldier to Blue's standing army or the saving if one is eliminated. In practical cases, great care has to be devoted to ensuring that these costs are correctly calculated. We shall assume, for simplicity, that the marginal cost and the marginal saving are equal. Changing the equation to represent a case in which they are not equal is an exercise in the problems in Appendix B.

FIG9-12.COS on the disk is set up to allow us to optimize with five of the costly parameters, which requires a cost equation:

```
n cost=(iblue-biblue)*costreg+(iblres-biblres)*costres
x +(bbmdel-bmdel)*delcst+(bmcap-bbmcap)*capcst
x +(brmcap-rmcap)*rmccst
c costreg=1.2
c costres=0.8⁵
c delcst=5000
c capcst=6
c rmccst=9
```

Finally, we must choose some optimization ranges for the chosen parameters, as shown below.

---

⁵ For simplicity we shall assume that the reserve troops are less expensive than those manning the frontier because it is easier to supply them with food when they are nearer their home base. We shall, however, suppose that they are as effective as the frontier force. If one wished to widen the scope of the model, one could assume, instead, that reserves are both less expensive and less effective than regulars. The reader may care to tackle that as a modelling project.

note   optimization ranges are:
note
note   iblue   5000–15 000
note   iblres   5000–30 000
note   bmcap   4000–6000
note   bmdel   3–6
note   rmcap   1500–2500

These ranges for five selected parameters allow IBLUE and IBLRES to vary independently; that is, they do not consider the pressure point of Total Force, which was shown in Fig. 9.10. In short, there may be more than one constraint in a problem; in this case it might be total money and total men. Amending the model to allow for two constraints is an exercise in Appendix B.

It should go without saying that you should, if possible, do some optimizations of your own, using different parameters and inventing your own cost values.

The necessary definitions are shown below.

d bbmcap=(blue/hour) base value for blue movement capacity
d bbmdel=(hour) base value for blue movement delay
d biblres=(blue) base value for initial blue reserves
d biblue=(blue) base value for initial blue forces in contact with red
d bpi=(arbunit) blue performance index to measure survival of blue's
x   forces
d brmcap=(red/hour) base value for red movement capacity
d budget=($) defence spending budget
d capcst=($/(blue/hour)) unit cost of increasing blue transport capacity
d cost=($) total cost of set of improvements being considered
d costreg=($/blue) cost of additional regular soldier for blue
d costres=($/blue) cost of additional reserve soldier for blue
d delcst=($/hour) unit cost of reducing blue movement delay
d iblres=(blue) initial blue reserves
d of=(arbunit) blue's objective function to be maximized
d rmccst=($/(red/hour)) unit cost of reducing red's movement capacity
d scale=(1) scaling factor to make objective function=100 for base
x   case

There are two options for optimizing the model: using only the five costly pressure points and using the five points plus the free parameter of the Blue Commander's Acceptable Force Ratio: the degree to which he is prepared to be outnumbered before committing reserves. It is, of course, a necessary check to add the computed constant, COST, to the output and check that the limit has not been exceeded. With the costly parameters, COST ends up at $23 954; in the other case it is $23 999.

Table 9.7 Table of final parameter values for the combat model

Parameter	Base case	Costly parameters	Costly and free parameters
BMDEL	6.00	5.06	5.60
BMCAP	4000	5644	5524
RMCAP	2500	1978	2065
IBLUE	10000	7260	10800
IBLRES	20000	30000	30000
BCAFR	2.00	2.00	1.14

These two cases produce rather different results for the two parameter sets, as shown in Table 9.7 and in Figs. 9.12A and 9.12B. In both cases, there is a resounding Blue victory, Red ending up by marching obligingly on to the battlefield to be destroyed by Blue's overwhelming firepower. In Fig. 9.12A, BLUE's final strength is 30980 and in Fig. 9.12B it is 35940. The commander's change of mind saves 5000 casualties!

The interesting insight is, however, in Table 9.7, showing final parameter values. In both cases, IBLRES is driven to its maximum value, because reserve troops are cheaper than the regulars who form the initial Blue Force in Contact with Red. However, the second optimization encourages the Blue commander to commit his reserves at an earlier stage, in which case it also finds that it will pay him to have more regular troops available at the beginning, even though he has less money to spend on reducing his movement delay and increasing his movement capacity. In the first case, in which he is unwilling to change his intentions, it pays him to reduce delay and increase transport capacity so as to be more easily able to make use of his increased number of reserves. At the risk of repetition, this is a very simple model and the results are neither realistic nor typical of those produced by more serious models.

It is important to try to explain optimization results by such verbal reasoning, rather than simply reading off tables of numbers from the optimizer printout.

Apart from explaining the results in verbal terms, the table repays a little more scrutiny. The key point is that the results for the costly parameters *are different in the two cases* of fixing and changing the free parameter. In policy terms, this is a most fundamental insight, regardless of whether the problem domain is military, corporate, or any other, strategy. The reason for the difference is that *the change of policy is only fully effective if the distribution of costly assets is changed.* Correspondingly, *redistributing assets only becomes effective if policies are changed.* Put another way, being flexible about policies makes it possible to use assets to the greatest effect, while acquiring the right assets makes it possible to adopt policies which will have the greatest effect. This is, of course, remarkably obvious after one has thought

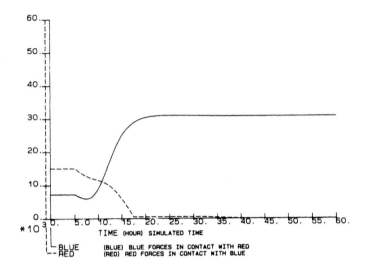

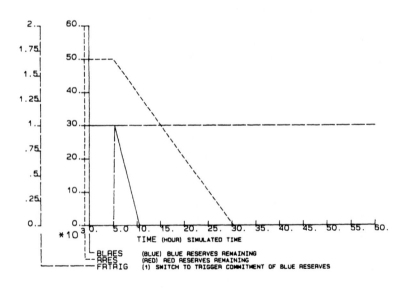

A - CONSTRAINED OPTIMISATION

Fig. 9.12 Optimizing the combat model.

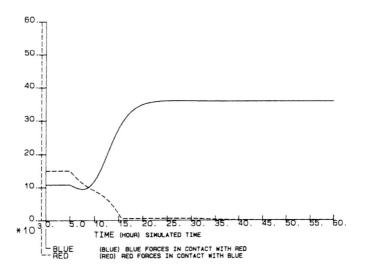

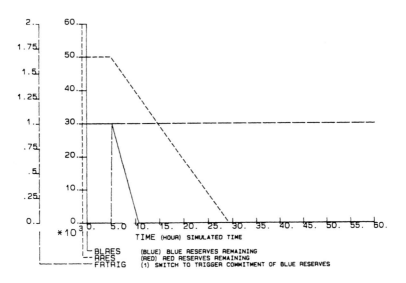

B - CONSTRAINED AND FREE OPTIMISATION

Fig. 9.12 *Continued*

of it. The role of optimization is to be able to detect the combinations of assets and policies which have the *greatest* effect.

A MATTER OF PRIORITIES

Optimization can be used in a rather different way to suggest broad priorities for investment. This is often illuminating, but it is also highly practical. In many strategic analysis problems, whether they be of military or corporate planning, the problem proprietors do not know how much money will be available to invest or, indeed, whether it is worth investing the maximum which might be available; it is possible that the marginal benefit may not justify the extra investment.

To illustrate this point, the combat model was optimized three more times. In the first case, the budget was set to $8000. The optimal values were then used as a new base case and the model was optimized again with a budget of $8000, effectively making $16000 in all. Finally, the process was repeated to give an effective investment of $24000. The models are on the disk as OPT8.COS, OPT16.COS and OPT24.COS. The idea is to show the effects of increasing amounts of money becoming available, perhaps over three successive budget years.

The optimal results are shown in Table 9.8, with the addition of the Total Blue Losses, but how are we to interpret them?

The first thing to notice is a very clear priority to cut the strength of the regular force by about 25% to release money for other purposes. The reader should use the cost values given above to work out how much is now spent on the other changes and should find that, when the budget is $8000, the total money available from the budget and from reducing the regular force is divided roughly equally between reducing BMDEL and increasing RMCAP and IBLRES. When the budget is increased

Table 9.8 Reoptimizing the combat model

Parameter	Base case values	Optimal values for			
		24000	8000	8000 + 8000	16000 + 8000
BMDEL	6.0	5.06	5.19	4.38	5.08
BMCAP	4000	5644	4144	4285	4784
RMCAP	2500	1978	2164	1864	1500
IBLUE	10000	7260	7360	7390	7457
IBLRES	20000	30000	23300	23551	29785
Final Blue strength		30980	22290	24620	31350
Total Blue losses		6280	8370	6321	5892

by a further $8000, the money is spread between reducing BMDEL and RMCAP still further, though there is a bias towards the former. Interestingly, if there are no technical means of reducing RMCAP, we may have identified a priority for Research and Development to find one. This is an example of the way in which system dynamics experiment and optimization sometimes triggers unexpected results in unanticipated areas.

Finally, when a total of $24 000 is available, the spread is between BMCAP, RMCAP and IBLRES, with a heavy bias towards the reserve forces. The move from $16 000 to $24 000 has involved allowing BMDEL to increase because the increased numbers of cheap reserves only have an effect if they can come into play. We are making the very unrealistic assumption that investments can easily be reversed, but, if they cannot, BMDEL would have to be excluded from the parameter plane in the 16 000 + 8000 optimization.

Before we leave the combat model, let us make two more comments, which are generally applicable. First, we observe that the final result in the 16 000 + 8000 case is rather better than when we simply optimized with $24 000. Specifically, the final blue strength is somewhat larger and the total blue losses are rather smaller. The shape of the response surface is normally so convoluted that more than one search strategy is usually worth trying. The second is that there are no *enormous* differences between the results from these different ways of spending $24 000. The nature of response surfaces makes it impossible to prove mathematically, but empirical evidence from the use of optimization suggests that the shape is much more like a corrugated roof than the peaks of the Alps. In other words, the differences between peaks are not all that large that a sub-optimal solution needs to be worried about too much, which also means that there is likely to be a large number of solutions all of which are pretty similar in terms of performance.

MAXIMUM PERFORMANCE OR MINIMUM COST?

In the previous section we maximized the performance which could be achieved from the simple combat model for the expenditure of a given amount of money. In many cases, however, one might wish to find the minimum amount of money, or other resource, required to achieve a given level of performance. A problem involving this appears in Appendix B.

MORE COMPLEX OBJECTIVE FUNCTIONS

In the DMC model, we used an objective function which recorded the extent to which four variables failed to meet their targets; for two of them, the target was obvious, and in the other cases we had to create a target, but the underlying theme was trying to make discrepancies vanish. For the combat model, we attempted to maximize a measure of Blue's ability to survive; we sought to get as far away from 0 as possible, that is we tried to increase a discrepancy. In short, the nature of the problem calls for some thought in creating an appropriate objective function and, to illustrate that style of thought, in this section we shall examine a case in which we need to account for two aspects of performance, the first being the extent to which an actual value is below its target and the second being the duration for which the actual falls below the target. Practical instances of this might be the extent to which hospital beds or fire engines are not available to meet possible emergency demands.

Consider Fig. 9.13 (FIG9-13.COS), which shows a variable called ACTUAL falling below TARGET at TIME=20 and continuing to fall until it reaches 60% of the target. After that nadir, ACTUAL recovers until it exceeds TARGET again at TIME=50. The fact that two targets, EIGHTY and SIXTY, are plotted is to suggest that greater failures to meet the target are more serious than lesser ones and should be penalized more heavily. To see how this can be handled, let us consider the following equations, the definitions appearing, as usual, after the equations.

note
note   target values
note

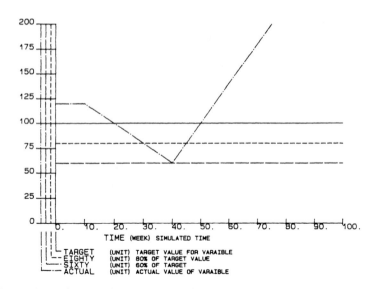

Fig. 9.13   Objective function for time and area.

```
a target.k=itarg
c itarg=100
a eighty.k=0.8*target.k
a sixty.k=0.6*target.k
```

These are some simple equations to set up a target and its 60% and 80% values. The approach works perfectly well with dynamic variables.

```
note
note    actual value
note
a actual.k=base−ramp(down,dtime)+ramp(up,uptime)
c base=120
c down=2
c dtime=10
c up=6
c uptime=40
```

Again, this is a simple driver for ACTUAL, with constants defined to allow other cases to be tested. The model on the disk tests ACTUAL under a range of cases and writing a small, simple, model such as this to test an objective function is an essential element of good practice.

```
note
note time storage
note
l below.k=below.j+dt*clip(1,0,target.j,actual.j+eps.j)
n below=0
a eps.k=0.00001*actual.k
```

First, we store the time which ACTUAL spends below TARGET. The CLIP function produces the value 1 whenever TARGET is greater than or equal to ACTUAL+EPS. EPS is a very small number and is used to ensure that the CLIP does not produce the value 1 when ACTUAL is equal to TARGET, because that would be a circumstance which we did not wish to penalize. As with all variables and functions in system dynamics models, the CLIP is evaluated every DT, so, for every occasion on which ACTUAL is below TARGET, DT is added to BELOW, thus recording the required total time spent below target. In the base case run, BELOW has the value 29.75, so the value of BELOW is DT too small, as the actual value of BELOW should be 30. The reason is that level variables always take 1 DT to be brought up to date, but the error is so small as to be negligible. It could be reduced by making DT smaller, but DT should be chosen to suit the delays in the system, not to make an objective function more 'precise'.

To compute an index of the extent to which ACTUAL falls below TARGET, consider the following rather complicated expression.

```
note
note    area storage
note
l areal.k=areal.j+(dt/dt)*nclip((target.j-actual.j)**2,0,target.j,actual.j
x +eps.j,actual.j,eighty.j)
n areal=0
```

COSMIC's NCLIP function takes the form:

$$V=NCLIP(A,B,C,D,E,F)$$

and gives V the value A if $C \geqq D$ *and* $E \geqq F$. In this case it produces the result (TARGET.J−ACTUAL.J)**2 if TARGET is greater than or equal to ACTUAL+EPS *and* ACTUAL is greater than or equal to EIGHTY. In other words it calculates a penalty if ACTUAL is within 80% of the target. The penalty is multiplied by DT/DT when added to AREA1. This ensures that the dimensions of AREA1 are [UNIT**2], rather than [UNIT**2*WEEK]. There is nothing to stop one taking the square root of the NCLIP, to give AREA1 dimensions of [UNIT]; it is largely a matter of choice and personal style.

It may seem strange to say, as we have on several occasions, that something is a matter of choice. The reader may expect there to be more of 'science' and 'the right answer' in a book on modelling. That would, however, be misleading. Modelling is much more an art than a science and the great skill is to model just enough of the problem to get a model which is well suited to its purpose, though the model must also be soundly constructed.

The line of thought used in writing the equation for AREA1 is also applied, in FIG9-13.COS, to AREA2 and AREA3, which calculate, respectively, penalties for being below EIGHTY but above SIXTY, and below SIXTY. The equations should be studied carefully. The difference is that these shortcomings are penalized more heavily by being raised to the fourth and sixth powers, respectively. Again, there is no magic to this. If one felt that being below EIGHTY was no more serious than being above it, there would be no need to use the higher power. If one felt that it was very much more serious, one might use an even higher power.

Using higher powers to penalize worse performance would not be valid if the discrepancies were calculated as *proportions* rather than as absolute values of the target. In such a case a suitable approach might be to use (ACTUAL.J/TARGET.J)**2, with the fourth or higher power used when the ratio exceeds, say, 1.2 and 1.4. The reader should rewrite the above equations to test this.

Clearly, care must be taken in formulating suitable objective functions, but the point is that objective functions can be created to meet any desired criteria for deficiencies in behaviour.

a total.k=area1.k+area2.k+area3.k

The three penalty terms are added together.

```
note
note    the objective function
note
a objfun.k=(belwt*below.k/scale1+areawt*total.k/scale2)/(belwt
x  +areawt)
```

The final objective function is now the sum of the area and duration penalties, suitably weighted and scaled to make each equal to 100 in the base case. The final division by BELWT plus AREAWT makes OBJFUN also equal to 100.

The reader should study the model on the disk, noting that all variables are printed out and that a series of test cases are used to verify that these equations work properly.

```
note
note    definitions
note
d actual=(unit) actual value of variable
d area1=(arbunit) penalty component if actual is less than target but
x               more then 80% of target
d area2=(arbunit) penalty component if actual is less than 80% but
x               more then 60% of target
d area3=(arbunit) penalty component if actual is less than 60% of
x               target
d areawt=(1) weighting for relative importance of being below target
d base=(unit) base level for actual value
d below=(week) time spent below target value
d belwt=(1) weighting for relative importance of time spent below
x               target value
d down=(unit/week) downward slope for actual value
d dt=(week) solution value
d dtime=(week) start time for slope down
d eighty=(unit) 80% of target value
d eps=(unit) small number to ensure correct comparisons between
x               target and actual
d itarg=(unit) target to be achieved
d length=(week) simulated duration
d objfun=(arbunit) value of objective function to be minimized
d pltper=(week) plotting interval
d prtper=(week) printing interval
d scale1=(1) scaling factor for time spent below to ensure objective
x               function=100 for base case
```

d scale2=(1) scaling factor for amount actual is below target to ensure
x   objective function=100 for base case
d sixty=(unit) 60% of target
d target=(unit) target value for variable
d time=(week) simulated time
d total=(arbunit) total of area components in objective function
d up=(unit/week) upward slope of actual value
d uptime=(week) time at which upward slope starts

FACTORS IN CHOOSING AN OBJECTIVE FUNCTION

The enormously wide range of applications of system dynamics makes it hard to give much prescriptive guidance on the selection of a suitable objective function other than

- to think carefully about the managerial problem
- to test the proposed equations to make sure that they work correctly
- not to be limited by the software

The need for thought is illustrated by the DMC model. A naïve analyst might be tempted to minimize Order Backlog, BLOG, but that would be to undermine the whole basis on which DMC deals with its customers. Similarly, minimizing Raw Material Stock seems like a good, cost-reducing, idea. Unfortunately, the minimum stock is zero, and, with that level of stock, production would have to stop. In both those instances, a little more thought makes it clear that keeping Backlog and Raw Material Stocks close to reasonable levels is what is required *by the nature of the managerial problem*, and it is that nature which the objective function must reflect.

Similarly, a little thought reveals why minimizing cost is rarely suitable as an objective function. In the DMC problem, one might well be able to calculate the costs of varying the Production Rate or of holding Raw Material Stock, and minimizing those costs would be well within the mainstream of management science thinking. If, however, one then ignored the importance of keeping Backlog close to its target, one would have missed out vital aspects of the managerial problem and it would be verging on fantasy to try to ascribe costs to the difference between Desired Backlog and Backlog. The essence of the matter is that the objective function should reflect what the problem proprietors think are the important attributes of system performance and, in management, cost is usually only part of their thinking. Similar considerations would apply to the simplistic maximization of profit, not the least of the difficulties being the difference between the short and long terms.

Minimizing cost might, however, be appropriate in constrained optimization. In the combat model, one might seek to minimize the cost of achieving a given level of Blue victory rather than trying to find the maximum Blue victory which can be achieved for a given amount of money (See Appendix B, problem 15).

Similar ideas apply to problems of corporate investment.

The importance of writing a simple model to test the workings of a proposed objective function with very simple inputs was demonstrated in the previous section. This is usually a very good strategy because the objective function is normally only formulated after the model has been developed, revised and experimented with. In practical analyses, one often has rather a large model, perhaps hundreds or even thousands of lines of equations and definitions, by that time. Attempting to use that as the basis for testing objective function, equations which, by their nature, ought to reflect some subtle ideas and, perhaps, embody some complex equations, is, at best, cumbersome and, at worst, prone to error. Having the patience and self-discipline to pause to develop a small model to test the objective function is good practice.

Finally, it is important to allow oneself full rein in writing equations which adequately capture the various factors in the objective function. There are no limitations on what can be modelled, so do not merely select some simple variable and simplistically optimize it.

For instance, had we chosen, in the DMC problem, to minimize the difference between BLOG and DBLOG and written a component of the objective function as:

a pen1.k=dblog.k−blog.k

we would have run into the problem that the minimum of PEN1 is when it is as negative as possible, which will happen when BLOG is very much larger than DBLOG, an absurd idea for this company. Using the absolute value:

a pen1.k=abs(dblog.k−blog.k)

would be no better because the software calculates the objective function only at the end of each iteration. That would mean minimizing PEN1 (and the sum of the other penalties) only when TIME=LENGTH, and paying no attention at all to the dynamic behaviour over the length of the run; equally absurd.

THE SIGNIFICANCE OF THE OBJECTIVE FUNCTION

Before moving on to some final technical points on optimization, this is an appropriate stage to understand the true nature of objective functions.

Towards the end of Chapter 4 we discussed the significance of DT and emphasized that DT is, in a sense, a figment of the modeller's imagination which is required only for the purpose of getting the model to run on a computer. It has nothing do to do with the real system, though its numerical value has to be chosen correctly.

Somewhat similar considerations apply to objective functions, in that they are an act of the analyst's imagination and are required in order to get the optimization software to run. Unlike DT, the objective function *does* have something to do with the real system, because it penalizes undesirable aspects of performance and rewards beneficial components.

It is, however, in the highest degree unlikely that, even with all the thought and care a skilful analyst bestows on formulating an objective function, it will *completely* capture the nuances and subtleties in the minds of the system's proprietors about what does and does not constitute good performance. In short, the objective function should not be accorded a status it does not deserve as the final arbiter of policy and structural options; that role belongs to human judgement of the dynamic behaviour those 'optimal' policies and structures produce. Put another way, the objective function should not be taken too seriously. Indeed, one should be prepared to abandon an objective function and create another if it triggers thinking about different, and better, ways of gauging performance. An objective function which does that will have served its purpose very well indeed because the underlying value of optimization is to stimulate deeper thinking about the problem than might otherwise have occurred.

NON-LINEAR OPTIMIZATION

Non-linear functions are common in system dynamics models and, as explained in Fig. 5.4, are normally represented as TABLE functions, or one of the variations thereof. As we saw there, a TABLE function takes a form such as:

T TFSRF=0/0.024/0.12/0.8/0.98/1.0

That particular table represents the effects of food supply on the longevity of foxes, but a similar construction is used in the pharmaceutical company's model in Appendix B to represent management's preference for allocating expenditure between two competing demands.

A TABLE statement is clearly a list of numbers written in sequence, in other words it is a special form of a constant. As such, it can perfectly well appear in the parameter plane if the optimization software is powerful enough to accept it. In order to provide the ranges within which the table is to be allowed to lie, one simply has to specify a pair of numbers for the upper and lower limits of each entry in the table. The COSMOS optimization software will then treat those as parameter limits and, in effect, design the shape of the curve which gives the optimal performance (in conjunction with any other ordinary parameters or tables which are in the parameter plane). Since this book is concerned with the principles of system dynamics rather than with any particular software support tool, we shall not discuss this any further, though the topic is explained in detail in the *COSMOS User Manual.*

FITTING MODELS TO DATA

In some cases, data are available about the historical dynamics of a problem, and some of the parameter values are uncertain. For instance, in the DMC problem one might have data about New Orders over a two-year period and the corresponding records for Production Rate over the same period. In such a case, one could set up two tables. That for NO would look like:

    a no.kl=tabhl(tno,time.k,0,24,1)
    t tno=120,145,163 . . .

Incidentally, the same method could be used to set up the Blue target, BTARG, mentioned above.

A similar table could be used for Recorded Production, RECPR. An objective function could then be:

    l pdiff.k=pdiff.j+dt*(pr.jk−recpr.j)**2

to minimize the difference between recorded production, RECPR, and simulated production, PR, when parameters such as TAPR, TABL etc. are allowed to enter the parameter plane. Care would have to be taken to get the initial backlog and raw material stocks correctly set to data values and to load up the delay in the raw material pipeline correctly, in that case by disaggregating the internal levels of DELAY3, as was discussed in Chapter 5.

The difficulty with this approach is that there may be more than one series of data; perhaps Raw Material Order Rate records also exist. Does one give more weight to matching the production data or the raw materials records? How does one know that the data are correct? Some of the parameters might be known with some confidence, in which case only those which are uncertain should lie within the parameter plane.

In short, there is nothing wrong with fitting models to historical data, though the problem is rarely as straightforward as might appear; the author's experience is that there have very rarely been occasions when one had enough confidence in the data to fit anything to them.

Above all, one must avoid thinking that one has proved, from historical data, that the model is 'valid', let alone 'right'. The reader should pause to review the comments made in Chapter 6 about the validation of models. Fitting a model to data does give some added confidence, but it is only part of a much deeper problem. One must also remember that system dynamics models are intended for policy design, and one is less concerned about what the parameter values now are than with what they should be to get good performance from the system. Most managed systems behave quite poorly, and there may be little point in carefully using optimization to discover the parameter values which give bad performance. It seems rather more constructive to use optimization to design policies and structures which give good performance.

SUMMARY

In this chapter we have sought to expose the reader to some of the range of powerful ideas in system dynamics optimization. We have discussed the vital role played by experimentation to find the range of options which might enter the optimization search and we have considered the formulation of objective functions. The ideas were illustrated by two cases of the different types of optimization. We laid great emphasis on simulation by repeated optimization, showing that optimizing the same problem in different ways could yield quite different results.

It is, however, absolutely essential to be clear in one's mind about what is being done. The aim is not the optimization of a model, it is the design of management policies and structures. Computations to search a parameter plane are really attempts to search the 'management plane'.

A major point was the idea that optimization is very powerful but, like all powerful tools, needs to be treated with care. Simply optimizing the obvious parameters without much thought is likely to lead to unimpressive results, no better than can be achieved by intelligent experiment. Optimization *after* experiment usually, however, leads to considerable further gains in performance.

In short, optimization is probably the most powerful development in system dynamics since Forrester's original stroke of brilliance in inventing the field. It is, however, neither a panacea nor a substitute for thought, and the objective function should never be allowed to be the final arbiter of system design; that role belongs to the judgement of analysts and managers, and more to the latter than the former.

CHAPTER 10

Advanced modelling

This Chapter deals with a number of special modelling techniques which will be required by the advanced student or the serious practitioner. For example, it explains how to model changes in a variable, including discrete changes. Achieving controls within limits are described, as is modelling of market share. Techniques for the modelling of forecasting are explained. A section is devoted to equations for financial accounts and balance sheets. The important subject of the correct use of multipliers is studied in detail. Several other techniques and examples are included. Additional advanced modelling techniques and cases are described in *Equations for Systems*, mentioned in Appendix C.

Such material logically falls into this part of the book but, to avoid distraction, the text of the chapter is included on the disk provided with the book. The models referred to are also on the disk, numbered as mentioned in the Chapter, as are the diagrams (in CorelDRAW format). For readers who do not have access to that excellent graphics package, the diagrams follow. In themselves, they are meaningless without the text, though they will give a flavour of what the text covers.

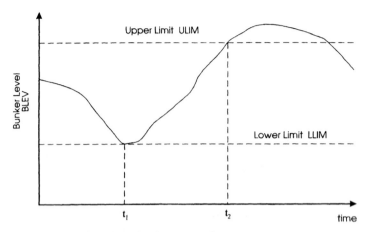

Fig. 10.1   The dynamics of production – case 2.

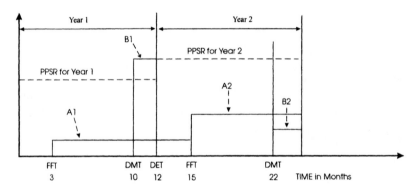

Fig. 10.2   Freezing of forecasts.

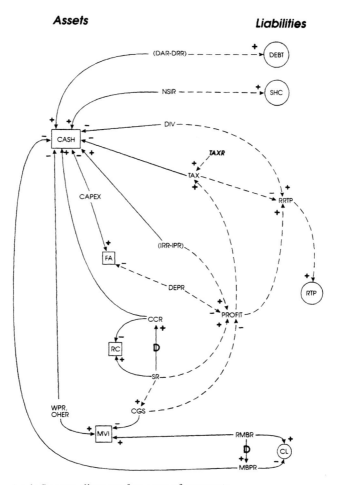

Fig. 10.3 An influence diagram for a set of accounts.

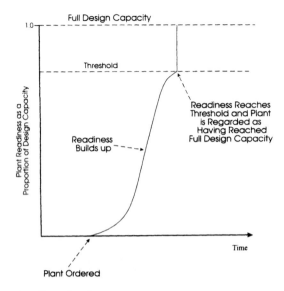

Fig. 10.4 Plant 'readiness' and output.

Table 10.1 Components of a balance sheet

TA	*Total Assets*	*TL*	*Total Liabilities*
CASH	Cash	SHC	Shareholders' Capital
+		+	
RC	Receivable	RTP	Retained Profit
+		+	
MVI	Monetary Value of Inventory	DEBT	Total Debt
+		+	
FA	Fixed Assets	CL	Current Liabilities

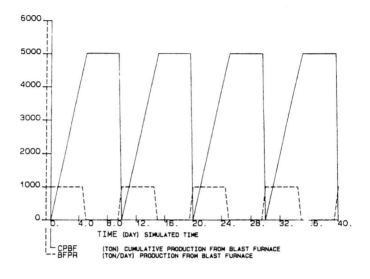

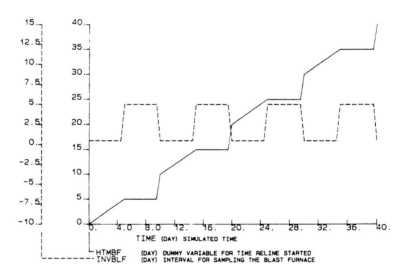

TRIAL OF MODEL

Fig. 10.5 Test of blast furnace equations.

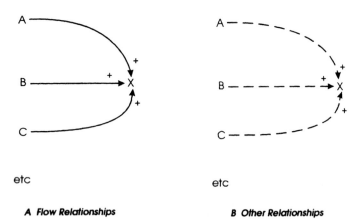

A Flow Relationships   **B Other Relationships**

Fig. 10.6   A simple influence diagram of multiple relationships.

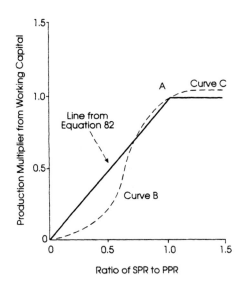

Fig. 10.7   Production multiplier from working capital.

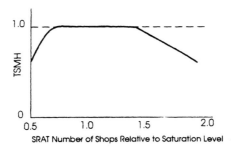

A The Saturation Multiplier

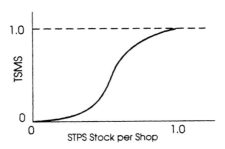

B The Stock Multiplier

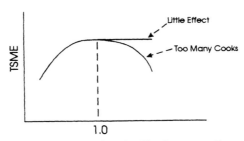

C The Employees Multiplier

Fig. 10.8 Three cases of multipliers.

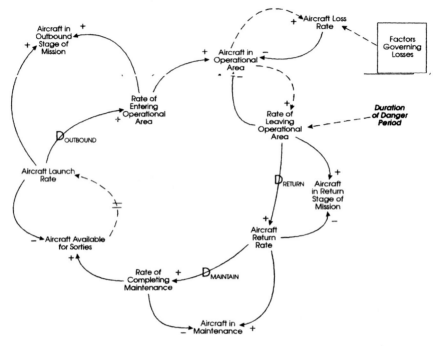

Fig. 10.9 Influence diagram for detailed approach to aircraft sortie problem.

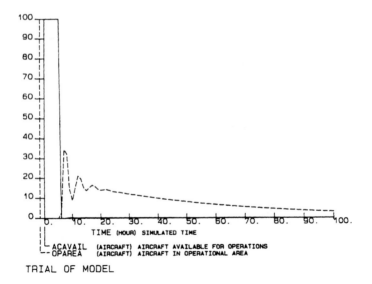

TRIAL OF MODEL

Fig. 10.10 Aircraft sortie model.

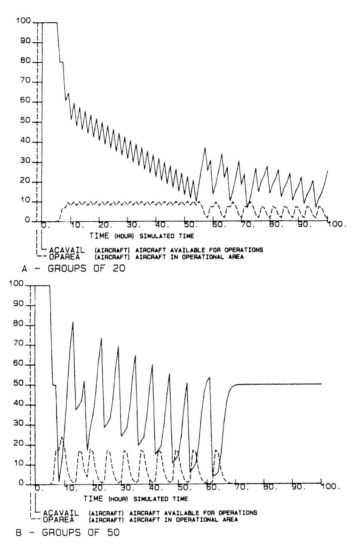

Fig. 10.11 Pulsed aircraft sortie model.

CHAPTER 11

The applications of system dynamics

INTRODUCTION

This chapter indicates the enormous range of applicability of system dynamics to help you to imagine how a problem might be analysed from a system dynamics viewpoint. We cannot hope to describe every single application or mention every worthwhile example; the latest bibliography of published work on system dynamics lists over 3000 references[1].

We start with a brief history of system dynamics. There follow a few case studies of system dynamics analysis. Finally, the defence applications of system dynamics, which have been particularly successful, are surveyed.

The disk at the end of the book contains listings of the models used as cases.

These models should be run to produce the graphs which are mentioned, but not shown, in the text (the diagrams are numbered to allow for this). This is vital practice in working with models and the student should not neglect it. In some cases, the model will offer interesting possibilities for optimization if the software being used supports that facility.

SOME SYSTEM DYNAMICS HISTORY

System dynamics originated at the Massachusetts Institute of Technology in the late 1950s thanks to Professor Jay W. Forrester.

Forrester's first book was entitled *Industrial Dynamics* (Forrester, 1961), though the wider term of system dynamics was soon coined. Work of such novelty led to the development of a number of models of industrial problems (see e.g. Roberts (1978)). These models are, however, completely closed unlike the DMC model, discussed in earlier chapters, in which the behaviour arises from the interplay between endogenous factors

---

[1] Available from The System Dynamics Society, 40 Bedford Road, Lincoln, Massachusetts 01773, USA. A search of that bibliography will produce numerous references in addition to those described here.

and **exogenous** driving forces (for some models of this type, see Coyle (1977)).

System dynamics, despite its origin in business management, was clearly applicable to social problems, and this has proved to be a fertile area of application.

In the late 1960s, American society was perceived to be in crisis and Forrester took up the challenge by presenting a model of the dynamics of a city.

Urban modelling was a strong theme in system dynamics for a few years. Refinements of the urban model were developed and applied to particular cities. Further publications were produced (see for example Alfeld (1975), but the application of system dynamics to urban problems seems to have fallen out of favour.

Unfortunately, that syndrome was repeated with the publication of models of the future of the world system in the 1970s. Supported by a group of concerned politicians and business leaders, The Club of Rome, Forrester produced *World Dynamics* (Forrester, 1971), the model being created during the flight from Boston to Rome! This was accomplished without the aid of laptop computers, perhaps suggesting that the real skill in model development is thought, not dexterity with software. The model extends to no more than four pages, including the definitions and dimensions of the variables. A second phase of the work was the WORLD2 model of Meadows *et al.* (1972). A full description of subsequent work on WORLD3 appears in Meadows *et al.*, and their influence diagram appears in Fig. 11.1 to indicate some of the characteristics of the world models.

The diagram reflects only what its authors saw as the most significant aspects of the model, so it is drawn without all the detailed conventions of influence diagrams[2]. To show the most striking assumptions, one being that Industrial Output per Capita decreases Fertility, whereas, on the other hand, total Pollution (not per capita) increases Mortality. Is this an inconsistency?

We do not aim to answer such questions, merely to show the reader how the careful study of a diagram can raise them. If one's assessment is that the diagram is correct, confidence in the model has been increased. If it is that the model is not correct, the model does *not* become useless. Either one has identified an area in which it can be improved or one has identified a limitation which can be taken into account when the model's policy implications are evaluated. The reader should study the subsequent case studies from this perspective. No model is 'right' in any absolute sense: the judgement is whether it is right enough to be useful.

The widespread emphasis in the 1970s on environmental problems naturally generated much further system dynamics work. An interesting exam-

---

[2] Users of causal loop diagrams do not adopt the dotted and solid line conventions.

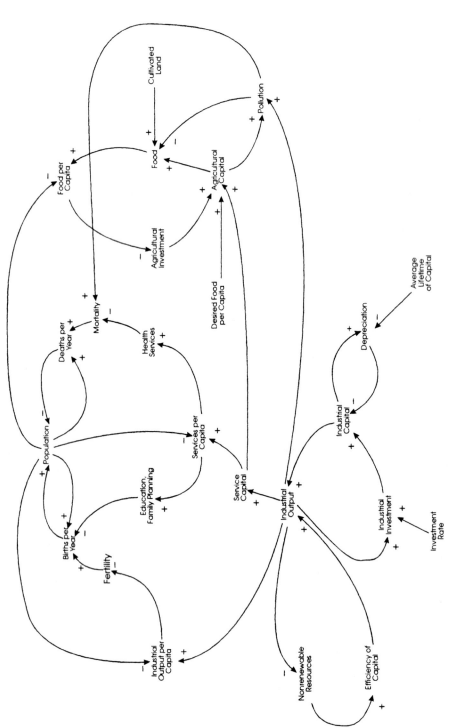

Fig. 11.1 Influence (causal loop) diagram of structure of WORLD3 (after Meadows *et al.* (1974)).

ple is a model of the transition from fossil fuels to sustainable energy sources (Naill, 1977).

In short, a number of themes had emerged from these early models. The first was that the concept of using the principles of control theory to analyse managed systems via a simulation tool had proved to be very practical. The second was that the fields of application were limited only by the wit and imagination of the analyst.

The dynamic behaviour of a wide range of systems could be modelled without much difficulty, and the policy implications were often striking and novel. Unfortunately, they were often also bitterly controversial. The source of controversy was that most people do not think in dynamic terms, and having someone else analyse 'their' problem from a completely novel standpoint was not always comfortable, especially so if apparent errors or misinterpretations of established theory could be pointed out[3].

Current work is covered in a refereed journal, *System Dynamics Review* and the *Proceedings* of the annual conferences of the System Dynamics Society[4].

We now turn to the more detailed examination of some system dynamics case studies.

THE NEW ZEALAND FORESTRY INDUSTRY[5]

Introduction

The economy of New Zealand is significantly dependent on the export of timber in various forms. Large areas had been planted during the 1960s and 1970s which meant that there would be a large crop during the 1980s and 1990s. Continued planting would increase that crop. However, there were risk factors, such as intensified competition in world markets. Consequently, in the early 1980s there was a need to design planting and cutting policies to cope with an uncertain future.

The suitability of system dynamics

How do we know that this is a problem for which system dynamics is more suitable than, say, linear programming, which has been applied to the optimization of land use? There are a number of factors:

---

[3] J. H. Plumb, Boxer (1965), writing of the need to make integrated sense of history, remarks that 'The majority of historians . . . have spent their leisure tearing to shreds the scholarship of anyone foolish enough to give the story of mankind a meaning and a purpose. Writers as diverse as Wells and Toynbee have been butchered with consummate skill. . . . the errors of interpretation have counted for everything; intention nothing'. One feels that could easily be rephrased to apply to the early workers who attempted to apply a new paradigm to the management of social and economic systems.

[4] For information on membership of the System Dynamics Society or subscriptions to *System Dynamics Review*, contact John Wiley & Sons Ltd, Baffins Lane, Chichester, West Sussex PO19 1UD, UK.

[5] For a fuller discussion see Cavana and Coyle (1984).

- There are exogenous shocks, such as changes in demand.
- There are long delays, trees taking about 30 years to grow to harvest size.
- There are clear feedbacks: the more trees are planted, the fewer will need to be planted to meet some anticipated future demand.
- There are significant policy issues for the long term, as opposed to short-term 'tactical' decisions.

In other words, a problem as vast as a nation's forestry policy can be interpreted in terms of the information/action/consequences paradigm shown in Fig. 1.1. In addition, the concerns of the nation's policy-makers are the four upper topics on the right-hand side of Fig. 1.5.

The main point is that an analyst should always justify the use of a methodology. To do so requires familiarity with other methodologies, and doing good system dynamics requires at least a working knowledge of other management science approaches.

Some simplifying assumptions

Any model is a simplification of reality, and the art is to ensure that there is nothing in the model which is not also in the real world and that the significant parts of the real world are in the model. Some modellers try to put as much as possible of the real world into their model. This leads to models which are intractable and difficult to interpret. The key step is to make some assumptions which are sensible for the problem and then to make sure that those assumptions are faithfully adhered to.

For this model there are two main assumptions.

The first is that trees can be grouped into four categories:

- Young forests less than 10 years old, which are non-productive and require attention for pruning and thinning.
- Immature forests (10–25 years old) which could be felled. However, New Zealand's plantation forests have usually been managed to produce logs from trees at least 25 years old, so the immature forests are treated as non-productive. Altering the model to vary this assumption would make an interesting project.
- Mature forests (25–40 years old) which contain wood that is at the desirable age for clearfelling.
- Overmature forests (more than 40 years old) containing slow-growing trees beyond the desirable age for clearfelling.

As a rule of thumb, the LENGTH for a model should be about two or three times the longest delay. If trees are in the system for up to 40 years, LENGTH will have to be about 100 years. This matches the reality that forest management is a very long-term business.

The second assumption refers to the price of wood. If demand for wood and the rate of felling are identical, the price is $10 per cubic metre. If the output of wood is less than the demand the increase in price will be small. If the output is greater than demand, the price will fall more sharply.

The physics of the system

We have emphasized the need to identify physical flows correctly, and the first assumption identifies the first physical flow of trees on a given piece of land moving from one age category to the next. This is, of course, exactly the same as for manpower planning models.

However, older trees contain more wood than younger ones, so there is a matching flow of *volumes* of wood.

Finally, what is sold is volume, which will determine the area to be clearfelled dependent on the age of the trees when they are felled. Thus, the third flow is land being planted, cleared, sold or purchased.

The influence diagram

You should now develop your own influence diagram. Having parsed, one might use the entity/state method for a clearer picture of the physics. The list extension method should also be used to think about control policies, and the two fragments should then be merged into a candidate diagram.

Figure 11.2 shows the resulting diagram at about level 3. The first and last two categories of trees have been grouped together to show the ageing of trees. This Area Transfer Rate creates the Volume Transfer Rate to feed the volume of wood which is available for cutting[6]. The volume is also fed by the Volume Increment, which is the continued growth of trees in the Mature and Overmature classes.

The need to ensure a balanced age structure in the forests implies the Ideal Clearfelling Rate (in hectares per year) and, through Average Volume per Hectare, gives the Ideal Cutting Rate (in cubic metres per year). However, Demand and Price also govern the actual Volume Cutting Rate.

The Planting Rate must replace felled areas and meet expected future demand. That leads to the land flow in the lower half of the diagram.

The reader should add a balance sheet, using the techniques discussed in Chapter 10.

The model equations

WOOD.COS should now be carefully studied, with the aid of a printout. It would also be useful, if the available software provides such features, to

---

[6] The volume of young and immature trees need not be considered, as they are deemed not to be capable of being harvested.

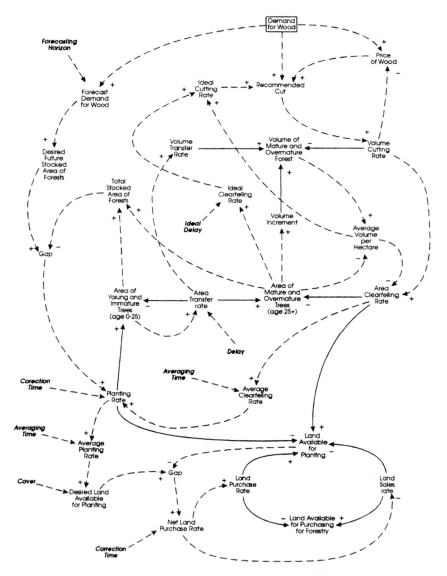

Fig. 11.2 Simplified influence diagram for the forestry model.

produce a printed map of the model's interactions, a documented set of
equations (see Chapter 5, p. 136) and a dimensional analysis.

The model has several of the characteristics of good practice, notably
comments within the equations, full definitions of all the variables, and
extensive print statements, providing evidence that the model has been
carefully debugged. Several plot statements have been suppressed, though
again they suggest thorough debugging, only those for the main aspects of
performance remaining.

Numerical data are from New Zealand Government statistics. The initial value for the Average Planting Rate reflects the previous policy of planting as much as possible. When the model starts, a more rational planting policy is introduced, with consequences which we shall see in the model's output.

The equations for profit and cash flow should be studied, as should the sector headed 'performance indices'. The numerical values for constants such as CWRFV have been found by running the model for the base case and finding, in this case, the value of Cumulative Wood Removed, CWR, at the end of the run. This is then used as a base value to assess the effects of other policies. No amount of further explanation by the author will help the reader to develop the practical skill of grasping what a modeller has done; you will have to work these equations out for yourself. The reader may wish to modify these performance indices, or to create others.

Another indicator of good modelling practice is the long list of sensitivity test experiments, suppressed by NOTE statements. The reader should reactivate some of these and see what they do.

Running the model

The reader should now run the model[7].

The base case, Fig. 11.3A, assumes medium demand growth at a constant rate of 0.3 million cubic metres per year. In these conditions, supply matches demand quite well in the long term. However, the heavy planting of the 1960s and 1970s comes into effect for 20 years from about 1995, when the available cut from the forests, ICUT, is much larger than demand. The pricing mechanism allows wood sales to increase somewhat, ACUT, but wood prices are depressed for about 25 years.

The previous planting history means that APLANT is much higher than PLANT, but eventually the two settle down. Land is sold during the 1980s, but land purchases stabilize towards the end of the run.

With lower demand, Fig. 11.3B, the lower prices last for about 50 years, and the sale of land continues for a long period. Even with high demand, Fig. 11.3C, there is still a period of price depression, but it is relatively short-lived. A problem looms towards the end of the run when land purchases cease because there is no more land left to purchase. Increase LENGTH to 150 years and explain what happens.

These runs are not, of course, forecasts; they are assessments of the future consequences of varying events. All show the serious implications of

---

[7] The reruns can, of course, be done interactively on a PC, but they have been left in the model for instructional purposes. Instructors on graduate or undergraduate courses will be able to develop some interesting individual student projects or class assignments using this model, or the others which appear elsewhere in the book.

the past policy of heavy planting. Even if high demand can be stimulated, a period of depressed prices is inevitable and the distant consequences of high demand are pretty unpalatable. Whatever happens, there are likely to be heavy costs from the previous policy of intensive planting which may well have serious knock-on consequences for the rest of the New Zealand economy.

The model forecasts demand for wood 30 years hence[8]. The three cases examined so far assume that demand is forecast perfectly (study the model at about line 23 to see how forecasting error can be modelled). Figure 11.3D shows the effects of forecasting errors, the pattern tested being pessimism when demand exceeds the actual felling rate, and vice versa. With the medium demand scenario, this introduces cycles of behaviour. For instance, the over-supply of wood during the 1990s causes managers to take a pessimistic view of the future and the planting rate drops. The eventual shortages of wood cause planting rate to rise, and it is not until towards the end of the run that the control policies restore stability.

Numerous policy experiments can be tested. Such as a rotation age of 37.5 years, and an alternative planting policy.

The performance indices mentioned earlier should be printed out and tabulated against the various runs as a way of assessing other effects of policy as well as what can be seen from the dynamic behaviour. (See also Coyle, 1981a.)

Summary

This model illustrates the process of turning a policy *problem* into a policy *model* and several aspects of modelling method. The reader should study the equations, and run the model, *carefully studying the output*[9].

What other assumptions have been made? What modelling simplifications have been made? Are there any other policy options?

Draw an influence diagram at Level 1 to reflect the insights derived after the model has been carefully studied. The reader who has access to optimization software will be in a position to explore this model very deeply, on the lines shown in Chapters 8 and 9.

Technically, this model is very classical system dynamics. It is essentially continuous in that land, tree age groups and timber volume do not change in sharp increments. We now turn to a policy problem for which those assumptions would not be reasonable.

[8] A map (p. 137) of the model will help immensely in tracing out these factors.
[9] This remark may seem superfluous. The reader will just have to take the author's word for it that he has seen many cases in which the modeller in question had done no more than glance at a few graphs on the screen and had completely failed to see errors which became obvious from even a cursory examination of the output and which undermined any credibility the model might have had.

DEVELOPING ARGENTINA'S ELECTRICITY CAPACITY[10]

Introduction

The purpose of system dynamics is the study of the evolution of a socio-economic system through time and this often requires the modelling of capacity expansion. National energy supply policies are clearly such a problem (see also Naill (1977)).

Argentina's energy policy in the late 1970s aimed to increase the use of hydroelectric generation. This programme was to be implemented over a long period, in the face of unpredictable changes in the price of oil and geared to meeting an increasing demand for electricity (for an interesting discussion of future prospects in Argentina, see Macrae (1994)).

Earlier system dynamics work on energy problems has often been based on models in which capacity can be added in continuous quantities. Hydro-electricity is, however, a case in which the generating capacity available from places where dams and power plants can be built varies enormously from one site to another.

In Argentina there are about 26 places at which hydroelectric power stations could be built, but their potential output varies from a few hundred to nearly 20 000 GWh (GWh) per year. The costs of development vary at least as much and the time required to build a station will also range between roughly 9 and about 12 years[11]. However, the output from the stations will be a continuous flow of electricity which will produce continuous flows of cash.

For credibility, each project must be treated individually, which requires the modelling of whether or not to order *a particular project* at a given time together with the continuous processes of the output of electricity and the flows of cash as revenue and expenditure during construction. Such complexity has often been assumed to be impossible within the framework of system dynamics software packages, but work on modelling coal production has suggested that the difficulties were not as severe as had been assumed (see *Equations for Systems*, described in Appendix C).

The reader should now think through the justification for using system dynamics for this problem.

---

[10] For a detailed discussion see Coyle and Rego (1982).

[11] The four river basins are the Plata, on the international border with Brazil, Uruguay and Paraguay, the Patagonian basin and two basins in the Andes, Cuyo and Neuquen. The Plata projects are large, with high construction costs and low unit output costs. Construction delays are about 12 years. The Cuyo system offers small projects, with relatively low construction costs but higher subsequent operating costs and construction delays averaging about 9 years. The main difference between the Patagonia and Neuquen basins is the likely construction delays, about 12 years for Patagonia, which is about 2000 km from the main electricity markets, and 9 years for Neuquen.

Capacity expansion policies

Figure 11.4 shows some possible policies. If capacity can be added in effectively continuous quantities, the idealized case in Fig. 11.4(a), projects will already be in the pipeline and demand will be forecast for the construction delay into the future, revealing a gap between Forecast Available Capacity, FAC, and Forecast Required capacity, FRC. Additional capacity can be ordered now to be completed by the required time, thus eliminating the gap. This process will roll indefinitely into the future and all should be well.

The physical realities with projects taking either 9 or 12 years to build, requires greater complexity. Figure 11.4(b) shows a policy of concentrating on meeting demand. Forecasts are made for 9 and 12 years ahead. The 9-year gap is filled first by choosing 9-year projects of the right size and additional 12-year projects are also ordered to try to fill any remaining gap at 12 years.

On the other hand, 12-year projects are likely to offer lower operating costs, and Fig. 11.4(c) shows a cost-orientated policy of selecting 12-year projects as far as possible and, as it were, filling in any 9-year corners with 9-year projects. Both of these policies depend on projects, or project combinations, of the right capacity being available, and, as we have pointed out, the projects vary considerably in their potential generating capacity.

The influence diagram

In Fig. 11.5 the system contains only negative feedback, the two loops which are apparently positive arising only as the drains from delays and being dominated by the true negative loops.

Figure 11.5 is a very aggregated view, and Fig. 11.6 indicates some of the complexity in the model, using the Cuyo basin for illustration, though the same ideas are used for the other basins.

The first idea is that the capacities of all 26 projects are stored in the Catalogue Table of Hydraulic Project Capacities. The second is that some of these projects, numbers 10 to 15, say, are projects on the Cuyo basin. A second table contains the preferred order in which these projects are to be taken: perhaps number 12 is first, 14 is second, and so on. This order could be varied to test, within the constraints of hydraulic engineering, alternative plans for Cuyo development.

Every time a project is selected from the Cuyo basin, the Cuyo Project Counter is increased by 1, but that only records the number of projects so far chosen, not which projects they were. The rest of the diagram shows a set of counters which record that Project N is moving from one stage to another. At the correct point, the actual capacity of Project N is picked up from the table and eventually added to the hydroelectric capacity.

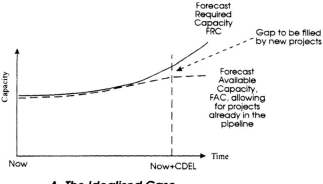

A The Idealised Case

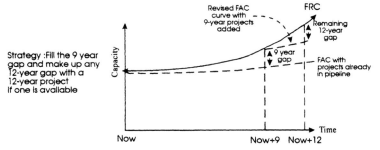

B The Demand Orientated Straegy

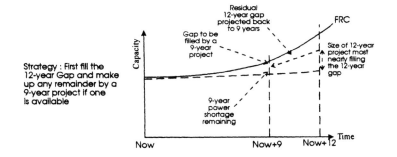

C The Cost Orientated Straegy

Fig. 11.4 Capacity expansion policies.

These ideas are expressed in equations of the following form.

First, the Cuyo Indicator Start Rate is generated from CYSTANDP, the next Cuyo project to be chosen, number 10, for instance, and CUYOPSW, a 0/1 switch to start a Cuyo Project:

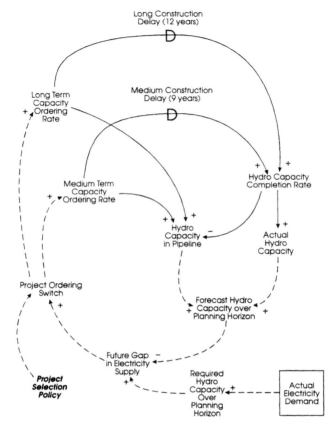

Fig. 11.5   Basic feedback loop structure controlling capacity growth.

r cyisr.kl=cuyopsw.k*(cystandp.k/dt)

Project number 10 is now stored for 1 DT:

l cycds.k=cycds.j+dt*(cyisr.jk−cyerase1.jk)
r cyerase1.kl=cycds.k/dt

which is just long enough for the capacity of Project 10 to be selected from the project capacity catalogue, TPROCAP:

a cycsds.k=tabsq(tprocap,cycds.k,0,26,1)

TABSQ is a special form of table function which gives the effect of a square table (see the COSMIC User Manual for more details). The 26 in TABSQ is the total number of projects in Argentina.

If this approach of using a control which lasts for only 1 DT is not used, the same project will be injected into the system over and over again.

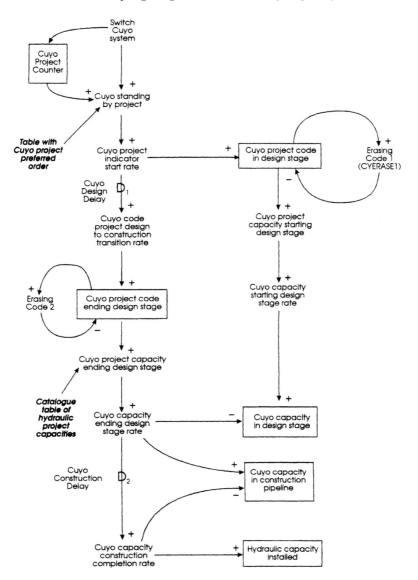

Fig. 11.6   Influence diagram of Cuyo capacity expansion.

The full detail of the problem is shown in Fig. 11.7.

The model

POWER1.COS. runs a reference scenario in which tables are used to define the proportion of electricity capacity from thermal and hydroelectric

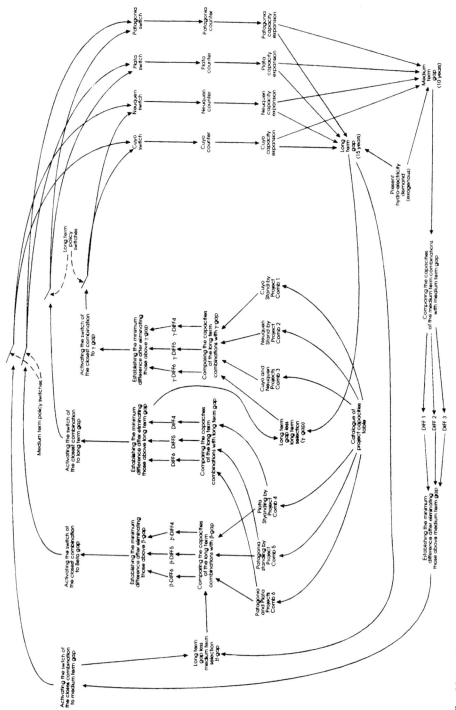

Fig. 11.7 Overall structure of the hydroelectric planning problem.

capacity. Prices are in 1978 US$ with a policy of maintaining constant real electricity prices. Note the extensive financial sector. Make some sensitivity tests. Add print statements and carefully check that the model works.

Figure 11.8 shows that the complicated equations for hydroelectric capacity addition produce the required discrete additions of capacity. The hydro sector produces growth in line with the intended plan. The trick of adding the capacities from the four hydroelectric regions gives a convenient display and the large steps in capacity are some of the very large projects from the Plata basin coming on line.

POWER2.COS, tests policy options with various scenarios of demand growth and oil price. With moderate growth in oil prices and low growth in electricity demand capacity, Fig. 11.9A, is added to provide at least the planned margin of safety, and return on investment is, if small, as least always positive. Depreciation cash flow and after tax profits always exceed the rate of repayment of long-term debt and the ratio of debt to assets is kept under control. In short, Argentina's national electricity company is solvent and can meet the demands imposed on it, under these particular conditions.

The lack of robustness in the policy of attempting to maintain constant real electricity prices and expand to meet demand forecasts in Fig. 11.9B, produces a financial catastrophe late in the 1990s. Other cases set up as reruns in POWER2.COS produce much the same result. In short, the official policies are not going to work, mainly because they were never designed from a dynamic point of view.

As with the New Zealand case, there is much that the student can do with this model. One policy option is suggested in Fig. 11.10, namely one of gearing the real price of electricity to the financial needs of the electricity company. Optimization will analyse this system in depth and design rich policy options for it.

Other options for improvement lie in the relationship between the electricity supply and wider national economic growth[12].

Summary

This model and the underlying problem have been described in somewhat less detail than the New Zealand case, mainly to stimulate you into investigating the model for yourself.

Technically, it is very complicated and shows how it is possible to use the concepts embedded in a sophisticated system dynamics software package to go beyond simple continuous models.

---

[12] System dynamics has been applied to the modelling of the growth of developing nations. Saaed (1994) is an excellent example. For an earlier study, see Forrester (1973).

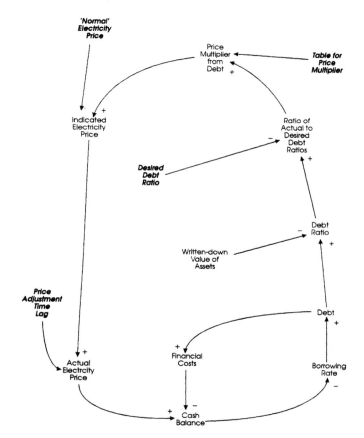

Fig. 11.10 Endogenous pricing policies.

COMPARISON OF THE NEW ZEALAND AND ARGENTINE CASES

These two models are very different and the reader should make a comparison between them. For instance, the New Zealand model is only about half the size of its Argentinian cousin. Its equations are simple, whereas the power problem is about the most complicated the author has ever worked with.

Is the Argentinian model 'better' because its techniques are more elaborate? Is the New Zealand model 'better' because it is easier to understand? It could be that the New Zealand model is 'worse' because it oversimplifies reality too much, or perhaps the Argentinian model is 'worse' because it has used overcomplicated techniques just to look impressive.

Such comparisons are virtually meaningless. The only way of really judging a model is *relative to its objectives*[13]. Review what was said about the two models and then try to evaluate them.

QUALITATIVE SYSTEM DYNAMICS

We now turn to some instances of the use of influence diagrams as models in their own right. The purpose is both to show the scope of qualitative analysis in and to provide opportunities for the student to think about problems and discuss them, perhaps in class or a syndicate. For that reason, some of the cases will have very limited explanation and you will have to work things out for yourself, seeking feedback loops and thinking about what they imply for the system's behaviour. The student may find some interesting opportunities for quantitative modelling in these studies, making assumptions about numerical parameters.

The case studies cover air traffic control, drug trafficking, the maintenance of business effectiveness during a reorganization and the criminal justice system in Britain.

As we work through these rather disparate problems, we shall not always adhere to the strict formalities described in Chapter 2. Influence diagrams are models, and all models are tools for thinking with. The emphasis is on the *thinking*, and the tool should be adapted to promote thought, not followed so strictly as to inhibit insight.

In studying these applications it is important to bear in mind that none of them required more than a few days of effort.

AN AIR TRAFFIC CONTROL SYSTEM

This problem (Fig. 11.11) is, at first sight, one of physical flow. The essence is that the more aircraft there are in the zone relative to the ability of the control system to handle them, the longer it will take for aircraft to get through the zone (we have all been in an aircraft which is 'stacked'). In practice, airlines negotiate with each other for landing slots, thus hoping to ensure that aircraft can enter the zone when they can be accommodated. The weather and problems at other airports ensure that the plans do not always work out.

There are two pressure points in the system. Buying more ATC computers, radars and people will increase Work Capacity, but there may be no point in doing so unless the Air Traffic Configuration is well designed. The $+/-$ sign on that link is to suggest that a given change to the configuration may make matters worse or better, depending on what it is. Obviously, no ATC planner *aims* to make matters worse by rearranging flight routes but

---

[13] These sections are derived from two Ph.D. theses, but describe only selected parts of those studies. One should not attempt to compare the two dissertations, merely these two models.

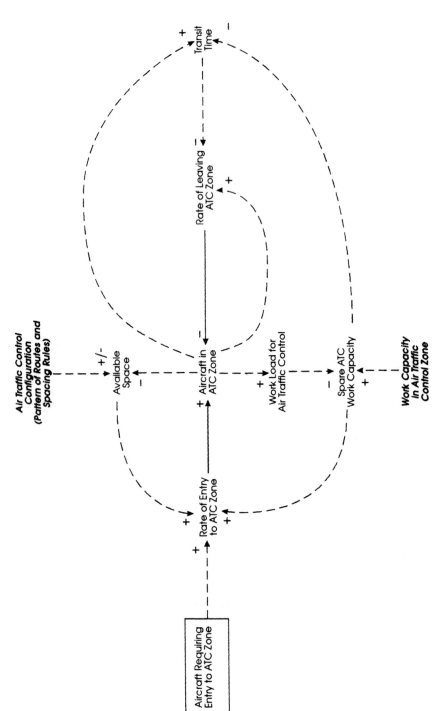

Fig. 11.11 Overall influence diagram for an air traffic control zone.

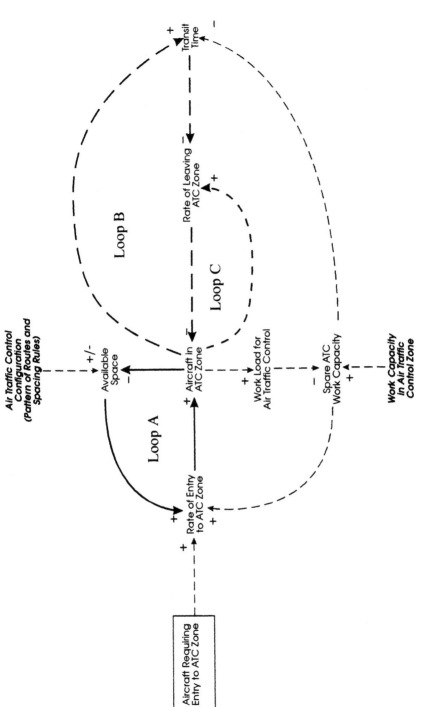

Fig. 11.12 Loops for an air traffic control zone.

dynamic systems can sometimes behave in counter-intuitive ways, so it is unlikely that all changes will be for the better, and careful design may be called for.

In Fig. 11.12, Loop A should regulate admission to what the system can handle, sometimes achieving its effect by delaying the departure of aircraft from, say, Paris so that the London ATC Zone should have capacity at the expected arrival time. Loop B, being positive, ensures that problems are likely to get worse once they arise, though Loop C, acting as a drain from the system, ensures that the problems will eventually dissipate.

Figures 11.11 and 11.12 are somewhat in the spirit of the DMC problem, in that the emphasis is on the physics of the system. Air Traffic Control is, however, but part of a system which also involves the airlines, the passengers, the airports and the aircraft manufacturers.

Some of this complexity is shown in Fig. 11.13, which is probably at about Level 2 or Level 3 in the cone. Numerous delays are indicated, of very different magnitudes. Since the problem is now essentially one of behaviour, no physical flows are shown, though they would be present in a more detailed diagram.

Ideally, the mismatch between supply of and demand for ATC services in the upper left-hand corner will trigger investment in ATC systems to increase the ability to handle flights and redesign of the configuration to get the most out of the available space, though the investment may be constrained by the vast cost of ATC systems. ATC charges may have to rise to recover the investment, the countervailing pressure being that charges can be spread over more flights. It also triggers efforts by the airlines and the ATC managers to rearrange flights to what the system can handle.

The economics of passenger and airline supply and demand are indicated in the lower left-hand corner, as are the effects of traffic growth and the changing size of aircraft.

Several pressure points are shown. One, overflights, is largely a matter of geography; the others are under the control of the ATC authorities, the airports, and the aircraft builders. Which is which, what do they mean and why do some have +/− effects?

THE DRUGS 'PROBLEM'

The use of drugs has been studied by several system dynamics workers. Homer (1993) developed a quantitative model of the dynamics of drug usage but to do so requires assumptions about the effects of, say, the supply of drugs on the rate at which people start to use them, Homer making diligent efforts to fit his model to available data.

Figure 11.14 attempts to capture the initial and tentative thoughts on the

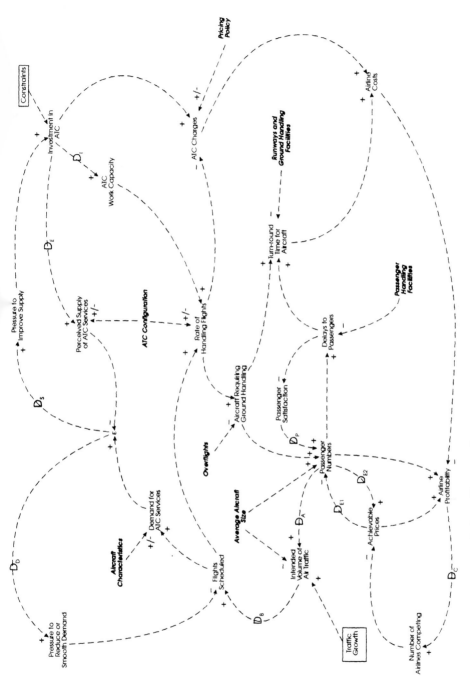

Fig. 11.13 The context of air traffic control.

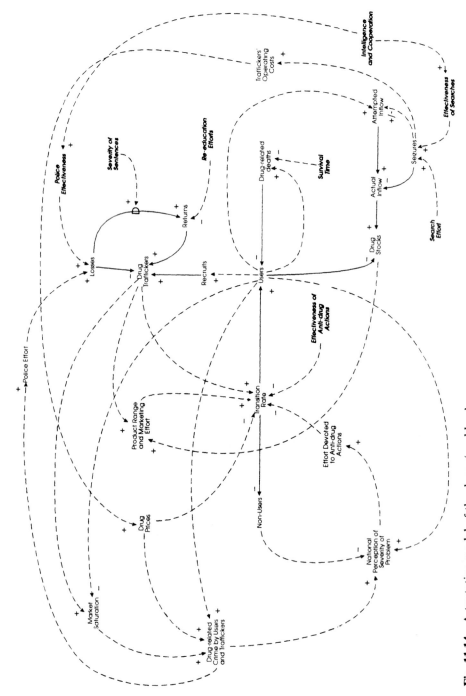

Fig. 11.14 A tentative model of the drugs 'problem'.

problem. The essence of building this diagram was the entity/state/transition method once one recognizes that there are three entities (at least): the ordinary population who are, or are not, drug users, and who can be viewed as moving between the two categories, the population of drug traffickers and, of course, the physical flows of drugs.

There are many areas in the model where questions are posed rather than answered. For instance, Homer argues that the price of drugs is determined by the traffickers' operating costs, rather than by more obvious forces of supply and demand. Perhaps excess supplies of drugs (the Drug Stocks in the diagram) stimulate traffickers to extend their product range. The influence of seizures of drugs on attempted inflow is shown by a +/− sign to suggest that up to a certain point, seizures increase attempted inflow. Presumably, the intention of the enforcement agencies is to seize so much that the sign turns negative and the trade becomes so unattractive that it diminishes. Perceptions by the non-using population that there is a problem may be a driving force in increased efforts against the trade.

An ideal way to make progress on a problem as complex and uncertain as this is by the study period method, involving discussions between interested parties and examined in the next chapter.

It is often useful to simplify a model to its essence by moving up the cone of diagrams. Figure 11.15 is a Level 1 interpretation. It suggests that the perceived severity of the problem triggers efforts to address it while, at the same time, other forces increase the drugs business. Two sorts of government action are shown: those to ameliorate the problem by social and educational means and those to destroy the value of the trade by preventive

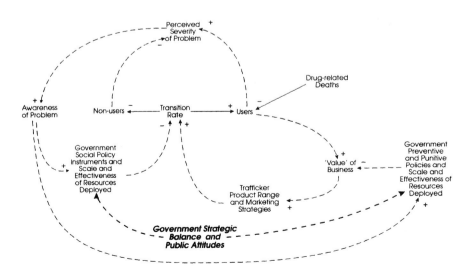

Fig. 11.15   The drugs 'problem' at level 1 in the cone of diagrams.

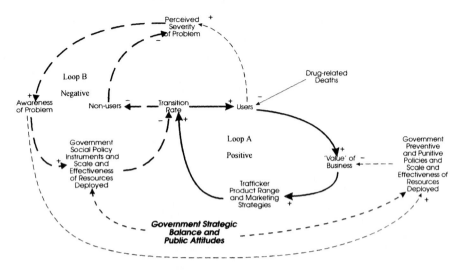

Fig. 11.16 Feedback loops at level 1 in the drugs 'problem'.

and punitive methods. The government must strike a balance between the resources allocated to the two approaches.

The effects of these mechanisms are highlighted in Fig. 11.16, which shows two of the loops. There are more and, armed with an understanding of them, the reader may be able to make more sense of Fig. 11.14. Certainly, an attempt to improve Fig. 11.14 (which is *not* the same as making it more detailed) would be a worthwhile project.

BUSINESS EFFECTIVENESS – THE CASE OF UNIVERSAL INDUSTRIES[14]

Universal Industries has expanded considerably over the last 10 years and its staff are scattered over offices in the south-east of England. This causes much inconvenience and makes it difficult to reorganize the activities and to take advantage of modern technology such as the 'paperless' office. Accordingly, Universal intends to move all its staff to a single large office in Shiretown in the English midlands.

The move will take a year to complete and there will be a need for staff to commute from London to Shiretown and vice versa for meetings. This effect will be largest when the transition is about half complete. While the move is going on it will be difficult even to know where people

---

[14] The case studies in this section and the next were provided by Jonathan M. Coyle. The author has somewhat simplified them and adapted them to the style of the book.

are (internal telephone directories are usually out of date before they are printed). As the move progresses, more time will be lost in travel from Shiretown to Universal's customers and suppliers who, by and large, are in London.

While individuals are moving home and office, they are effectively un-available for work. To that effect must be added the fact that there are a number of people who do not want to move at all. They may be expected to increase wastage from the London staff and to add to the time lost due to illness. Naturally, wastage must be replaced, which will add to time lost in recruitment and training.

In Fig. 11.17, the Work Profile represents a base load with fluctuations. The Actual Work Output is based on the number of effective staff and a normal working week, though the introduction of the 'paperless office' at the new site will have the effect of increasing the work output.

A significant factor is the nature and extent of the departmental reorgani-zation. This is shown with a ? sign (another example of being flexible with the conventions of influence diagrams) to show that the effect on the ability to locate people may or may not be beneficial, and will depend on the nature and timing of the reorganization.

The model is based on a very simple idea of it. 'Business effectiveness' can be a very vague concept, in this case it has been represented as the fraction of the work output which can be achieved against that which is required. To that, of course, could be added the financial costs and benefits of the move. The object is to come up with a Movement Plan to minimize penalties and maximize paybacks.

A STUDY OF POLICE FORCE MEASURES OF PERFORMANCE

This study is an example of how the qualitative strengths of system dynam-ics can be applied to address a problem in which the boundaries and issues are far from clear. The diagrams presented here were intended to be purely illustrative and no claims can be made about their comprehensiveness. The diagrams do not represent United Kingdom government policy or opinion in any way.

The study was conducted in two iterative stages:

1. A short period of initial fact finding followed by the creation of a draft influence diagram.
2. Presentation of that diagram to the client followed by revision of it.

This study also shows the flexibility of influence diagrams. In the drugs problem, the major diagram was presented first, leading to two higher level portrayals of the major elements of the problem. In this case, the approach is to build the diagram up in stages, so that each can be understood before the next is encountered. Dotted lines are what was on the preceding

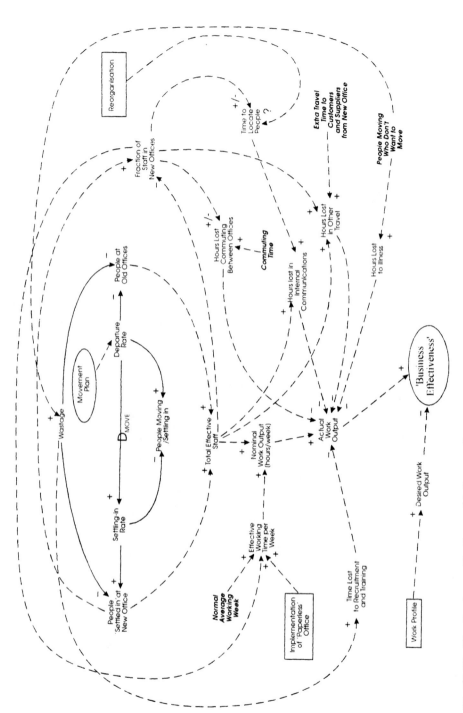

Fig. 11.17 The 'business effectiveness' influence diagram.

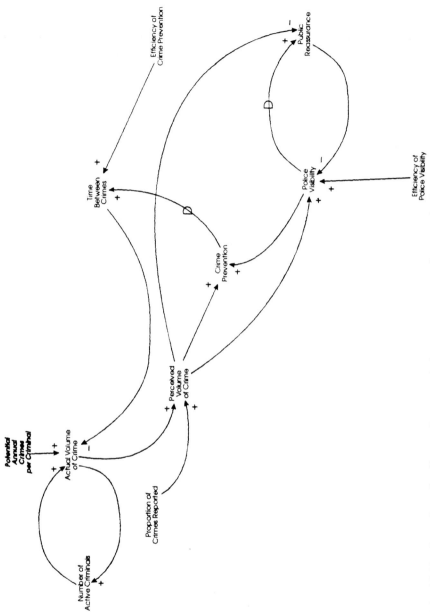

Fig. 11.18 Overview of police performance (crime occurrence and prevention).

diagram and solid lines show what has been added. Factors which might appear to be parameters in one diagram will be variables in the next. The true parameters will be shown in bold italic type in the last diagram, Fig. 11.21.

Figure 11.18 represents the first element of the diagram.

Volume of Crime represents a function of the number of active criminals in the population, the number of crimes each criminal could commit if there were no police, and the time between actual crimes. It attempts to capture the idea that, since a car thief's 'productivity', in crimes per month, may well be higher than someone attempting to commit a complex fraud. However, the Perceived Volume of Crime is a function of the Proportion of Crimes Reported as well as of the actual value. This means that the complex fraud described above might be seen by the public to be more significant or worrying than car crime if a higher proportion of fraud cases than car crimes were to be reported. In the late 1980s this was indeed the case, though, in the early 1990s, car crime has higher visibility than fraud.

Perceived Volume of Crime causes an increase in demand for resources to be allocated to Police Visibility (police officers on foot patrol, as opposed to less visible car patrols) and also decreases Public Reassurance. In this way, increases in the Proportion of Crimes Reported have two effects:

- They cause Perceived Volume of Crime to equate more closely to the actual value and so allow the resources allocated to, say, crime prevention, to match the real requirement.
- They increase public concern about crime, leading to a greater demand for resources to be allocated to Police Visibility, which might not be the most beneficial use of police resources.

Figure 11.19 expands the previous diagram to cover crime detection and the Crown Prosecution Service (CPS). It suggests that increases in Perceived Volume of Crime and decreases in Public Reassurance will also cause more resources to be allocated to Criminal Identification. After a delay, the Time to Prosecute, which is the time taken to prepare a case, and is constrained by the Capacity of the CPS, this will increase Successful Prosecutions which increase Public Reassurance about crime, thus forming a large and significant negative feedback loop, running from Police Visibility via Response Time to the volume of crime variables. The Proportion of Crimes Reported, as well as affecting Perceived Volume of Crime, may also cause a decrease in the Time to Identify Criminals and so increase the rate at which they are identified. The Time to Identify Criminals is also a function of the Probability of Arrest, which may increase as the Number of Crimes Reported rises. Decreasing Police Response Time should increase the Probability of Arrest and so enhance Public Reassurance. Unfortunately, Police Response Time

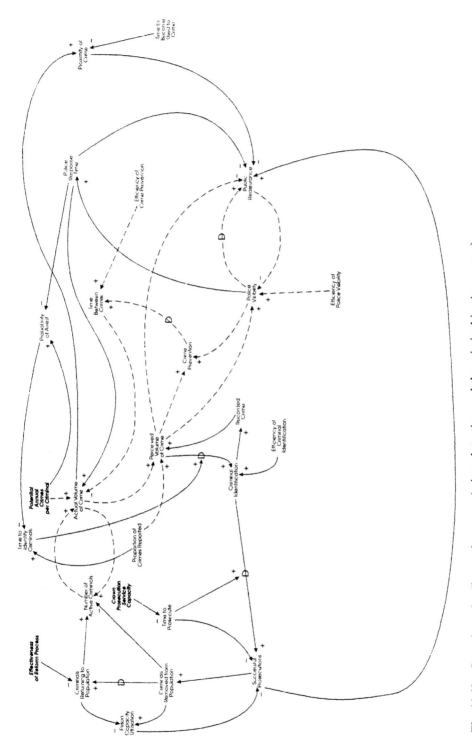

Fig. 11.19 Overview of police performance (crime detection and the criminal justice system).

may be increased by diverting officers to Police Visibility. In this way, we see that there may be a conflict between two of the control mechanisms in the system (the police visibility loop and the response time loop)[15].

This is a good example of one of the major strengths of the influence diagram approach. Police Visibility and Response Times are two variables that the public often use to measure the performance of their local police. However, the diagram shows that these measures conflict with each other. This highlights a potential weakness in treating them as independent measures.

Figure 11.19 also shows that an increase in Successful Prosecutions will lead to an increase in Criminals Removed from the Population. However, after serving their sentences, these criminals will return to the population, unless the reform process is effective. The diagram shows that the criminal identification process is a negative control mechanism, but it requires investment in prison resources to keep Prison Capacity Utilization to an acceptable level in order to sustain the process. However, the link from Successful Prosecutions to Public Reassurance may well be quite a strong one.

Figure 11.19 shows that increases in Actual Volume of Crime increase the Proximity of Crime, the feeling that crime takes place 'close to home', thus causing Public Reassurance to fall. This may have a stronger effect on Public Reassurance than Perceived Volume of Crime. This last variable is also affected by the rate crime is recorded. Two variables in this system, Recorded Crime and Proportion of Crimes Reported, potentially distort the Perceived Volume of Crime. Given that this last variable is the driver for many of the negative loops in the system, it is clearly important that it is as close as possible to the Actual Volume of Crime.

Figure 11.20 examines community relations, police morale, behaviour and training. The parameters are now shown in bold italic type. Some of them, such as Time to Become Used to Crime, are characteristics of human behaviour. Others, such as Police Training and Crown Prosecution Service Capacity, are pressure points which might improve the system's performance. It ought to be possible to apply the techniques of constrained and free optimization to deduce broad priorities for improvement strategies, as was done in Chapter 9.

Improvements in Police Morale and Police Training will increase the Efficiency of Criminal Identification, while Police Training will increase the Efficiency of Criminal Identification, Police Visibility, and Crime Prevention. These links are not negative: making morale worse or having less well trained police officers is unlikely to improve their performance. Police Morale is likely to be enhanced by Successful Prosecutions and reduced by Complaints Against the Police and by those that are upheld. Complaints

---

[15] You will be wasting your time in studying this diagram unless you are also finding and assessing the effects of these loops.

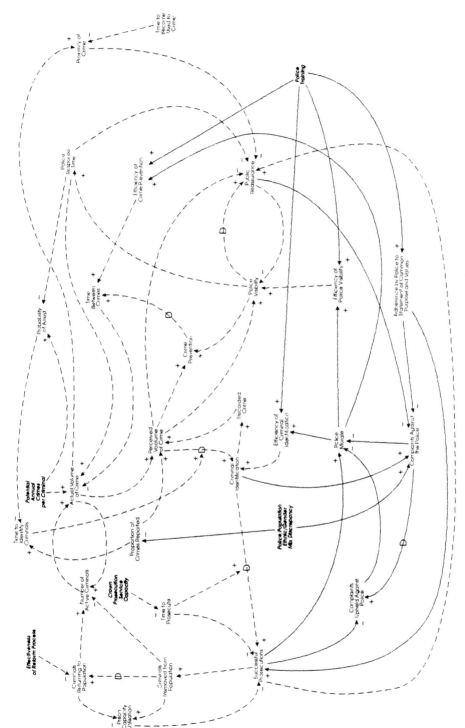

Fig. 11.20 Overview of police performance (community relations and police morale, behaviour and training).

Against the Police will increase if there is an imbalance between the ethnic or gender mix of the force and its local population. This variable will also cause a decrease in the Proportion of Crimes Reported, which has already been identified as a significant variable. Other factors which affect the number of complaints against the police include Adherence by Police to Statement of Common Purpose and Values (which presumably could be increased by further police training) and the number of criminals identified (thus the police could be victims of their own success in some circumstances). Finally, falls in Public Reassurance could lead to increases in the number of complaints against the police.

The analysis of influence diagrams is sometimes helped by laying Venn diagrams over them to emphasize sectors of a model and to identify sector interactions. Figure 11.21 does this for some of the measures of effectiveness proposed by the Audit Commission (1992). Each sector is identified by a name in bold Roman type and an arrow to the boundary of the sector.

It can be seen from Fig. 11.21 that there is considerable overlap between some of the measures of effectiveness. For example, those connected with complaints against the police are completely overlapped by other measures. This may be a desirable feature in that measures which overlap can provide checks on, or alternative insights into, other measures.

The diagram addresses the issue of the quality of service provided by the police in a number of indirect ways. However, the value of using complaints against the police as a measure of quality is that the volume of complaints could increase as the police become more 'successful'. For instance, an increase in the number of criminals identified could in itself lead to a corresponding increase in complaints.

The feedback loops that deal with crime prevention, criminal identification and police visibility are all affected by how 'efficient' the police are at that particular task. In other words, the Time Between Crimes is affected not only by the volume of resources allocated to crime prevention but also by the Efficiency of Crime Prevention. This logic can of course be applied in reverse; if it was possible to measure the Time Between Crimes (perhaps by interviewing known criminals at the time of arrest or conviction) and the level of resources allocated to crime prevention, the analyst would start to gain insights into the efficiency of a particular police force in the role of crime prevention.

In summary, the diagrams illustrate how the technique could be used to structure debate about, and research into, this issue, perhaps through the medium of the study periods discussed in Chapter 12.

THE COLLAPSE OF A COMPANY

In an earlier chapter we examined the case of the collapse of the Mayan civilization. Similar dramas occur in business life and, to encourage the reader to search the management literature we refer him to some seminal

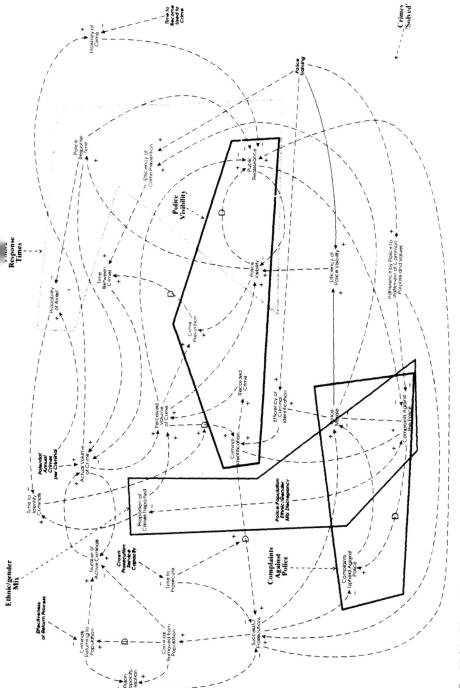

Fig. 11.21 Venn diagram of sectors in police measures of effectiveness.

work by Hall (1976) on the collapse of the magazine *The Saturday Evening Post*. Apart from its business interest, this case shows the power of a small model and the use of other management science techniques, in this case regression analysis, in a system dynamics model.

SYSTEM DYNAMICS APPLIED TO DEFENCE ANALYSIS

Introduction

This section uses the open literature to describe the work which has been done and to suggest reasons why system dynamics has been a useful tool for defence analysis.

Coyle (1981b) shows many of the ideas which have emerged in this body of work. It was suggested by General Sir John Hackett's hypothetical scenario, *The Third World War* (Hackett *et al.*, 1978). The model proved to be surprisingly easy to develop and to 'fit' to Hackett's data. When the initial conditions are changed, the model also fits a different scenario (Bidwell *et al.*, 1978).

The model has 300 variables and requires a few seconds to run on a PC, whereas some combat simulations consist of 200 000 lines of code, and require major effort to run. This difference in size and tractability will be explored later.

WW3.COS can be run to produce Hackett's scenario of war after a period of warning and Bidwell's scenario of a surprise attack.

Qualitative analysis of influence diagrams has been very useful in the defence arena. A paper written *before* the Falklands War analysed the problems of sea control over a group of islands (Coyle, 1983).

Qualitative modelling was further applied (Coyle, 1984) to the relationships between East and West in the context of the balance of interest between Northern and Southern nations which had been suggested by the report of the Brandt Commission, mainly with the intention of posing research questions. Influence diagrams have also been used to model counter-insurgency warfare (Coyle, 1985). The main insight was that the government forces should emphasize preventing insurgents from forming parties large enough to attack sources of government power, combined with a policy of controlling the availability of supplies to them.

Wolstenholme (1986) describes a quantitative model of a Red force attacking a Blue defender; an extension of system dynamics modelling from strategic problems to more detailed analyses of Blue's fire discipline and Red's attack tactics. He comments that:

> In general the role of system dynamics [is to create] initial insights into problems prior to detailed conventional modelling. This is . . . an important role.

System dynamics analyses the influence of information feedback processes on the behaviour of systems. It was, therefore, natural to apply the method to military command and control (C2) systems. Coyle (1987) describes models which cover two aspects of the problem. The first is an influence diagram of the 'theory' of command and control systems. The second is a quantitative model of the ability of given C2 assets to handle specified volumes of information. The output indicates how quickly a C2 system can respond to sudden events, and, indeed, whether the system will ever recover, how long it takes the system to cope with extra demands, and how much query traffic would be generated by a sluggish response.

Wolstenholme *et al.* (1992) describe the use of system dynamics for the design of information systems in general, using two military examples. The first shows how a proposed information system can be used to regulate the movement of ammunition more effectively. The authors make the very profound point that

> the proposed information system must be evaluated in terms, not only of its immediate impact, but also in terms of its potential to support more efficient policies.

The qualitative sea control model was developed into quantitative form by Coyle and Gardiner (1993), in a model of submarine patrols. The work dealt, firstly, with the construction of a fleet of submarines and their recycling throughout their lives through major overhauls, and, secondly, with the ability of a fleet of vessels to maintain a given level of patrols. The output from the first model indicates how frequently a fleet of a given size would be able to put a certain number of boats on patrol over a life-cycle of 20 years. For example, with a fleet of four vessels, there will never be four available, whereas with a much more costly fleet of 10, four will be available about two thirds of the time. The second model represented submarines being despatched on patrol but having to return if there is a mechanical breakdown.

Another maritime model (Coyle, 1992a) deals with a fleet of aircraft carriers attacking enemy bases while surviving against enemy attack. The technical interest lay in treating carriers as integer objects during deployment and combat. These discrete variables must be mixed with continuous processes of carrier sorties, damage infliction and self-repair, to some extent, by the carrier crews. A copy of the model appears on the disk as CARRIER.COS[16].

---

[16] This model has some apparent dimensional inconsistencies: for instance CACE is the total number of aircraft available and is the sum of bombers and fighters. Dimensional validity is a vital ingredient of a successful model, but it should be used flexibly where *exceptional circumstances* arise. The author, who has a lot of experience in modelling, spent many hours carefully studying the printed output to make sure the model was working correctly.

Unconstrained optimization methods were applied by Wolstenholme and Al-Alusi (1987) to the Red attack versus Blue defence model described in Wolstenholme (1986), a most interesting aspect being the 'design' of the non-linear policy curves used by Red to control his attack. The optimum curves turn out to be very different from the curves intuitively chosen in the simulation of the same problem and between optimizations using different objective functions.

Constrained optimization is a theme in an analysis of defence expenditure (Coyle, 1992b) using an aggregated model of land, sea and air warfare in a hypothetical country. Several different levels of defence budget are optimized separately, and the results indicate the priorities for spending.

Clark (1986a) modelled the dynamic effects on finance of the expansion of the US Navy during the 1980s. He draws on Cobb–Douglas functions to estimate operating costs.

Clark (1986b) studied the effects on the stability of Central America of US arms transfer policies. His model simulates the acquisition of military capability and perceptions of threat in four nations.

Clark (1993) has studied the economic context of defence procurement. He discusses a model of the effects of cut-backs in the forces on the procurement of assets and, in turn, on procurement by manufacturers.

System dynamics has been applied to a considerable number of real problems by defence analysis civil servants and consultants in several countries. Topics have included combat modelling, manpower planning, equipment design, equipment support management and many others.

Some of this work is in the process of appearing. For example, Coyle and Richardson (1993) have used a system dynamics model to evaluate lessons from the Gulf War.

System dynamics has enjoyed a considerable degree of application to defence problems. The reasons are not difficult to see.

Warfare is fundamentally a dynamic process governed by the initial forces and weapons (parameter values), attrition and reinforcement (physical flows), command and control (information feedback) and heavily influenced by non-linearities. That is precisely the paradigm which system dynamics addresses, so it is not surprising that the approach fits the problem domain extremely well.

Further, system dynamics, by its insistence on capturing the principal mechanisms in a system, is consistent with a top-down, strategic view of problems, which accords well with the expectations and viewpoints of senior decision-makers. The clarity and simplicity of a well-constructed influence diagram appeals to their sense of logic and does not demand too much of the time of very busy people, which engenders a feeling of control and problem ownership.

A factor in the success of system dynamics has been the willingness of system dynamics modellers to go outside their own tradition of rather broad, continuous models and to tackle problems involving discrete events and stochastic processes, while still aiming at the holistic paradigm of system dynamics. The flexibility of the software has been a powerful tool in achieving that aim.

Wolstenholme was quoted earlier as remarking that a system dynamics model might be seen as a precursor to a detailed conventional model. While that is certainly true, one feels that it is not the whole story. There is also a role for system dynamics as a tool for guiding modelling strategy. If, for example, a system dynamics model showed that the investment in a new weapon system could not be beneficial for any reasonable range of aggregate parameter values, there would be little point in any further detailed conventional modelling. One would therefore revise Wolstenholme's remark to read

the role of system dynamics is to create strategic insights into defence problems as a stimulus to, and guide for, any further work which might, *or might not*, be required.

That applies to every area to which system dynamics has been applied.

SUMMARY

This chapter has explored the application of system dynamics to real problems. The aim has been to stimulate the reader's insight into the capabilities of the discipline and to show how problems of great complexity *can* dealt with. The reader will get most value from the chapter by running the models, studying their output, becoming familiar with how they work, and then experimenting with them. Optimization will deepen understanding even more.

The examples, and the instance cited in the system dynamics bibliography (see footnote 1 in this chapter) cover such an enormous range of topics that it is hard to identify problems to which system dynamics cannot be applied. That is not the same as saying that system dynamics *should* be applied to those problems, and, in the final chapter we shall consider some criteria for assessing the circumstances in which system dynamics might be an appropriate methodology.

REFERENCES

Alfeld, L. (1975) *Introduction to Urban Dynamics*, Wright-Allen Press.
Audit Commission (1992) *Consultation Paper: Citizen's Charter – Performance Indicators*, September.
Bidwell, S. *et al.* (1978) *World War 3*, Hamlyn, London.
Boxer, C.R. (1965) *The Dutch Seaborne Empire*, Hutchinson, London.
Cavana, R.Y. and Coyle, R.G. (1984) A policy analysis model for New Zealand's plantation forestry system, *New Zealand Journal of Forestry*, **29**(1), 24.50.

Clark, R. (1986a) Operating and support costs of the US Navy: some analytic facts, *Budgeting for Sustainability*, Military Applications Section, Operations Research Society of America.

Clark, R.H. (1993) The dynamics of force reduction and reconstitution, *Defence Analysis*, **9**(1), 51–68.

Clark, T. (1986b) A management systems analysis of the arms transfer process, *Decision Sciences*, **17**(4), Fall.

Coyle, J.M. and Richardson, H.D. (1993) A recreation of the Gulf War using the CARCO system dynamics model, presented at International Symposium on Military Operational Analysis, Royal Military College of Science, Shrivenham. (Copy available from Coyle at HVR Consulting Services, Selborne House, Mill Lane, Alton, Hampshire GU34 2QJ, UK.)

Coyle, R.G. (1977) *Management System Dynamics*, John Wiley & Sons, Chichester.

Coyle, R.G. (1981a) Modelling the future of mining groups, *Transactions of the Institution of Mining and Metallurgy*, **90**, April, A81–A88.

Coyle, R.G. (1981b) The dynamics of the Third World War, *Journal of the Operational Research Society*, **32**, 755–65.

Coyle, R.G. (1983) Who rules the waves – a case study in system description, *Journal of the Operational Research Society*, **34**(9), 885–98.

Coyle, R.G. (1984) East and West and North and South: a preliminary model, *Futures*, December, 594–609.

Coyle, R.G. (1985) A system description of counter-insurgency warfare, *Policy Sciences*, **18**, 55–78.

Coyle, R.G. (1987) A model for assessing the work processing capabilities of military command and control systems, *European Journal of Operational Research*, **28**, 27–43.

Coyle, R.G. (1992a) A system dynamics model of aircraft carrier survivability, *System Dynamics Review*, **8**(3), Fall, 193–212.

Coyle, R.G. (1992b) The optimization of defence expenditure, *European Journal of Operational Research*, **56**, 304–18.

Coyle, R.G. and Gardiner, P.A. (1991) A system dynamics model of submarine operations and maintenance schedules, *Journal of the Operational Research Society*, **42**(6), 453–62.

Coyle, R.G. and Rego, J.C. (1982) Modelling the discrete ordering of hydro-electric projects – the Argentinian case, *Dynamica*, **8**(1), Summer.

Forrester, J.W. (1961) *Industrial Dynamics*, MIT Press.

Forrester, J.W. (1969) *Urban Dynamics*, MIT Press.

Forrester, J.W. (1971) *World Dynamics*, Wright-Allen Press.

Forrester, N. (1973) *The Life Cycle of Economic Development*, Wright-Allen Press.

Hackett, J. *et al.* (1978) *The Third World War*, Sidgwick & Jackson, London.

Hall, R.I. (1976) A system pathology of an organization: the rise and fall of the old *Saturday Evening Post*, *Administrative Science Quarterly*, **21**, June.

Homer, J. (1993) A system dynamics model of national cocaine prevalence', *System Dynamics Review*, **9**(1), Winter, 49–78.

Macrae, H. (1994) *The World in 2020*, HarperCollins, London.

Meadows, D.H. *et al.* (1972) *The Limits to Growth*, Potomac Associates.

Meadows, D.L. *et al.* (1974) *Dynamics of Growth in a Finite World*, Wright-Allen Press.

Naill, R.F. (1977) *Managing the Energy Transition*, Ballinger Publishing Company.

Roberts, E.B. (ed.) (1978) *Managerial Applications of System Dynamics*, Wright-Allen Press/MIT Press.

Saaed, K. (1994) *Development Planning and Policy Design*, Avebury Press.

Wolstenholme, E.F. (1986) Defence operational analysis using system dynamics, *European Journal of Operational Research*, **34**, 10–18.

Wolstenholme, E.F. and Al-Alusi, A.-S. (1987) System dynamics and heuristic optimization in defence analysis, *System Dynamics Review*, **3**(2), Summer.

Wolstenholme, E.F., Gavine, A., Watts, K. and Henderson, S. (1992) The design, application and evaluation of a system dynamics based methodology for the assessment of computerised information systems, *European Journal of Information Systems*, **1**(5), 341–50.

CHAPTER 12

Summary and conclusion

INTRODUCTION

We have followed a long and hard road to study the principles, techniques and applications of system dynamics and have now reached the stage of trying to draw the threads together. Accordingly, this chapter restates the essence of system dynamics, and discusses problems of applying it in practice. That leads to some considerations of the assessment of the quality of models. Approaches to the teaching of the subject are explored and some further reading is suggested. Finally, the author indulges himself with some personal comments.

THE ESSENCE OF SYSTEM DYNAMICS

Many aspects of system dynamics have been described in this book. The techniques of influence diagrams and examples of their use were studied. The principles of simulation were explained. Modelling techniques and the formulation of models have been explored and the depths of optimization were plumbed. In the preceding chapter we tried to bring all those topics together in a survey of the applications of system dynamics, including some rather detailed case study models. Those topics are all very well and the reader will have expended much effort in mastering the material, but what is the essence of it all?

To see the essence, recall that an aircraft flying from, say, London to Hong Kong does so almost entirely under automatic control. During the flight, the aircraft's weight changes considerably as fuel is used, it flies over high mountains and is exposed to head and side winds which are largely unpredictable. This standard of performance is achieved safely and economically every day, and practically regardless of the weather, because the designers of feedback control systems for large aircraft and many other engineering systems know their business. They are helped considerably by having access to a large body of accepted and tested theory of, say, aerodynamics and some powerful mathematical methods of control system design.

Socio-economic systems must also be, so to speak, piloted through time, and we know from the practical experience of daily life that they are all too frequently far from successful in that enterprise. The most dramatic examples are the decay of cities, the changes in social behaviour, the cycles of economic expansion and recession and the rise and fall of nations. Less dramatic cases are the growth and decline of business firms and the daily efforts of managers to meet targets for production, to recruit staff or to complete projects on time and to cost. Many other instances have been mentioned in the book; even more will occur to the reader.

The essential aim of system dynamics is to achieve in socio-economic systems the standards of controllability and dynamic behaviour which are commonplace in engineering systems. The decades of work in system dynamics which were summarized all too briefly in Chapter 11 offer ample evidence that these targets are achievable in a very wide range of socio-economic systems. The paradigm, or fundamental point of view, is that dynamic behaviour is produced by feedback mechanisms and that the quality of the design of the *policies* which regulate those mechanisms and the *structure* of their interactions are the factors which produce good or bad performance.

To be sure, the system dynamicist dealing with a managed system faces more difficult problems than the control engineer. The first is often a lack of any accepted theory of the problem in question. The saving grace is that all socio-economic systems contain physical flows of something, and getting the physics right is the key step to understanding a problem and getting a good model. The injunction to *think physics* was emphasized when we discussed influence diagrams, and the principle of physical flow is, perhaps, the nearest thing system dynamics has to a general theory for structuring and understanding problems.

Further difficulties arise because socio-economic systems usually contain significant non-linearities, are often larger than engineering problems and cannot be subjected to laboratory testing under controlled conditions. Even more than that, socio-economic systems may well involve positive feedback loops, or badly behaved negative loops, capable of driving the system far outside the normal operating range within which an engineering system can be made to remain.

Even further, socio-economic systems cannot always be represented as objects which are separate from the environment that causes the shocks. A Boeing 747 does not create the headwind with which its automatic pilot must cope; a business firm, on the other hand, may very well create or destroy the demand for its products to which it is trying to react. One consequence is that system dynamics models of socio-economic systems quite frequently produce counter-intuitive behaviour, in which the result of a policy or structural change is different from, or even opposite to, what logic might initially have predicted. Thinking more deeply, with the aid of the insights from the feedback structure, should reveal the reason for the counter-intuitive response, and that ought to enrich one's understanding of the policy problem the model represents.

To add to all that, socio-economic problems, because of their non-linearities, positive feedback and sheer size are simply not amenable to the rigorous mathematics of control theory. Indeed, one of Forrester's powerful insights was to see that simulation, supported by software with a suitable syntax, was the only way in which socio-economic problems could even be represented. The techniques of optimization of system dynamics models take that idea a valuable step further.

If socio-economic systems analysis was simply replete with difficulties it would be an impossible undertaking. Fortunately, socio-economic problems typically offer a multitude of design options, as we saw, for instance with the DMC problem (Fig. 7.10 in particular). There are usually few laws, human or scientific, to prohibit the system's *structure* from being changed. The *policies* by which it is managed can be whatever its managers choose to adopt, or can be persuaded to apply when the results of analysis have been explained to them. This extensive freedom to redesign a system's structure and policies makes system dynamics not only a practical proposition but also a discipline capable of providing useful policy guidelines for problems of great practical and human significance.

Designing good behaviour into socio-economic systems is the essence of the matter, and the tools and techniques are simply the servants of that aim. Even apparently abstract research, such as the model of the decline of the Mayan empire has, at its roots, a policy orientation, even though the 'managers' concerned are long since dead. The policy emphasis is still present in models of the dynamics of ecological systems, in which human management may not be overtly represented. You should now review Chapter 1 and write down your own ideas in your own words about the fundamental nature and purpose of system dynamics.

IN WHAT CIRCUMSTANCES?

Chapter 11 described the enormous range of the potential applicability of system dynamics, but the fact that a methodology *can* be applied to a particular problem does not mean that the circumstances are necessarily propitious to a *successful* application. We therefore need to consider some of the factors which might make for success or failure.

The first point we need to establish is that the purpose of any management science investigation is to change a situation. This is clearly true of practical work in a business or in a government department. The object is usually to support the making of a decision and, whether the study is simply to confirm whether a given decision is sound or whether it is to help in choosing the decision, the end result will be to change the system. This factor of change is still true even when the investigation is into, say, the processes which led to the collapse of the Maya. The object now becomes to influence the ways in which scholars have interpreted and understood that problem or to help them to organize their insights, and hence generate new ones, through the construction of an influence diagram, but the net effect will still be change.

To bring about change to a system, a situation or a point of view is essentially a *political* act, not a scientific one. To be sure, it is a good idea if the technicalities are correctly carried out, which is why we have laid so much emphasis on checking models and dimensional analysis, but the technicalities of modelling are subordinate to the political realities.

This line of reasoning suggests that the first consideration in a successful application is identifying the potential client for the work and discovering the reasons why that person or group should want to be a client. Unless the study can be conducted in such a way as to satisfy their demands, its results are, at best, likely to be ignored.

Use of the term 'client' conjures up images of fee-earning work for industry or government or of directly sponsored research on a defined problem. That is, however, too restrictive, and it is helpful to keep the idea of a client in mind even for abstract academic research on a problem which excites one's interest[1]. The client will be the academic community in, say, economics. Economists have their own agenda of research, their own understanding of problems and their own paradigms, and one ignores that at the risk of wasting one's time on work which will be neglected or derided. At the same time, it is pointless to pretend to be, for this illustration, an economist, on the strength of having read a couple of books and audited a few classes. Truly expert advice is needed, and getting on the same wavelength as the client is as vital in academic research as in consultancy practice.

Market identification is rarely easy, but it is essential. The snag is that clients may not realize that they should be interested in a system dynamics approach. Overselling is rarely a good idea. Being able to point to successful applications elsewhere may help, but the fact remains that this stage can be very difficult. Time, effort and thought must be expended. The author is a great believer in what he calls 80/20 modelling, which means that 80% of the time when one's brain and insight are working at full throttle (not the same thing as 80% of one's working day) should be spent on thinking about why this problem is important, who thinks it is and why they care about it, the other 20% being available for clever programming. He has seen rather too much work which is more like 2/98 modelling.

The second factor is to be able to identify dynamic behaviour in a system, which is usually not too difficult, and to be able to imagine ways in which it might be improved, through policy changes and/or structural modifications.

---

[1] The author's research interest at the time of writing is, in collaboration with Professor Piatelli of the Italian National Research Council, a study of the growth of sea power in the Classical Greek city states. On the face of it, this is an abstract problem, though one which may have considerable implications for modern politics and defence strategy. Who might the clients be for such work?

If the system cannot be changed, it is either perfect or paralysed and, in either case, the study would be a waste of time.

The third step in deciding whether or not to try to apply system dynamics is to have some idea of the time by which results would have to be available in order to be able to meet client requirements. Can anything useful be done within that time and with the resources available?

SYSTEM DYNAMICS IN PRACTICE

System dynamics is a fascinating and challenging approach to policy design. At a technical level it is easy; at a deeper level it is very difficult because of what we shall call the 'blank paper problem'. This means that one does not know in advance what the model should look like. By contrast, the analyst knows in advance that a linear programming (LP) problem will consist of a set of linear constraints and a linear objective function. Building an LP model involves steps which are distinctly non-trivial, but the fundamental mathematics are well understood and the form of the solution is known in advance. By analogy with the carver who starts with a piece of wood and cuts away all the bits which do not look like a duck, the analyst starts with a problem and cuts away anything which does not look like a linear equation. In other words, the problem is forced to fit a linear structure, which may or may not be legitimate, depending on the problem[2]. Much the same is true for most other management science methods; the form of the solution is known and it is a matter of judgement and knowledge as to how well that form fits the given problem.

System dynamics, in common with other systems methodologies, such as the soft systems methodology (Checkland and Scholes, 1990), has no fixed form and the analyst is faced with a blank piece of paper on which a model must be created. The essential trick is to realize that one cannot do this alone: one must work with the client or problem proprietor, and preferably with a group of three or four people who are involved in the problem.

The real subtlety is that the clarity of influence diagrams, especially at the higher levels of the cone, where the conventions of dotted and solid lines can be dispensed with, is so great that clients will be able to work with the analyst on model formulation. It is noteworthy that the Mayan problem involved one system dynamicist and three historians working together.

One of the exciting things which can happen is to have a member of the client group say 'that diagram is wrong because ...'. From that basis, progress can be made with the clients heavily involved in what is going on

---

[2] The author is very well aware that there is a good deal more to LP than that and that there are integer and non-linear programming methods.

rather than them having the feeling that they are detached, or divorced, from progress and that they will have to leave it to the analyst to come back later and tell them the answer. That is never going to lead to accepted results.

Diagram development is best done with a whiteboard and room to walk around. The group dynamics are such that people will wish to argue, debate, think and physically stand back and look at what has been done. A good practical trick is to use the kind of whiteboard that makes a print of what has been drawn before it is amended too much, or the use of a Polaroid camera. Using graphical software too soon is bad practice, as it is hard for more than about two or three people to see what is on the screen, only a small part of the diagram can be seen at any time, the person controlling the mouse may come to dominate the thinking and the rest of the group are likely to lose interest. Software which projects onto a large screen is usually not helpful because the person running the program cannot be involved in the discussion and may be perceived as being too much in charge of the debate, and there is the risk of delays as the diagrams are built and projected. The essence is to keep everyone involved and active.

An important aspect of a good project is to feed results back as they arise. Again, one of the attributes of system dynamics techniques is their rapidity; initial results, whether from a diagram or a model, can often emerge within days or even hours and showing the client what is happening is a good way of keeping up his or her interest. This is especially important in academic work. A given project may lead to a paper based on the diagram, comment on which may be helpful in revising the diagram and improving the subsequent model, and it also establishes the project as an accepted part of the scenery, so to speak. When final results are produced, they are less of a surprise to the academic community and hence may be more likely to be taken seriously, rather than being 'butchered with consummate skill' (p. 309, footnote 3).

This ability to produce early results quickly has a more profound implication. It is a truism of analysis that the first model of a problem is always wrong, usually because the client's need for it has not been properly identified or the client has been inadequately educated as to what his or her needs are. Being able to produce an initial model from a few hours or days of effort means that so little time, money and personal prestige have been invested that one can abandon the initial model and try again with the benefit of the improved understanding. Any method which requires person-months of effort before any results emerge will produce models which are, psychologically and politically, almost impossible to amend, even if they profoundly miss the point of the problem.

Sufficient has been said in earlier chapters about the importance of not making the simulation model over-complicated, and we have also explained that there is nothing in most of the software packages to prevent one from modelling exactly as much complexity as is needed. In other words, the problem can be modelled just as it needs to be represented and there is no need whatever to force the problem to fit a given format.

The fact that even quite complicated models will run in a matter of seconds on a PC means that client involvement can be maintained and exploited through the medium of study periods. These are intensive working sessions of about one or two days, based on groups of three or four client staff, or academic colleagues, each assisted by someone familiar with the model and adept at running it. The results which emerge are perceived by the client as belonging to the client, and this sense of ownership of the answer to the problem of which they are also the proprietors does wonders for acceptance of the value of a study. Until one has taken part in a study period one will not realize the benefits they produce[3]. Study periods with academic colleagues from other backgrounds are a valuable part of interdisciplinary academic research.

Study periods are particularly valuable as a precursor to optimization. As we have seen, optimization of models is a very powerful approach, but, as with all strong medicine, it needs to be taken cautiously; it is vital to optimize the right model and to optimize the right aspects of the problem. Careful attention to the discussions in the study period will usually throw a good deal of light on what the problem proprietors consider to be important in the system's behaviour. That leads to more acceptable objective functions. It sometimes happens that there are two or more objective functions which represent the aims of competing interest groups within a problem. The fact that a thorough optimization search can be done within a few seconds[4] allows one to run study periods devoted to optimization of candidate, or competing, objective functions with the aim of finding policy options which satisfy everyone.

These are all positive aspects of practical work. The negative side is that there is an acute danger that the model will take on a life of its own. Indeed, the phrase 'in reality the model shows . . .' sometimes occurs in published work. This is nonsense. The model is not reality; it is an artefact and a simplification and it is essential not to lose sight of that.

Presentation of results is an important aspect of practical work and has three aspects. The first is the influence diagram. It is usually disastrous to show the complete diagram as a single illustration in a report or paper. Building the diagram up through several stages is often helpful (see, for example, the police effectiveness case study in Chapter 11 and Coyle (1992)). Using different diagrams from the various levels of the cone can be useful. Do not assume that the diagram is self explanatory; a picture does

---

[3] The author once led a study period at which a rather tasty chicken dish was served at lunch on the first day, but not on the second. When the participants completed their appraisal forms, the only improvement suggested was that chicken should be served on both days. There spoke some happy customers!
[4] This depends heavily on the capabilities of different software packages.

not save 1000 words – it may well require them. Secondly, presentation of the equations requires some thought. A full listing of a model is not always helpful, and documented printout of the most significant equations, supported by a few sentences of explanation, may be more appropriate. It convinces clients of the quality of the work, without overwhelming them with detail. Finally, the recommendations for policy or structural change need to be handled with care. It is often a good idea to present the results in such a way as to show clients two or three things about the system which they already knew or suspected as well as the novel insights and recommendations which have emerged.

The essence of good system dynamics practice is to bear in mind that the model represents a policy problem for a client and it is the policy which is the object of study, not the model.

ASSESSING THE QUALITY OF A SYSTEM DYNAMICS MODEL

A critical aspect of good system dynamics practice, whether it be in consultancy or in academic research, is to ask a colleague to review one's model. This exposes the model to, hopefully constructive, criticism and is an essential step in quality assurance. The review should be done more than once during a project. The following are some of the criteria which can be applied.

Purpose: Has a clear purpose been defined? Is it a worthwhile aim, likely to help in the solution of a real problem?

Suitability: Does the model meet its purpose? This is rarely clear-cut. It depends on some of the following criteria, and the question ought really to be answered at the end of the review. Posing it at an early stage is a useful way of creating the right frame of mind for the remaining questions.

Basis: Is the model soundly based on accepted theory of the processes in the problem, if there is any, or does it contain sweeping generalizations? The question ought to be stated in this provocative form if later, and probably much more hostile, questioning by the client is to be avoided. Failing accepted theory, what assumptions and simplifications have been overtly stated, preferably in writing, and what others have been made implicitly. The model builder often does not realize what has been assumed.

Credibility: Does the sponsor attach value to its results? If the results are unexpected or provocative, how can we ensure that they are seen as interesting and valuable, as opposed to controversial and unacceptable?

Creativity: Does it provide new insights into problems and suggest new solutions? A model which did no more than confirm accepted wisdom would verge on the sycophantic and would be a most unusual event in system dynamics work.

Simplicity: Has the real world been simplified to just the right extent? This is rarely a technical matter and often has much more to do with showing a particular interest group that 'their' part of the problem has been included. The difficulty is that it can lead to intractable models. The real subtlety is that the model must be simple enough to be tractable while ranging widely enough to contain the possible solutions to the problem and to ensure *creativity*. The reader might like to review the case studies in Chapter 11 to see how well they meet this test, bearing in mind that case studies in management education are never intended to show examples of good or bad practice; they are intended to make the student think.

Redundancy: Can parts of the model be removed without affecting its behaviour?[5] This does not necessarily mean that the model has not been simplified enough. It often means that aspects of the problem which everyone knew to be important are not significant. In the psychology and politics of studies, this may cause some parts of the client group to heave a sigh of relief when they can prove that the problem was not their fault. On the other hand it may upset them deeply to discover that they are not as important as they thought they were. Careful handling will be called for, and it is as well to discover this kind of result early enough for precautions to be taken.

Transparency: Are the model's assumptions, simplifications and equations well understood by analyst and sponsor? If not, the analyst is working blind and the sponsor has to take the results on trust. Study periods should have ensured that the model and its results are the shared property of analyst and client.

Flexibility: Can the model deal with new inputs and parameter values?

Generality: Can the model represent other problems, perhaps when parameter values are changed? It is often poor practice to try to make a model too general. The value of a model depends fundamentally on how well it is suited to its purpose and a general model may be trying to satisfy too many purposes. System dynamics models are typically quick and cheap to develop and it is generally better to develop a new model for a new purpose.

Sensitivity: How severely does it react to changes to input data, parameters or initial conditions? Testing a model under extreme conditions is an essential step to verifying that it is technically correct.

Soundness: Does it satisfy technical criteria such as dimensional consistency, accuracy of mass balances, correct use of DT and legitimate use of rate-dependent rates? In other words, within the framework of the assumptions and simplifications, does it do the same things as the real system and for the same reasons?

---

[5] The COSMOS optimization software includes a simplification technique designed to show which parts of a model can be switched off without degrading the final value of the objective function by more than X%.

Productivity: Is it easy to use for experimentation and optimization?
Cost: Is it expensive or slow to develop?
Promotion: Are the claims made for the model's results reasonable or exaggerated?

A thorough review of one's work can be a deeply uncomfortable process. It is rather akin to having one's cherished research butchered by a journal referee who shelters behind his or her anonymity and abuses it to show how clever the referee is and how he or she would have done the work. Nonetheless, the experience must be endured partly because it is an inevitable component of true professionalism, but also because it may well avoid a lot more grief later at the hands of the client. This is especially true when the work is abstract research and the client is some other part of the academic community.

TEACHING SYSTEM DYNAMICS

System dynamics is difficult to teach; at any rate the author has found it to be so, despite many years of trying. To be sure, the basic techniques of influence diagrams and simulation are easily understood, and most students have no difficulty in following models developed during class periods, running them and doing policy experiments and optimization. The difficulty arises when they subsequently try to tackle a problem of their own, without guidance. Forrester has always argued that to become a good system dynamicist one must spend a period of apprenticeship with an experienced practitioner but, while this works well for the student who can spend a year or more on a higher degree, it is a counsel of perfection, as most students cannot do that.

One of the purposes of textbooks such as this, and especially the problems and cases it contains, is to provide students with synthetic experience and reference material to help them over this hurdle. The essence of the teaching problem is, however, the same as the difficulty in practising system dynamics, discussed earlier, which is that one has to start with a blank piece of paper, and many people find that to be very hard. The only thing the instructor can do is to be aware of the problem and to try to frame the teaching to deal with it. The purpose of this section is to try to pass on some experience and suggestions to those who may wish to embark on their own teaching of system dynamics.

Although helping students to develop a deep understanding of system dynamics is difficult, it is easy to construct a course of lectures and practice to give at least initial competence in the discipline. What follows is an outline of a one-week intensive course intended to start at about 11 a.m. on Monday and finish at lunchtime on Friday, a common pattern for short courses. Students are expected to undertake evening reading.

Figure 12.1 shows a timetable based on periods of 50 minutes, as students from industry are not used to sitting in a classroom and frequent breaks to stretch their legs and their brains are essential.

The course is geared to modelling practice so that, at the end of the week, the student has built, debugged, analysed and optimized two models. The software used is irrelevant, as the focus is on the principles and fundamentals of system dynamics, not on a particular syntax and graphical front end. The production model is the Domestic Manufacturing Company's problem, studied in earlier chapters. Since most of the students come from a defence background, the second model is the simple combat case study. Obviously, this is easily varied to suit different student groups. Readings can be assigned from chapters of this book or from the suggestions made below for further reading in system dynamics.

The author has experimented with many approaches to equation writing with students of various backgrounds in different universities. The approach he now uses may seem old-fashioned, but it does work[6]. The method is to write the equations on a whiteboard, with the students copying. Three boards are used, one for equations, another for definitions and dimensions and the third for explanations of, say, the smoothing equation. The author's handwriting is less than perfect and he may make mistakes (even deliberately). Students are likely to make copying errors and will probably make more when they type the model into the computer.

The end result is that the model is practically guaranteed not to work, the point being that students will learn far more about system dynamics by debugging the model than they ever will by being given a perfect version to start with (an instructor solution is issued afterwards, of course). To save time, the author once issued each student with a personal version of a model containing deliberate mistakes, and the reaction was universally that the students felt that they would never make such silly errors. He now leaves them to find for themselves just how silly are the mistakes they can make. Dimensional analysis is a key part of this process. Although COS-MIC has a graphical model-building front end it is not used in the instruction as the emphasis is on modelling principles, not on a particular package.

This approach works well with class sizes of up to 30, but would be unlikely to be successful with larger groups.

The timetable is easily adapted to give a course of about two periods per week over a term or semester. Parts of the material can be used to give a more condensed treatment of system dynamics within a management science syllabus, though teaching system dynamics as part of a course in simulation techniques is missing the point; simulation is only a tool

---

[6] At the end of the course, the students, who are usually experienced modellers, complete a 'happiness sheet' in which they rate each topic according to the technical level at which it was presented (too low, about right, too high) and the time spent (too long, about right, too short). The course is normally given a practically perfect score.

Time	Monday	Tuesday	Wednesday	Thursday	Friday
08:50 09:40		Principles of simulation	Modelling practical	Development of a combat model	Advanced aspects of SD modelling II
09:50 10:40					System dynamics case study II: optimization of defence expenditure
11:10 12:00	Introduction to system dynamics	Formulation of production model	Optimization in SD	Guest speaker: SD in practice	Applications of SD
12:10 13:00	Introduction to influence diagrams		Optimization of production model		Review of SD software and course discussion
14:10 15:00	Develop IDs for production model and combat model	Load, debug and run production model	Advanced aspects of SD modelling I	Running of the combat model	Optional modelling practical
15:10 16:00					
16:10 17:00	Influence diagramming problems	Policy design in production model	System dynamics case study I: qualitative analysis	Optimization of the combat model	
17:10 18:00	Influence diagram work	Experiments with production model	Modelling practical	Modelling practical	

Fig. 12.1 Outline program for a system dynamics course.

by which the underlying thought patterns of system dynamics are implemented.

Apart from teaching system dynamics modelling *per se*, the format of Fig. 12.1 is also used to teach information system modelling to students on a masters degree in information systems. The DMC model is still used, but with the emphasis on the information structure, its interaction with policy, and the consequences these have for system behaviour. Only one model is built, and other sessions are used to consider applications of system dynamics to other problems. Such a course occupies about 25 lectures.

Students on Master of Business Administration courses may take system dynamics as a course in a management science or strategic management option, essentially using the timetable of Fig. 12.1, though with different models as the case studies. For students not doing such options a course of about eight lectures and cases can be structured around influence diagrams under the title of 'Organizing Complexity'. The theme is that management problems are inherently complex and dynamic and that understanding the mechanisms of dynamic behaviour is an important aspect of strategic thinking. The diagrams used are for business firms, socio-economic problems and industrial structures. The last couple of lectures are used to show what a simulation model can do and the effects of policy and structural optimization.

FURTHER READING

Any book reflects its author's prejudices and experience, and a subject, especially one as subtle as system dynamics, can only be deeply understood, as opposed to technically grasped, by reading of other work and the absorption of other views and prejudices. This section outlines some suggestions, in chronological order of publication.

The author first read Forrester's masterpiece, *Industrial Dynamics* (Forrester, 1961) about 30 years ago and he still re-reads it from time to time, always to his profit. Since system dynamics was established, numerous authors have written texts on the subject.

Richardson and Pugh's *Introduction to System Dynamics Modeling with DYNAMO* (Richardson and Pugh, 1981) is a good example of this literature. It provides a number of examples of simple models and is, to some extent, a supplement to the DYNAMO User Manual (Pugh did a great service to system dynamics by developing the original software).

Roberts *et al.* (1983) cover the techniques of using the DYNAMO software and offer a wide range of examples, most of the models being rather simple.

The common themes to both these books are that only DYNAMO is mentioned and that nearly all of the models they contain are completely closed in that the dynamics arise solely from the structure of the model, its parameter values and initial conditions, and not from any exogenous influences. This might be termed the American school of system dynamics, as all the authors were for a long time associated with Forrester's group at MIT. The view adopted in this book is that both closed and driven systems exist (the combat model and DMC are, respectively, examples) and that one needs to know how to deal with both.

Both books are now some 12 years old and a more recent offering is Wolstenholme's *System Enquiry* (Wolstenholme, 1990). The examples it contains are from business, coal mining and defence analysis.

A useful comparison of system dynamics and event-based simulation is provided by Pidd (1992).

Finally, the System Dynamics Society's journal, *System Dynamics Review* (see p. 309, footnote 4) provides an invaluable picture of current work in the field.

DEFINITIONS OF SYSTEM DYNAMICS

In Chapter 2 several definitions of system dynamics were put forward. You have, we hope, pondered these as you have worked and you should now understand this fascinating and wide-ranging subject fairly thoroughly. Perhaps, as a culminating task, all that knowledge and study can be put to use to formulate your own definition of the subject. The author would be delighted to hear from you with your suggestions.

FINALLY, ON A PERSONAL NOTE

I first encountered system dynamics in 1967 at the feet of a master, Professor Jay Forrester. I had read his book before that, but the subject really came alive at his hands and I owe him a debt.

Since then, most of my working life and, indeed, many of my waking hours have been spent doing and thinking about system dynamics. Apart from providing me with an academic's typically modest living it has given me an immense amount of intellectual and practical enjoyment. The intellectual part came from dealing with some challenging problems and working with many excellent students, many of whom subsequently became respected colleagues and good friends. Practical enjoyment came from tackling some demanding consultancy assignments and feeling that I was doing something useful in the real world.

A great personal pleasure is that my son, Jonathan, who wrote part of Chapter 11, also enjoys system dynamics and is very good at it.

This almost reads like a premature obituary, but I hope to continue in the same vein for many years yet, God willing. In fact, the labour of writing this book has been a rewarding chance to draw together thoughts which have been developing during the 15 years since my first book on system dynamics (Coyle, 1977).

My final word is that I hope that the hard work you have put in to studying this book will help you to enjoy system dynamics as much as I do.

REFERENCES

Checkland, P. and Scholes, J. (1990) *Soft Systems Methodology in Action*, John Wiley & Sons, Chichester.

Coyle, R.G. (1977) *Management System Dynamics*, John Wiley & Sons, Chichester.

Coyle, R.G. (1992) The optimisation of defence expenditure, *European Journal of Operational Research*, **56**, 304–18.

Forrester, J.W. (1961) *Industrial Dynamics*, MIT Press.

Pidd, M. (1992) *Computer Simulation in Management Science*, 3rd edn, John Wiley & Sons, Chichester.

Richardson, G.P. and Pugh, A.L. (1981) *Introduction to System Dynamics Modeling with DYNAMO*, MIT Press (reissued by Productivity Press, undated).

Roberts, N. *et al.* (1983) *Introduction to Computer Simulation: A System Dynamics Modeling Approach*, Addison-Wesley, Reading MA.

Wolstenholme, E.F. (1990) *System Enquiry: A System Dynamics Modelling Approach*, John Wiley & Sons, Chichester.

System dynamics software packages

INTRODUCTION

System dynamics models can be written, with some effort, using spreadsheet packages or general-purpose languages such as FORTRAN but there are several commercially-available software products specifically for system dynamics modelling. The author has, however, seen some remarkably good system dynamics models written, laboriously, in FORTRAN and some quite unbelievably bad models written, very easily, using some of the packages described here. In general, good quality system dynamics is more to do with intelligent and diligently careful use of the package one has available, and less to do with having any particular one, though the packages differ substantially in the tools they provide.

This book was written using those parts of COSMIC which are consistent with traditional formats for system dynamics models, but COSMIC has many other features, as have the other packages. In an attempt to be fully comprehensive, this Appendix provides a brief summary of the available packages.

The Appendix falls into three parts. The first is an outline of the historical pedigree of the different packages, indicating the degrees of consistency and difference between them.

The second part is composed of summaries of the capabilities of the several packages, provided by the companies concerned. The author has eliminated qualitative assertions, such as that Software X is 'easy to use' so as to provide a factual summary, though the author accepts no responsibility for the accuracy of the information.

Finally, the author offers his own suggestions, based on many years experience of doing system dynamics, of some factors to be considered when selecting a software package.

HISTORICAL PEDIGREE

To put the individual packages into context a brief discussion of their historical pedigree may be useful. Figure A.1 shows the lines of evolution.

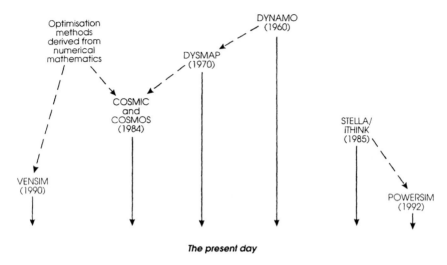

Fig. A.1   The evolutionary lines of system dynamics software.

The solid vertical lines show that the package is still available, but does not show the several stages of development that each has gone through. For instance, the DYNAMO which was available in 1960 bears little relationship to what is now produced, DYNAMO, DYSMAP2 and COSMIC, though originally developed for mainframe computers, all now exist in PC versions.

The dotted lines indicate that one package is very similar in operation and syntax to its 'predecessor', though not, of course, plagiarised from it. For instance, *ithink* and Powersim are both based on Forrester's flow diagrams. It should be understood, however, that although Package Y may be 'descended from' Package X it is very likely to contain features which Package X did not have and, of course, since the evolutionary split X itself may have acquired facilities which Y does not possess. Indeed, in some cases, facilities which were new to the 'descendant' have since appeared in the 'predecessor'.

In short, the only reliable guide to what a given package does, and does not, support is careful reading of the relevant manual and, where possible, **close** comparison with other manuals. Even better is to use the packages or, failing that, to talk to users.

The three groups of packages are very different from each other. The DYNAMO/DYSMAP/COSMIC evolution is derived from the concept of writing equations using a text editor, though COSMIC also has a graphical modelling support system. The STELLA/Powersim line is geared to the use of symbolic icons for drawing level/rate diagrams, from which *some* of the equations are written automatically. Vensim is rather different from both these groups, being derived from the text editor approach but also supporting icon-based modelling.

SUPPLIER SUMMARIES OF THE PACKAGES

Introduction

The following paragraphs were provided by the software suppliers. The descriptions are in alphabetical order of the product name. All the product names and italicized features are trade marks of the respective suppliers and all trade marks are acknowledged.

To emphasise that this is what the suppliers chose to say about their own products, in response to the author's invitation, the text is shown in quotation marks.

The *COSMIC* and *COSMOS* software

'COSMIC and COSMOS are an integrated environment for the development, analysis and optimisation of system dynamics models. The software suite is based on a Workbench concept and can be run on any PC. It can run under DOS or WINDOWS.

Model building is supported by a graphical development environment in which the user builds Influence Diagrams on the screen, creating and defining new variables as required. The Influence Diagram data are stored and can be used for model building. Level equations are written automatically; rates and auxiliaries are created on the screen by the modeller using a mouse and the appropriate part of the Influence Diagram. As equations are created, a fully-documented model is stored.

COSMIC has a total of 44 built-in functions and interactive guidance is provided on their correct use. TABLE functions are created by drawing graphs on the screen. The graphical environment also provides facilities for editing diagrams and equations while maintaining the consistency between them, for tracing paths and loops in models, and for browsing equations and the associated Influence Diagrams.

Alternatively, the modeller uses the text editor to type in the equations in the traditional manner. The modeller can change from the graphical workbench to the text editor at any stage in model development.

Debugging and running of the model is supported by extensive diagnostic software including dimensional analysis, detection of feedback loops, logging of simulation runs and statistics of function usage. Output can be produced as tables or graphs, the latter being available on the screen (with hard copy options), on the printer, or on a flat-bed plotter with user control of line pattern and pen colour. The model can be saved as a FORTRAN program.

COSMOS supports four methods of optimization. Direct optimization uses a hill-climbing algorithm to design policy options or to fit a model to parametric or historical data. Base vector optimization tests the sensitivity of the model to subsets of parameters which are chosen from candidate sets provided by the user. Simplification is used to test which parts of a model have the most effect on behaviour. The Planning Horizon method can be used to study the ability of a model of a system to adapt itself to changing events.

For more information contact The COSMIC Holding Co, 8 Cleycourt Road, Shrivenham, Swindon SN6 8BN, UK. Telephone +44 (0)1793 782817, or use the order form at the end of this book.'

The *DYNAMO* software

'DYNAMO allows the user to create large models (via a text editor) using arrays, macros (either built-in, such as DELAY, or definable) and functions (either built-in, such as TABLE, or definable in C). Models can be documented to collate all uses of a variable. Compilation errors are shown via an integrated editor. Simulation results can be stored and DYNAMO can integrate with spreadsheets for data input or output.

To support ownership of models, the DYNAMO *RepW* scripting language supports templates, FOR loops, macro actions to permit 'what if', or a gaming system under Windows 3.1. This system permits up to 4 views (Reports, Plots or Flowcharts) on a single screen. Reports analyse or explain simulation results and prompt the user to change parameters with on-line help. Plots displays curves. Flowcharts or other diagrams can be imported for display.

For further information contact: Pugh-Roberts Associates, 41 Linskey Way, Cambridge MA 02142, USA. Telephone +1 717 864 8880, Fax 864-8884.'

The *DYSMAP2* software

'DYSMAP2 accepts models from a source file. Equations may be written in any order and variable names may be up to 32 characters. Syntax errors are reported to the screen and to an optional file and run-time errors are reported. The software supports dimensional analysis and includes the usual logical, time-related and delay functions. Delays of orders 1, 2 and 3 and a pipeline delay are provided.

After execution of a model the user is placed in an Interactive Command Environment (ICE) with all variables saved for subsequent plotting. Comparative and scatter plots can be performed. Zooming on the time axis is permitted.

Up to 10 runs can be performed in any one session and new runs can be launched interactively from the ICE. These usually follow the Change or

Insert commands which allow the user to respecify any equation, table or constant in the model, or introduce a new variable, table or constant.

Input and output includes reading table data direct from file and output of numerical data in tabulated or ASCII free-format.

For further information contact Salford Software Ltd, Adelphi House, Adelphi St, Salford M3 6EN, attention Dr Olga Vapenikova. Telephone: +44 (0)161 834 2454, Fax: +44 (0)161 834 2148.'

The *ithink* software[1]

'The *ithink* software comes in two versions: Core and Authoring. The Core version is designed to enable people to construct system dynamics models by piecing together stocks, flows, convertors and connectors on an electronic tableau. In addition to the standard Reservoir, the software includes three additional, specialized, stock types: Conveyors, Queues and Ovens. Other facilities include automatic cycle-time calculations and statistics, drill-down sub-model capability, high-level mapping capability, unit-conversion within flows and space-compression objects (for managing visual complexity). The package supports multi-variable sensitivity analysis and run-by-sector capability. The documentation comprises: *An Introduction to Systems Thinking, The Process Improvement Module, The Business Applications Guide* and *The Technical Reference Manual*.

The Authoring version of the software enables people to create run-time accessible, stand-alone Learning Environments, or Management Flight Simulators. No knowledge of programming is required to create these high-level interfaces, which can include *QuickTime* movies, slider inputs, graphical function editors, on-condition message posting and numeric readouts. Authors can regulate how much access to allow end-users to the underlying model structure which they have created.

For more information contact High Performance Systems Customer Service at +1 603 643 9636 (9 am–5 pm Eastern Standard Time, USA), Fax +1 603 643 9502.'

The *Powersim* software

'Powersim is a Windows-based graphical system dynamics modelling package. It supports both flow diagrams and causal loop diagrams. The underlying equations can be entered when the model's structure is defined. Equations are created with the aid of a library of built-in functions and a visual editor for drawing graph functions.

Simulation results are presented as animation, numbers or graphs. Powersim also offers dynamic objects which display selected variables as time graphs, scatter graphs, time tables, bars and numbers.

---

[1] The **ithink** software also exists as STELLA II, which is marketed to academic institutions.

Powersim supports the creation of games and "management flight simu-lators". Parameters can be changed using slider buttons. Push buttons, radio buttons and check boxes can be added to the gaming interface and data can be imported and exported using additional modules.

Powersim supports multidimensional arrays. The results of a simulation can be saved and compared with other results. Sensitivity analysis can be performed by setting up multiple simulations with varied inputs. Separate models can be connected and simulations run in parallel.

For more information contact ModellData AS, Zachariasbryggen, PO Box 642, N-5001 Bergen, Norway. Telephone +47 55 31 32 38, Fax +47 55 31 31 36.'

The *Vensim* software

'Vensim is an integrated environment for the development, analysis and application of system dynamics models. It supports causal loop diagrams and stock and flow diagrams. Information from the diagrams is stored to allow the creation of model equations. It provides *Document, Loops, Outline* and *Tree* tools to generate different representations of model structure and includes a dimensional analysis facility.

Any number of named runs or scenarios can be created and there are Bar Graph, Gantt Chart, Graph, Stats, Strip Graph and Table tools for the analysis of data and simulation results. Vensim's tools support *Causal Tracing*, facilitating the review of simulation results and the tracing of underlying causes.

Reality Check will test families of models under a variety of assumptions.

Optimization and data handling tools support calibration of a model against historical data and the selection of policy interventions.

The scripted application language can be used to develop learning environments, game and decision support systems with access to the analysis tools. The software supports subscripts, special variable types and user defined macros. An interface can be used to create special purpose functions.

For more information contact Ventana Systems Inc., 60 Jacob Gates Road, Harvard, MA 01451, USA, Telephone +1 508 456 3069, Fax +1 508 456 9063, Email vensim − world.std.com.'

FACTORS IN SELECTING A PACKAGE

Introduction

The newcomer to system dynamics will need software support and, as we have seen, there is a fairly wide choice. Even the established practitioner of system dynamics can find it useful to have more than one package available. The price differences between packages are not large and, in any case, the price is trivial relative to the power they confer. How is one to choose?

In this section, we shall attempt a summary of the features typically found in system dynamics software, working from the available information, but reasoning from first principles. Where it is necessary to mention particular packages by name, alphabetical order is used.

The views expressed are very much the author's opinions. They are based on much experience of system dynamics, but others might disagree with his views. Recall also, that all the packages are undergoing continual development so, if software X is criticized for lacking feature Z it may well be that Z will appear in X in due course. If software X is mentioned as having facility Z, it is simply to illustrate the argument and does not necessarily imply that other packages do not have an equivalent of Z.

In general, a system dynamics software package needs to be assessed according to

- its basis in fundamental system dynamics theory;
- the ease with which it can be used;
- the support it gives to model building;
- the extent to which models can be documented and explained to a sponsor;
- the facilities it has for debugging a model;
- the ease of making experiments and producing output;
- the scope of its facilities for policy design.

Speed of running a model is nice to have but the critical feature is the support of thought, not the crunching of numbers.

Fundamental theory

All the packages except one are based on the fundamental Level, Rate and Auxiliary variables which were discussed in Chapter 4, especially Figure 4.2. The exception is Vensim which only recognizes Level and Auxiliary variables in its text editor, though allowing icons for rates in its graphical tool.

DYNAMO, DYSMAP2 and COSMIC identify different types of variable by initial letters in the equations, such as L, R and A. *ithink* and Powersim identify them from the icon used to draw that diagram; using a box symbol will create a level variable. Vensim identifies variables by the form of the equation defining the variable and optionally allows the use of icons and built-in typographical conventions to highlight variable names; the name for a constant, for instance, will be displayed in UPPER CASE.

Ease of use

DYNAMO and DYSMAP2 depend on models being typed into a file from a text editor. COSMIC, *ithink*, Powersim, and Vensim have Graphic User Interfaces (GUI) which allow one to draw diagrams on the screen as a precursor to equation writing. In all cases, a mouse is used to link the appropriate symbols into diagrams. Powersim and Vensim also provide 'causal loop diagrams' (see page 43).

The advantage of drawing a diagram on the screen is considerable; the assumed influence of one variable on another is abundantly clear, as it is in an Influence Diagram. In both cases, however, the **form** of the influence is not clear. The disadvantage with *ithink* and Powersim is that the diagram must correspond exactly and in every detail with the equations which are to be created, because, in those packages, the GUI is the portal to the creation of equations. If the model is to become larger, so must the diagram. If the model is to have very complicated logic, then the diagram must have a very complicated structure, though Powersim's array feature can reduce diagram complexity and *ithink*'s space compression object allows complexity to be shown or hidden as required.

COSMIC and Vensim, on the other hand, while supporting GUIs (though the two are very different), allow one also to enter equations using a text editor. This allows the user to take advantage of the distinction between Level 3 and Level 4 in the cone of Influence Diagrams in Figure 2.16. Level 3 is drawn to show the main features of the problem. Level 4 need not necessarily be drawn, as it is implicit in the equations, and the connection between the two levels is carried in the analyst's mind or can be inferred from the code if needed.

All the GUIs will, to some extent, write some of the equations automatically, and it is sometimes believed that these packages 'write the model automatically'. This, however, **is not the case**. The most that happens is that the Level equations are written automatically, but the Rate and Auxiliary equation still have to be created by the user. Creating initial conditions for levels must also be done by the user. This can be an important part of modelling and it is worth checking how much flexibility a given package offers.

In the author's opinion, heavy reliance on a graphical user interface can sometimes lead the modeller into very bad habits. It is simply too tempting to rush ahead with putting equations together and then looking at graphs on the screen rather than carefully checking that the equations are correct and **do the same things as the real system and for the same reasons**.

In summary, all the packages are easy to use, once one has mastered the appropriate procedures, though the beauty of a GUI is, perhaps, very much in the eye of the beholder. What to one person might seem elegant and powerful may be over-complicated and confusing to another.

Support to modelling

A package will support modelling by the extent and diversity of its built-in functions. For example, delays are a common and significant feature of dynamic problems so a system dynamics package which did not support functions for modelling delays would be next to useless. In fact, all the packages do have greater or lesser numbers of various forms of delay function. COSMIC has 7 versions of delay functions; *ithink* has special functions called 'Conveyors, Queues and Ovens', which produce discrete, event-orientated delay processes, though a standard DELAY3 would have to be written as a sub-model.

In general, however, all the packages are fairly comprehensively equipped with standard functions though, before selecting a package, the reader should check exactly what functions a given package does and does not support.

Model documentation

The process of describing what a model does, what it contains and how it works is called 'documentation'. It is vitally important to the production of good models as it enables both the creator of the model and the customers for it (whether they be fee-paying clients or the academic community) to understand the model. Without understanding, the model will not be accepted as a basis for management action or scholarly progress and will therefore be a waste of effort. The quality of a software package is, therefore, more profoundly affected by the extent to which this vital process is supported than by the elegance and sophistication of its GUI.

The original DYNAMO concept provided for 'maps' of the connections between all the variables and for documentation of the equations. This concept was followed almost exactly in COSMIC (DYSMAP2 and Vensim achieve a corresponding effect in a different way) and has been illustrated in this book whenever individual equations have been discussed. All the variables in an equation are defined and any constants used in the equation are searched for and shown, together with their definitions, even though the constant may be scores or hundreds of lines away from the equation being documented. See, for example, page 116. The other three packages have similar features.

Facilities for debugging

Debugging a model means the process of clearing it of errors of syntax and logic.

All the packages provide syntax error messages if, say, a closing parenthesis has been omitted. All of them ensure that the model will not run if variables are used without having been created. In short, all the packages do a thorough job of checking for errors of syntax.

The extent of displaying the error messages is rather more variable between packages. The DYNAMO concept, reflected in COSMIC and DYSMAP2, is to provide a printout of all the error messages, which, combined with the model map, enables one to detect related errors in equations which may be hundreds of lines apart in the model. The other packages depend more heavily on showing error messages on the screen, which can make it difficult to see where the true source of an error lies, apart from such obvious cases as missing parentheses. Printout of error messages is, however, also available.

Checking a model for errors of logic is more difficult and is a function both of the steps the modeller has taken to build checks into her model and the features the software supports.

Built-in checks have been illustrated in several places in this book. By far the most important is to include mass-balance checks for the physical flows in the model.

Software support is rather variable. The author's experience, based on hundreds of models, is that it is completely impossible to debug a model simply by looking at a few graphs on a screen. Recall the discussion of the simple combat model in Chapter 6, page 178, in which the model's dynamics looked right even though the model was still incorrect. Even if one graphed all the variables, only about six can be shown on the screen at any one time and the precision of the graphs is too low for fine details to be seen. In short, the only way to be sure that a model is doing the same things as the real system and for the same reasons is the careful examination of printed tables of the values of variables against time. This was the original DYNAMO concept and it is unquestionably the correct one. All the other packages support printed output.

Apart from printing out values of model variables, the most powerful check is dimensional analysis of model equations. The importance of dimensional analysis cannot be overstated. The author has worked on some hundreds of system dynamics models, experience which has proved to him that he is very good at making mistakes and that software for dimensional analysis saves any amount of worry over the quality of his models.

Ease of making experiments

The concept employed in the original DYNAMO was to allow the user to define a set of experiments by providing new values of constants or tables followed by a run command. This procedure could be repeated a certain number of times (up to 50 in COSMIC) to provide a set of experiments which would be executed in one run of the model. This approach was essential with Mainframe computers, though it is still essential with PCs, provided one has a suitable printer. COSMIC extends its value by providing a run logging option which prints out exactly the experimental conditions defined by the successive changes of constants and tables. Although there is some risk of becoming buried under paper, the batch experiment, followed by careful study of the tabulated values of model variables under different run conditions is an invaluable way to debug a model, and experiment with it in the fashion described in Chapter 7.

All the software packages also provide for interactive experiments in which one changes parameters as required and runs the model again. The packages vary a little in what they offer. Powersim, for instance, allows parameters to be changed during a simulation run.

All the packages provide some means of saving the results of model experiments. For example, COSMIC allows the parameter values to be printed and the screen graph to be copied to a laser or colour printer.

In summary, all the packages make it fairly easy to perform model experiments though, as remarked above, it is the author's experience that heavy reliance on graphical interaction can result in serious errors. To develop quality models it is vital to review equations carefully and get printed output documenting the model and giving detailed simulation results.

Policy design facilities

All the packages make it easy to perform experiments, none of them hamper the analyst's intuition and all can be said to possess good policy design facilities.

We have, however, attempted to demonstrate that optimization is potentially a very powerful policy design tool.

COSMOS allows the user to choose the objective function so that it can be maximized or minimized as the nature of the problem requires, typically finding the optimum of a large model in about 30 iterations, taking some 15 seconds on a 486DX2 PC running at 33 Megahertz. There are also facilities for three additional forms of optimization, beyond the simple hill-climbing, or direct optimization, which was described in the earlier chapters.

SUMMARY

All of the packages are very impressive and required much effort to develop, but, before selecting one, try to probe into what facilities it actually has. A few criteria are given below.

The essential distinction is between the COSMIC/DYSMAP2/ DYNAMO grouping, which offer the traditional ability to produce extensive output on paper for debugging and documentation, as well as differing degrees of graphical capability, and the *ithink*/Powersim/ Vensim concepts which are rather more heavily geared to the use of screen graphics for building and running models, though they can produce differing degrees of paper-based output. Some people like the idea that the model's diagram (though not its equations) can be seen on the screen, though, in practice, for even a moderately sized model, only a small portion will be visible on the screen at any one time.

All the packages are reasonably well equipped with built-in functions but they are **by no manner of means identically equipped**. Exactly what functions does the package support and exactly what do they do? How cumbersome and time-consuming would it be to have to write one's way round a missing feature?

All the packages have to translate the model into another language. Some packages **interpret** the model into, say, C, which is quite quick but produces a model which runs relatively slowly. Other packages have an option to **compile** into, perhaps, C or FORTRAN, which takes a few seconds but produces models which run much faster than interpreted models, which will be important if many runs are to be made or optimization searches are to be done. Is the cost of the interpretation or compilation software, and a valid licence, included in the price? If not, which compiler is recommended, and can it be guaranteed always to work efficiently, especially regardless of the size of the model?

Some suppliers will provide a demonstration disk which will give some insight into the product. A much better source is the manual for the package and it would be reasonable to ask the supplier to sell an inspection copy, deducting the cost from a subsequent purchase (it would be unreasonable to expect suppliers to give free copies of expensive publications to people who only might buy the software). Studying the manual before making a commitment may save a lot of subsequent grief and the cost of buying a few manuals would, in the long run, be trivial.

When studying a manual do not simply browse quickly through it to see what features the product supports and what it does not, though that is the first important test. The second test is to read at least parts of the manual carefully to see if what is described is **clearly** described in a style you feel comfortable with. Writing manuals (and books) is very difficult and it is hard for authors to find a style which will appeal to everyone. In other words, try to assess whether you will be able to understand the package in question well enough to use it **effectively**. Search for warnings about things the software does not do; you also need to be able to use the software **correctly**. The third test is to see if the manual is self-contained in the sense that, if you understand the principles and techniques of system dynamics, you will be able to understand the manual **without any additional knowledge**. A package which required knowledge which is not in books such as this or in its manual would be very difficult to use effectively and correctly.

Nothing in the foregoing paragraph implies that software vendors are careless or inaccurate in writing their manuals. It is simply a recognition of the fact that writing manuals is very difficult. The manual's author **knows** the software deeply and intimately and it is hard for him to avoid all the difficulties which might be encountered by someone trying to learn how to use the package.

Studying the manual is important but, if possible, talk to people who have used the package you are considering and, best of all, talk to someone who has experience in system dynamics modelling and has used more than one package.

In general, choose carefully before buying. Consider buying more than one package so that you can get exactly the mix of strengths you need. Once you have developed skill in system dynamics by using one package, you will soon learn to work in another style, switching easily to the one most appropriate to the purpose of the particular model you are building.

Once you have acquired the software of your dreams, do not, **on any account**, rush into building big models. Work up slowly to using all the facilities it contains, making sure you understand each one before you use it in a serious model. For instance, suppose software X supports built-in function Y. It would be foolish indeed simply to use Y without first having written a small model, perhaps only a few lines long, which shows what Y does under a variety of conditions, carefully checking **by printed output**, probably at intervals as small as DT, that what Y does is what you expected it to do. If it does not, it is extremely unlikely that the fault lies in the software; it lies in your lack of understanding of Y and, if you don't understand Y well enough to use it correctly and in the circumstances in which it is intended to be used, your model is very unlikely to do **the same things as the real system and for the same reasons**, and is, therefore, practically valueless, if not downright misleading to your clients.

System dynamics problems and solutions

INTRODUCTION

This appendix is intended to help the reader to develop skill by practice on problems. The problems are numbered from 1, but are also arranged according to the chapters of the book. All the models appear on the disk as SN.COS, where N is the problem number. They are all in the traditional DYNAMO-like format of COSMIC, and tutors may have to translate them into the format of whatever software is available. The student should, of course, do at least some of the problems before proceeding to the next chapter.

In most cases, the solution to a problem will be found on the model disk, so the solutions section of this appendix is sometimes no more than a few words of explanation. Some of the problems have no solution.

Diagrams in this appendix are numbered according to the problem.

PROBLEMS

Chapter 1

Problem 1

Write notes to explain to another person the purpose of system dynamics and the type of problem which it does, and does not, address.

Chapter 2

Problem 2

In a manufacturing company, the inventory of parts is depleted by consumption. Consumption is determined by exogenous factors. Inventory is replenished by the production of new parts.

The rate of ordering new parts is determined by the difference between desired and actual parts inventory, with the rate of ordering being such as to correct the discrepancy over TCI weeks. The desired inventory is the consumption rate, averaged over a period of TAC weeks, and multiplied by the number of weeks for which average consumption is to be covered by inventory, TTCAS. The rate of ordering new parts is also influenced by the average consumption rate itself.

The rate of ordering new parts feeds into a parts manufacturing backlog and the manufacturing department set the production rate so as to eliminate the backlog in TEBLOG weeks.

Using the list extension method to get started, draw an influence diagram for this problem.

Problem 3

Oil companies import large amounts of crude oil into Europe by means of a fleet of oil tankers. Most companies own some tankers, but the great part of their needs is met by chartering vessels on the open market. A charter usually lasts for between 3 and 5 years, during which time the chartered ship fetches cargoes, mainly from the Middle East producers, as required by the company.

The demand for refined products varies with the business cycle and is strongly seasonal. Demand is met from refined stocks, which are replenished by refinery production. Producing a ton of refined products requires more than a ton of crude oil. For simplicity, assume that target product stocks are constant. The rate of production in the refinery is planned to correct any discrepancies between target and actual product stocks and to match current average consumption.

Assuming, again, that target crude stocks are constant, the company determines its required rate of inflow of crude oil so as to correct the crude stock discrepancy over a suitable time period and to make up for current average consumption of crude oil. The amount of oil which the tanker fleet can carry at any one time is measured in deadweight tons (DWT), and the required shipping capacity is determined by the rate at which crude oil should be flowing into the refineries.

Tankers are chartered so as to correct the discrepancy between the required and actual carrying capacities over a suitable period.

Using the entity/state/transition method, produce an influence diagram.

Chapter 3

Problem 4

In the United Kingdom, a building society is a finance company which specializes in accepting deposits from savers and providing mortgages to house buyers. The following is a very simplified description of the mechanisms operating in large building societies.

A building society is affected by three exogenous factors: the general level of deposit interest rates in other financial sectors; the normal rate of inflow of new deposits from small savers, and the normal demand for mortgage lending. In practice, these three factors are not mutually independent, but we shall assume that they are.

Funds flow into a society from several sources: the repayment of capital from the mortgages it holds, interest payments on mortgages and the inflow of new deposits. The last is lower when building society deposit rates have for some time been low relative to the general level of interest rates and vice versa. Funds flow out of a society as it lends money on new mortgages and as it pays interest to its depositors.

The rate at which the society is willing to grant mortgages depends, in part, on the amount of money currently held in the society. However, the society will not be willing to lend that all at once, but will try to spread the available money over a certain period of time; a 'cover period'.

The same 'cover period' of the normal demand for mortgages governs the funds the society thinks it desirable to have, and the difference between that and the actual funds drives the society's choice of the interest rate to be offered to depositors.

The interest rate charged on mortgages is usually a multiple of that offered to depositors. The normal demand for mortgage lending is affected by the average mortgage interest rate and thus determines the current demand for mortgages. The rate at which mortgages are actually made depends on this current demand for mortgages and the rate at which the society is prepared to lend money.

Starting with list extension, draw an influence diagram for this problem.

Problem 5

(a) Find some negative feedback loops in the solution to Problem 4 and explain how they affect the dynamics of a building society.
(b) Repeat (a) for a positive feedback loop.

Chapter 4

Problem 6

Using your own words, explain the differences between level, rate and auxiliary variables.

Problem 7

Write notes to explain in your own words, starting from the basic principles of dynamic simulation, why the equations in system dynamics models can be written in any order.

Chapter 5

Many of the following problems ask you to formulate equations. Always run a simple model to test your equations and, within the limitations of your package, thoroughly test the model by printing out all variables, and plotting graphs of a selected set of variables. If possible, check the dimensions of your equations by software; failing that, do it by hand. Ensure that your equations do the same thing as the real system *and for the same reasons.*

Problem 8

This problem is designed to give some practice in writing some very simple equations; in some cases a few equations may be needed. Do not neglect it; much experience shows that one of the greatest hurdles to overcome is writing one's own equations for the first time.

Write equations for the following:

(a) Consumption Rate starts at 100 units per week and increases to 150 at the start of week 15. It falls back to 25 units per week at week 25.

(b) Order Backlog is increased by the Rate of Receipt of New Orders and is depleted by the Rate of Starting Production. The rate of Starting Production is the Order Backlog divided by a time constant called the Backlog Elimination Time.

(c) Average Consumption is the Consumption rate averaged over a period called the Consumption Averaging Time, which is 6 weeks.

(d) The Production Start Rate depends on the difference between Desired Stock and Actual Stock, with the discrepancy being eliminated over a Stock Correction Time of 8 weeks.

(e) Production Completion Rate is a third-order delay of Production Start Rate, the delay being 6 weeks. How would the equations be altered if it was decided that a sixth-order delay of the same total duration was a better model of this process?

Chapter 6

Problem 9

Using the Consumption Rate equation from Problem 8(a), write a complete working model for Problem 2. Set TEBLOG, TAC, TTCAS and TCI all to 6 and use a time unit of weeks.

Problem 10

Develop a working model for the tanker chartering system in Problem 3.
Assume that product consumption has a base value of 10×10^6 tons per

month and oscillates with an amplitude of 1×10^6 tons per month. Target Product Stocks are 30×10^6. Making a ton of refined product requires 1.2 tons of crude oil so Target Crude Stocks are set accordingly. The value of CONVF is 0.4266 and all time constants are 3 months. The duration of tanker charters can reasonably be modelled as a third-order delay with an average magnitude of 48 months.

Chapter 7

No particular problems are set for this chapter. You already have numerous models to work with and you should attempt to experiment with some of them, as suggested in the chapter.

Chapter 8

Problem 11

Write an essay explaining in your own words the principles of system dynamics optimization, considering carefully the limitations and advantages of the approach.

Chapter 9

You will be able to do some of these problems even if you do not have access to optimization software.

Problem 12

Amend the equation for COST in the combat model optimization (p. 281) to allow for different marginal costs of adding and subtracting resources.

Problem 13

Making whatever assumptions you wish about the marginal costs of increasing and decreasing assets, optimize the model again.

Problem 14

Modify the objective function for the simple combat optimization to allow for two constraints: money and manpower.

Problem 15

Optimize the simple combat model to achieve the minimum cost for a given level of performance. Assume that the aim is for Blue to have a certain number of men surviving at the end of the combat.

Chapter 10

Problem 16

Amend the equations for discrete production planning to allow for production to change only at certain intervals.

Problem 17

Chapter 10 referred to the problem of making samples happen at variable intervals, though the example given there used only two possible values of the sampling interval. Write and test a model which takes as input a sine wave with a base value of 100 units, an amplitude of 30 units and a period of 25 weeks. The sampling interval has a base value of 2 weeks but varies with random noise between 1 and 3 weeks. Such techniques are required when one is testing the sensitivity of a system to random effects.

A technical hint is that the standard SAMPLE function takes a sample at intervals of INTV weeks. When each sample is taken, the value of INTV at that time is stored and a count is kept against TIME to determine when the next sample is due. This is true whether INTV is, as usual, a constant or, as required in this problem, a variable.

A second hint is that INTV, or whatever you choose to call the sampling interval, must be treated as a level to act as memory for the sampling process.

This is a nasty little problem which is intended to give disciplined practice in thinking in terms of time as well as developing a useful technique. The reader will gain no benefit at all by simply looking at the solution, and the only way to approach it is to write and test equations, observing *very carefully indeed* exactly what happens to the variables every DT in the run. If your software does not make it easy to have printed output it is unlikely that you will solve the problem correctly.

Problem 18

The bidirectional multiplier has the form:

a desfam.k=table(ttdfam,avinc.k,arg1,arg2,arg3)
d desfam=(children/family) desired family size
d avinc=($/year/family) average family income
d ttdfam=(children/family) table of value for desfam as affected by
x   average family income

and ARG1, ARG2 and ARG3 are numerical arguments for the range and step size in TTDFAM.

This formulation has been used in some system dynamics models of economic development models to make families have fewer children as

380 Appendix B: System dynamics problems and solutions

their income increases, but it makes family size increase again if they subsequently become poorer.

Amend the equations to ensure that, once family size has fallen due to increased income, it will not subsequently increase, no matter how far incomes subsequently fall.

Problem 19

Amend the air sorties model to allow for raids to take place every 2 hours or as soon as aircraft are available and thereafter every two hours.

Problem 20

Formulate a model for the pharmaceutical company problem shown in Fig. 3.12. Assume the following numerical data[1]:

Average Cost of Product Research	61 $/product
Delay in Product Research	20 months
Shelf Life of a Developed Product	10 months
Delay in Product Introductions	10 months
Average Cost of Product Introduction	560 $/product
Lifetime of a Product at Market	20 months
Disposable Revenue per Product:	

60 $/month at age 8 months or less
37 $/month at age 10 months
20 $/month at age 12 months or more.

Company policies are to:

- adjust product research effort over 6 months
- adjust discrepancies in products at market over 3 months
- as the average product age rises above 10 months there is a slight bias towards allocating money to product introductions and conversely if average product age is less than 10 months

This problem will require the use of many of the techniques covered in the book, particularly multipliers, discussed in Chapter 5 and considered in detail in Chapter 10. The problem is also typical of real cases in that not all of the policies are explicitly identified.

To derive initial conditions, assume that the flows of products in the firm are balanced at 15 products per month.

One major difficulty is the development of a formula for Average Product Age, APA. One approach is

APA=PY/PAM

[1] The data are completely fictitious and the problem is heavily adapted from a real case. Assume all monetary values to be in tens of thousands.

where PY is product years of total age, or the number of products at a certain age, multiplied by that age and summed over all products, and PAM is the number of products at the market. Differentiating with respect to time (represented by a dot over the variable) yields

$$A\dot{P}A = \dot{P}Y/PAM - \dot{P}Y*PAM/PAM^2$$

Now, $\dot{P}Y = PAM - PWR*PLT$, where PWR is Product Withdrawal Rate at the Product Lifetime, PLT, and

$$\dot{P}AM = (PIR - PWR)$$

where PIR is the Product Introduction Rate. Hence

$$A\dot{P}A = 1 - (PWR*PLT + (PIR - PWR)*APA)/PAM$$

All that remains is to translate this into system dynamics syntax.

Chapter 11

Problem 21

Choosing a problem in which you are interested, carefully identify who you think might be the clients for a study of it. Draw an influence diagram of the main features of the problem. Explain how you would apply system dynamics to it, indicating what you would anticipate to be the most difficult aspects of developing a quantitative model. What would you expect the qualitative and quantitative models to tell you about the problem that you did not already know?

Chapter 12

Problem 22

Explain the main stages of the system dynamics methodology, drawing your own diagrams to explain the points you make.

Problem 23

Summarize the limitations and strengths of system dynamics as an approach to understanding complex managed systems.

SOLUTIONS

Solution 2

This problem is intended to give some practice in basic influence diagram techniques. The solution in Fig. S.2 is fairly straightforward, provided one

---

[2] COSMIC has the option of plotting a line to mark the zero point on an axis, which will make it obvious if a variable has gone negative.

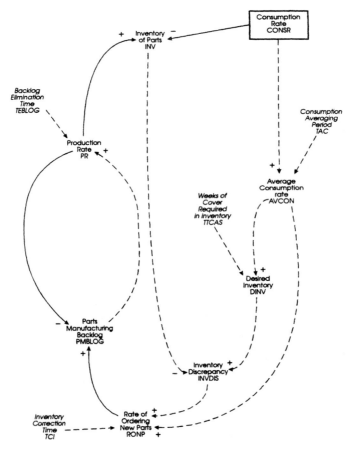

Fig. S.2   Influence diagram for problem 2.

carefully parsed the narrative. Check your solution against this to make sure you correctly used solid and dotted lines, that your signs are right and that parameters have been properly identified.

Solution 3

This is rather more difficult than the previous problem. The first trick is to recognize that there are *three* entities, refined oil, crude oil and oil tankers. The first two involve only simple levels, the third is a delayed flow. The solution appears in Fig. S.3.

The second key to this problem is to recognize the need for a conversion factor between the amount of carrying capacity in the tanker fleet, measured in DWT, and the rate at which oil should, and does, flow in, measured in Crude Tons/Month. Using round figures, the return trip between Europe

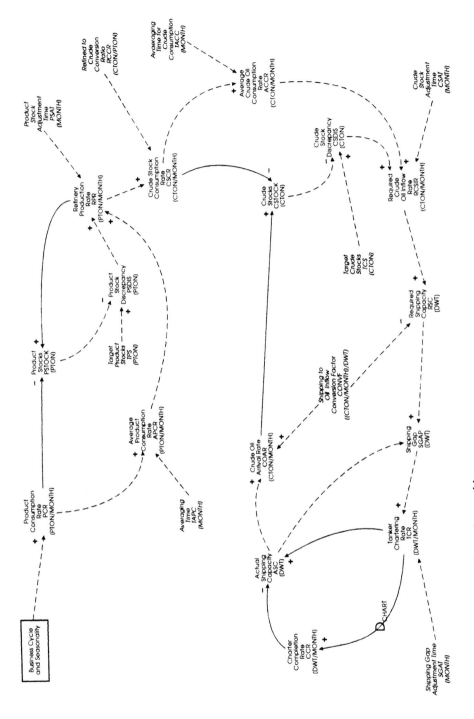

Fig. S.3 The oil tanker chartering problem.

and the Middle East takes 2 months, so a ship of, say, 250 000 DWT brings in 250 000 tons of crude oil every other month. Averaging over many ships, the rate of inflow of crude oil is half the tonnage on charter. The exact figure is 0.4266, so it is clearly important whether one multiplies or divides by that number, so we have shown + and − signs on the links from that parameter.

Solution 4

This problem is very much more difficult than problems 2 and 3. One of the main sources of difficulty is the terminology. The 'rate' of interest is not the same thing as a rate variable in system dynamics. To add to the confusion, phrases such as 'the level of interest rates' are used and sound like nonsense to a system dynamics ear. In fact, what is meant is the 'average interest rate'.

The solution is shown in Fig. S.4. As always, the main trick is to 'think physics' and identify the flows in and out of the pool of funds currently held by the society. After, that, it is a matter of mapping the influences that those flows exert and which are exerted on them. Thus, the Rate of Inflow of New Deposits is the Normal Rate, scaled up or down, according to whether the society's deposit rate is, on average, higher or lower than general deposit interest rates.

A similar idea is used for the Actual Rate of Lending. The Current Demand for Mortgage Lending is the Normal Demand, modified by the effects of the Mortgage Rate. The smaller of that demand and the potential supply from the society gives the actual rate of lending.

The parameter for the duration of a standard mortgage was not mentioned in the narrative, but is clearly necessary.

The most interesting idea is that the society bases its Desirable Level of Funds on the Normal Rate of Lending, which is a variable in the general economy and which might be rather hard to determine, rather than on the Current Demand, which could probably be measured reasonably accurately.

Solution 5

Figure S.5A shows four negative loops. The numbering is arbitrary and it is not implied that Loop 1 is the most significant. The reader should work out where these loops share common paths.

Loop 1 is negative and acts to raise the deposit rate to correct a shortage of funds. Loop 2 inevitably reinforces whatever Loop 1 does, so that if, for example, deposit rates are raised to bring in more cash, the mortgage rate also rises, bringing in yet more money as mortgagees pay more interest. When society funds are low, Loop 3 ensures that less lending takes place, while Loop 4, eventually, reduces demand for mortgages.

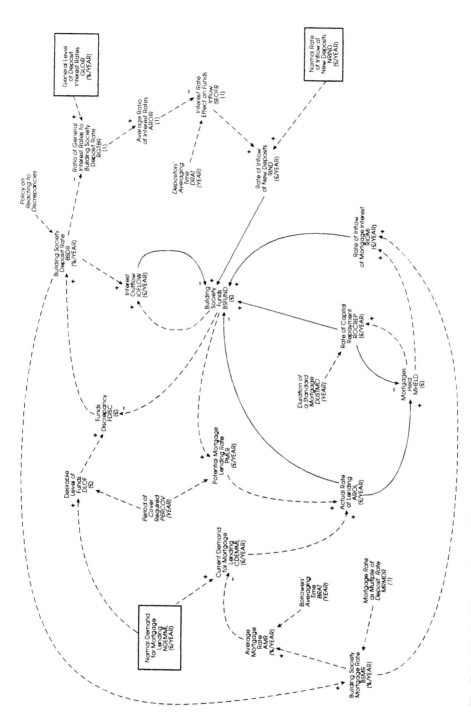

Fig. S.4 The building society problem.

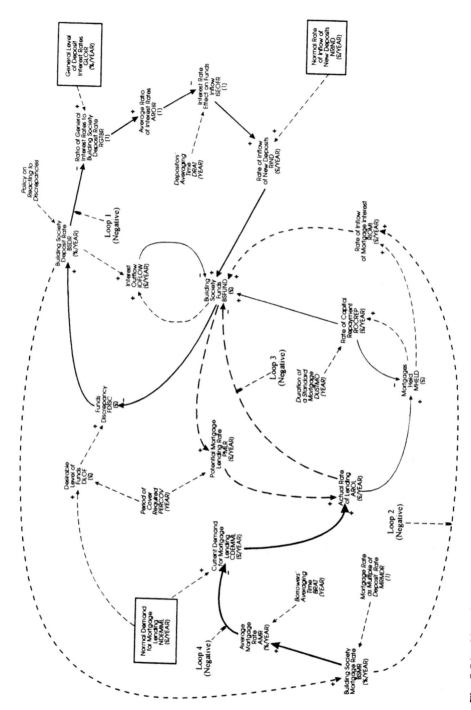

Fig. S.5A Negative loops in the building society problem.

It could well be that these four effects over-compensate for the shortage of funds, at which point they will act so as to reduce excessive funds. One suspects that, since the delays in the system are relatively long, the effect may be to reinforce any cyclical effects, such as the business cycle, in the exogenous inputs.

Figure S.5B shows a positive loop which acts to reinforce shortages or surpluses of funds. Whenever rates rise because funds are low, more interest has to be paid to depositors, thus driving funds further down. The society hopes that this will be more than compensated for by the influx of new funds through Loop 1, and the other negative effects in the system.

This positive loop shows how very careful one has to be in formulating models. Does all the cash in the society belong to depositors, or is some of it 'profit' due to the mortgage rate being higher than the deposit rate? It may be that the cash funds in the society are less than the amount belonging to depositors because their funds have been lent out as mortgages. Legally, all the assets of the society belong to its members, but both borrowers and depositors are members, and assets may be more than the funds on deposit. Care would be needed in building a quantitative model!

Are there any more positive loops?

Solution 8

(a)

```
CONSR.KL=BASE+STEP(STEP1,STPTM1)−STEP(STEP2,STPTM2)
C BASE=100
C STEP1=50
C STPTM1=15
C STEP2=−125
C STPTM2=25
D CONSR=(UNIT/WEEK) CONSUMPTION RATE
D BASE=(UNIT/WEEK) BASE VALUE FOR CONSUMPTION
D STEP1=(UNITS/WEEK) HEIGHT OF FIRST STEP
D STPTM1=(WEEK) TIME OF FIRST STEP
D STEP2=(UNITS/WEEK) HEIGHT OF SECOND STEP
D STPTM2=(WEEK) TIME OF SECOND STEP
```

In general, is better practice, if slightly more effort, to define all constants explicitly than to write:

```
R CONSR.KL=100+STEP(50,15)+ ...
```

The explicit form allows more flexibility for model experiments and ensures that all factors have been defined in words, which allows their dimensions to be more thoroughly checked.

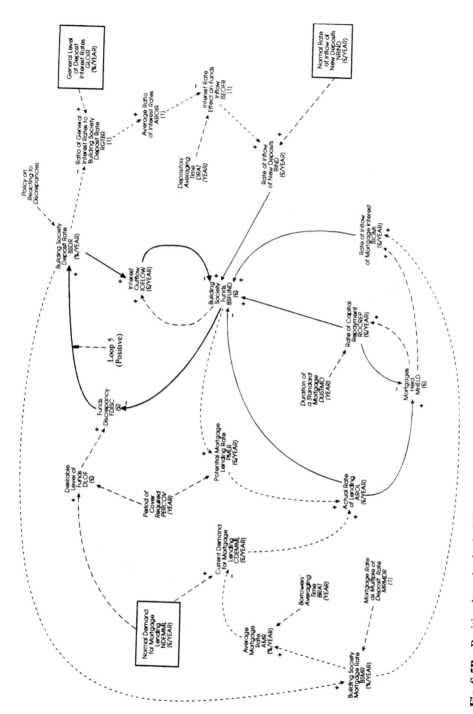

Fig. S.5B Positive loop in the building society problem.

(b)

```
L OBLOG.K=OBLOG.J+DT*(RRNO.JK−RSP.JK)
N OBLOG= ...
R RSP.KL=OBLOG.K/BET
C BET= ...
D OBLOG=(UNIT) CURRENT BACKLOG OF ORDERS
D RRNO=(UNIT/WEEK) RATE OF RECEIPT OF NEW ORDERS
D RSP=(UNIT/WEEK) RATE OF STARTING PRODUCTION
D BET=(WEEK) TIME CONSTANT FOR WORKING OFF
X              BACKLOG OF NEW ORDERS
```

A suitable initial condition would be required for OBLOG and a numerical value for BET.

(c)

```
L ACON.K=ACON.J+(DT/CAT)(CONSR.JK−ACON.J)
N ACON=CONSR
C CAT=6
D ACON=(UNIT/WEEK) AVERAGE CONSUMPTION RATE
D CONSR=(UNIT/WEEK) CURRENT CONSUMPTION RATE
D CAT=(WEEK) TIME CONSTANT FOR AVERAGING
X              CONSUMPTION RATE
```

(d)

```
R PSR.KL=SDIFF.K/SCT
A SDIFF.K=DESSTK.K−ASTOCK.K
C SCT=8
D PSR=(UNIT/WEEK) PRODUCTION START RATE
D SDIFF=(UNIT) DIFFERENCE BETWEEN DESIRED AND
X              ACTUAL STOCKS
D DESSTK=(UNIT) DESIRED STOCK
D ASTOCK=(UNIT) ACTUAL STOCK
D SCT=(WEEK) STOCK CORRECTION TIME
```

Note the use of the auxiliary variable, SDIFF, the value of which could be printed or plotted as a check that the model was working correctly. Obviously, these simple equations are going to work correctly, but they may reflect incorrect operation of some other part of the model and giving oneself as much diagnostic help as possible is a good way to avoid a lot of trouble.

(e)

```
R PCR.KL=DELAY3(PSR.JK,PDEL)
C PDEL=6
D PCR=(UNIT/WEEK) PRODUCTION COMPLETION RATE
D PSR=(UNIT/WEEK) PRODUCTION START RATE
D PDEL=(WEEK) PRODUCTION DELAY
```

The sixth-order delay can be represented by introducing a dummy rate, DUMR, and writing:

    R PCR.KL=DELAY3(DUMR.JK,PDEL/2)
    R DUMR.KL=DELAY3(PSR.JK,PDEL/2)

In all the above we have referred to the entity involved as 'UNIT'. That is acceptable for simple practice problems but, in real models, it is far better to call the entity what it is, such as WMS for washing machines, or STTONS for tons of steel.

From this point, you will have to run the solution models to produce the results. This is to encourage you to gain the invaluable practice of working with models, as opposed to looking at ouput diagrams. For example, Figure S.9 is referred to in the solution, but you will have to run the model to produce it.

Solution 9

The solution appears on the disk as S9.COS, though the base case behaviour in the upper half of Fig. S.9 should be studied first. Clearly, something is very wrong, as the backlog of parts, PMBLOG, becomes negative at about TIME=35, and a negative backlog is clearly ludicrous. The reason is that non-negativity constraints were not built in, and the system's policies are not able to handle the extreme shock of the drop from a consumption rate of 150 parts/week to 25.

Again, this illustrates how *very* careful one has to be with models. It might well have been decided that the only variables of interest were CONSR and INV, so if PMBLOG had not been plotted the graphs of those two variables would look plausible. It is only when one studies the printed output that errors such as this can be found.

The reader should experiment to find a value of STEP2 which does not cause negative PMBLOG. That will be the value which is the limit of the robustness of the stated policies.

Turning to the model itself, none of the equations should present any difficulty. Note that *all* variables are to be printed. The equation for PR is set up, for explanatory purposes, to allow one to switch between the stated production policy and one which includes a non-negativity constraint. When the model is run with the latter, the graph in the lower half of Fig. S.9 is produced. If, of course, STEP2 is such that a correct response is obtained, it does not matter whether MODSW is 0 or 1.

Study of the printed values shows that INV and PMBLOG settle at a value of 125 at the end of the run. Is that correct?

Solution 10

Most of the equations in S10.COS are fairly straightforward, providing one works carefully. A good example of the mistakes one can make in modelling is that, in developing this model, the author made the typing errors:

L ASC.K=ASC.J+DT*(TCR.JK−**CSCR**.JK)
N ASC=IASC
N IASC=ACCR*CONVF

Compare with the model to find the correct equations. In both cases the error was found by COSMIC's dimensional analysis software. It could have been found by hand checking, though that would have been more laborious.

Note the inclusion of a mass balance check for tankers chartered and develop similar check equations for refined products and crude oil. When the model is run, the value of TCHECK is about 200, which is near enough to zero when it is realized that cumulative charters amount to some 450×10^6 DWT over the 200 month length of the simulation.

Is 200 months too long? Probably. The period of the oscillation in consumption is 51 months, so a LENGTH of about 100 or 150 should be enough to show the main dynamics.

The behaviour in Fig. S.10 is for two cases; the base case in the main model and an alternative of charter durations of only 12 months and the company attempting to close its shipping gap over 12 months instead of 3. This is obviously wrong again, as CSTOCK goes negative, reaching a minimum of about -8×10^6 tons.

The reader should try to work out why this happens before reading the next paragraph. Close study of the graphs, comparing the times at which peaks and troughs occur, will help.

The reason for the erroneous behaviour is that large amounts of shipping reach the end of their charters during the first few months of the run. The company's slower reaction to placing new charters means that shipping does not reach its peaks until about 10 months after the peaks in the base case; the oscillations in tanker capacity have been shifted in phase relative to those in refined stocks (and therefore relative to those in crude stock consumption). For this reason, the oscillations in crude stock are much larger than in the base case and the slower chartering means that the large fall in crude stocks in the early months is never made up. When SGAT is left at 3 months, with the shorter charters, the behaviour becomes acceptable.

Clearly, negative crude stocks should not happen, and it is tempting to add a non-negativity constraint of the type:

A IPR.K=PSDIS.K/PSAT+APCR.K
R RPR.KL=MIN(IPR.K,CSTOCK.K/DT)
D IPR=(PTON/MONTH) INDICATED PRODUCTION RATE

IPR is the rate at which production *should* take place and RPR will be that rate or the available stock, whichever is less.

In practice, we should adopt the strategy of regarding the negative stock as simply a phenomenon of a catastrophic policy which should never be adopted.

Solution 12

For the men, the required change is:

n cost=max(0,iblue−biblue)*upcost+max(0,biblue−iblue)*downcost

If the value of IBLUE chosen in a given iteration exceeds the value, BIBLUE, from the base case, then the first MAX will operate, and vice versa. The extension to the rest of the equation is obvious.

Solution 14

This is a relatively trivial problem.
    The original objective function was:

a of.k=bpi.k/scale−10e06*max(0,cost−budget)

Simply add another term:

−10e06*max(0,men−menlim)

where MEN is the total of IBLUE and IBLRES, parameters which will be varied in the optimization search, and MENLIM is the total number of men available.
    If there are two manpower limits, those available for the regular force and those available for reserves, the penalty terms become:

−10e06*max(0,iblue−reglim)−10e06*max(0,iblres−reslim)

This simple approach can be extended indefinitely; the scope of optimization is restricted only by the analyst's insight!

Solution 15

The technique is to define a target to be achieved, such as Blue having 20000 men left at the end of the battle. The objective function then becomes:

a of.k=cost+10e06*max(0,tblue−blue.k)
c tblue=20000

The equation for COST will need to be changed so that it is now of the form:

n cost=iblue*costreg+iblres*costres

and so forth, as all we are concerned with is the minimum cost of a given performance, not the maximum performance for a given budget. The method of the previous solution can be used to avoid overspending.

The objective function, OF, is now to be *minimized*, as the problem is to achieve a given performance for minimum cost. If the optimal solution is to reduce current levels of spending, COST will be negative, which will be ideal. If BLUE.K, the value of Blue's force at the end of the simulation, is less than TBLUE, a huge penalty will be added to OF and the minimization will not be achieved. On the other hand, if BLUE.K is greater than TBLUE, no penalty will be added.

That could, of course, leave Blue with more than 20 000 men at the end of the battle, which, in human terms would be fine. In the brutal world of optimization, one might aim to be as close as possible to 20 000, in which case one might use:

a of.k=cost+10e06*(tblue−blue.k)**2

The method can be varied to make BLUE follow some desired profile throughout the battle and penalizing variations below, or from, that profile.

There is, of course, no magic about the use of 10e06 as a penalty weight. If, for example, COST is of the order of 10e09, then the penalty weight needs to be 10e12.

Although these two solutions are in terms of the simple combat model, the ideas apply to any optimization problem.

Solution 16

This is very straightforward. Instead of using:

r pr.kl=opr.k+pchan*int((dpr.k−opr.k+0.5*pchan)/pchan)

we write:

a dum.k=opr.k+pchan*int((dpr.k−opr.k+0.5*pchan)/pchan)

so that DUM can be printed out as a check that the equations are working properly and then use:

r pr.kl=sample(dum.k,prodper,dum.k)
d prodper=(week) interval at which production planning decisions
x   are made

Solution 17

* Figure s.17 variable sampling times
note
note   file named s17.cos
note

```
note    sine wave to be sampled
note
a input.k=base+amp*sin(6.283*time.k/perd)
c base=100
c amp=30
c perd=25
```

A standard sine wave for INPUT, the variable to be sampled.

```
note
note    sampled value of input
note
a output.k=sample(input.k,intv.k,input.k)
```

The resultant sample is OUTPUT.

```
note
note    sampling interval sector
note    =============
note
note    noise in sampling interval
note
a samvar.k=int((basint*(1+noise(seed))+0.5*dt)/dt)*dt
c basint=2
c seed=19
```

SAMVAR is expressed as an exact number of DTs. There is no need to take random numbers at set intervals, as was done in Chapter 10, as all we need is to store the appropriate value of SAMVAR when INTV is changed.

```
note
note    saving of random value until a sample is due
note
l save.k=save.j+(dt/dt)*samvar.j*clip(1,0,save.j−time.j,dt)
x *clip(1,0,dt,save.j−time.j)
n save=basint
```

This is the key sector of the model. The sampled value is saved to monitor the passage of time. Since, however, the sampling interval will only be changed when a sample is taken, we require to store the value of SAMVAR 1 DT before the sample is due to be taken. This will then be held as the sampling interval to determine when the next sample is due.

The two CLIPs ensure that the change only takes place when the difference between SAVE and TIME is exactly 1 DT. Run the model again, removing one or other of the CLIPs, and explain what happens.

```
note
note    saving the sampling interval at the appropriate time
note
```

```
l intv.k=intv.j+(dt/dt)(samvar.j−intv.j)*clip(1,0,save.j−time.j,dt)
x *clip(1,0,dt,save.j−time.j)
n intv=basint
```

The value of INTV is updated whenever the comparison between SAVE and TIME makes TIME 1 DT before SAVE.

```
note
note   model control and output
note
c dt=0.25
c pltper=0.25 for precision graphs
c prtper=0.25 printing every dt to check model !!!!
c length=40
print 1)input
print 2)output
print 3)samvar
print 4)save
print 5)intv
plot input=a,output=b(0,200)/intv=c(0,5)
```

This sector is absolutely critical. Note that each variable is printed at intervals of DT to ensure that the right things are happening at the right times. Only a few variables are plotted, though simply plotting INPUT and OUTPUT might suggest that the model is correct. Figure S.17 shows that, as required, INTV changes exactly when the samples are taken, except where the randomness has produced the same value for INTV twice in succession.

```
note
note   model tests
note
run A trial of equations
c seed=3
run B a different series of random numbers
```

Testing the model with at least two sequences of pseudo-random numbers increases confidence that it is working correctly.

This procedure of testing the tricky parts of models by writing a mini-model and closely studying what is happening is an essential method of developing anything other than a purely trivial case.

```
note
note   definitions of variables
note
d amp=(unit) amplitude of sine wave for input
d base=(unit) base for sine wave for input
d basint=(week) base value for noise
```

d dt=(week) time step in simulation
d input=(unit) variable to be sampled
d intv=(week) sampling interval
d length=(week) simulated period
d output=(unit) sampled value of variable to be sampled
d perd=(week) period of sine wave for input
d pltper=(week) plotting period
d prtper=(week) printing period
d samvar=(week) values for random sampling interval
d save=(week) cumulatively saved random sampling intervals
d seed=(1) seed for random number generation
d time=(week) time within simulation

Solution 18

It is not possible to write, as one might in FORTRAN, C or Pascal,

a desfam.k=min(desfam.k,tabhl (. . .))

as that requires DESFAM to depend on itself, which prevents the package sorting the equations into the correct computational sequence.
    A possible solution is to store the previous value of DESFAM:

l odesfam.k=odesfam.j+(dt/dt)(desfam.j−odesfam.j)
d odesfam=(children/family) value of desfam one dt previously

and then to use:

a desfam.k=min(odesfam.k,table(. . .))

Solution 19

* Figure s.19 revised aircraft sortie model
note
note   file named s19.cos
note

Most of the model is unchanged from FIG10-11.COS, so only the new equations need to be explained.

note
note   launch rate
note
note   the policy switch, POLSW, allows us to change from
note   one decision rule to another
note
c polsw=1

To demonstrate the difference between the old policies and the new, a policy switch is introduced.

```
r launch.kl=polsw*clip(sgroup/dt,0,acavail.k,sgroup)
x *pulse(1,wstart,abintvl.k)+(1−polsw)
x *pulse(clip(sgroup/dt,0,acavail.k,sgroup),wstart,stsint)
```

The first two lines are the new policy of launching every two hours or as soon as the required number of aircraft become available. The CLIP tests for aircraft availability and the PULSE sends a series of 1s to allow the CLIP to work (or not). The frequency of the 1s is shown below and is either the standard interval, STSINT, of 2 hours, or DT, which has the effect of a continued monitoring of aircraft availability when there were too few to launch at the last two hour point.

The third and fourth lines are the old policy from Chapter 10.

```
a abintvl.k=clip(stsint,dt,acavail.k,sgroup)
c stsint=2
```

The launch interval, ABINTVL, is either 2 hours when aircraft are available or a continuous check every DT if they are not.

```
c sgroup=40
```

The groups have been set to 40 in order to get some clear comparisons between the two policies. The reader should, of course, experiment with other group sizes.

```
note
note   record time at which launches took place
note
l ltime.k=ltime.j+(dt/dt)(time.j−ltime.j)*clip(1,0,launch.jk,sgroup)
n ltime=0
```

An equation to record the times at which launches took place.

```
note
note   model control and output
note
c dt=0.015625
c length=40
c pltper=0.125 for precision plots
```

With such a small value of DT, PLTPER must be chosen by experiment to get reasonable graphs. Run the model with PLTPER=1 and observe the difference in *apparent* behaviour. This is yet another demonstration of the inability to debug a model properly from the plotted graphs.

```
print 8)abintvl,ltime
```

The new variables are added to the print commands for debugging.

```
plot acavail=a,oparea=b(0,100)/ltime=c(0,50)
```

The launch time variable, LTIME, is added to the plots.

```
run A Launch every 2 hours or when 40 available
c polsw=0
run B launch every 2 hours if 40 are available
```

Experiments to compare the two policies.

```
note
note    definitions of variables
note
d abintvl=(hour) minimum interval between launch of groups of
x                        aircraft on sorties
d launch=(aircraft/hour) rate of launch on missions
d ltime=(hour) time at which launches occurred
d polsw=(1) switch to choose alternative policies
d stsint=(hour) standard interval for launching sorties
```

Definitions are added for new or revised variables.

The behaviour with the two policies is shown in Fig. S.19. The behaviour is quite significantly different, the old policy giving 10 launches over the 40 hours while the new is nearly ready for a twelfth. With the old policy, there are some quite clear cases where rather more than 30 aircraft are available, but nothing happens until the next two hour point is reached.

The value of developing models to include such fine detail will depend on the case. There is no virtue in complicating a model simply to show one's modelling virtuosity. However, policy nearly always makes a significant difference to system behaviour, so correct modelling of policy is usually vital.

It is an open question whether some of the packages described in Appendix A would be capable of this degree of sophistication.

Solution 20

```
* Figure s.20 the pharmaceutical company model
note
note    file s20.cos
note
note    research and product availability
note    ================
note
a rrsr.k=((dpafm.k−pafm.k)/prat+apir.k)*acpr
```

The aim is to bring Products Available for Marketing up to a target level and to replace products being introduced, the effect being expressed as a desired rate of spending.

```
c acpr=61
c prat=6
a aspr.k=max(0,drflow.k−aspi.k)
```

Actual spending may be less than the desired level. The model makes no provision for the spending of accumulated balances, an area which the reader may wish to explore.

```
note
note    product research starts and completions
note
r prsr.kl=aspr.k/acpr
r prcr.kl=delay3(prsr.jk,dres)
note
note    products available for marketing
note
l pafm.k=pafm.j+dt*(prcr.jk−rspi.jk−par.jk)
n pafm=base*dres
c base=15
```

The pool of Products Available for Marketing is initialized to the base flow of 15 products per month.

```
c dres=20
a dpafm.k=apir.k*dres
```

The target for Products Available for Marketing.

```
r par.kl=pafm.k/sl
c sl=10
```

Products are lost due to the shelf life effect.

```
note
note    product introductions
note    ============
note
a rrspi.k=((apwr.k*dlife−pam.k)/mpat+apwr.k)*acpi
c mpat=3
```

A similar concept to that use for Products Available for Marketing.

```
a aspi.k=min((pafm.k/dt)*acpi,min(1,rsr.k*pisf.k)*drflow.k)
```

The Actual Spending may be increased by a factor affected by Average Product Age, PISF (see below).

```
r rspi.kl=aspi.k/acpi
n rspi=base
r rcpi.kl=delay3(rspi.jk,dintro)
c dintro=10
c acpi=560
l apir.k=apir.j+(dt/tapi)(rcpi.jk−apir.j)
n apir=rcpi
c tapi=2
```

Simple equations of flow and averaging.

```
note
note    products at market
note
l pam.k=pam.j+dt*(rcpi.jk−pwr.jk)
n pam=base*dlife
note
note    product withdrawals
note
r pwr.kl=delay3(rcpi.jk,dlife)
c dlife=20
l apwr.k=apwr.j+(dt/tapwr)(pwr.jk−apwr.j)
n apwr=pwr
c tapwr=2
```

Again, a simple flow of Products at market.

```
note
note    average product age
note    ==========
note
l apa.k=apa.j+dt*(1−(pwr.jk*dlife+(rcpi.jk-pwr.jk)*apa.j)/pam.j)
n apa=dlife/2
```

The mathematical derivation gives the rate of change of APA and therefore appears in a level equation as shown.

```
note
note    revenue flows
note    =======
note
a drfpp.k=tabhl(trl,apa.k,8,12,2)
t trl=60/37/20
a drflow.k=drfpp.k*pam.k
l cumearn.k=cumearn.j+dt*drflow.j
n cumearn=0
note
note    product development policy
note    ===============
note
s tns.k=rrsr.k+rrspi.k
```

A supplementary variable, solely for output.

```
a rsr.k=rrspi.k/(rrsr.k+rrspi.k)
```

The ratio of the two spending requirements, which was used to determine the Actual Rate of Product Introductions.

```
a pisf.k=tabhl(tpisf,apa.k,9,11,1)
t tpisf=0.9/1.0/1.1
```

The actual ratio is scaled up or down depending on Average Product Age.

```
a tots.k=aspr.k+aspi.k
a factor.k=tots.k/drflow.k
```

Some additional output variables.

```
note
note    output and control
note    ==========
note
c dt=0.25
c length=100
c pltper=1
a prtper.k=1+step(9,10.5)
print 1)prsr,prcr,factor
print 2)pafm,par,dpafm
print 3)rspi,rcpi
print 4)pam,pwr
print 5)apa,drfpp,cumearn
print 6)drflow,tns,pisf
print 7)rrsr,aspr,rsr
print 8)aspi,aspr,rsr
plot pam=a,pafm=b(0,800)/apa=c(8,12)
```

As always, all the variables have been printed and the output closely examined.

```
note
note    simulation experiments
note
run A Trial of model
note t tpisf=0.0/1.0/2.0
note run very steep tpisf
note t tpisf=0.9/1.0/1.1
c prat=12
run B Policy of smoother reactions
```

Two experiments are carried out, though many more are suggested.

```
note
note    definitions of variables
note
d acpi=($/product) average cost of product introductions
d acpr=($/product) average cost of product research
```

d apa=(month) average product age
d apir=(product/month) average product introduction rate
d apwr=(product/month) average product withdrawal rate
d aspi=($/month) actual spending on product introductions
d aspr=($/month) actual spending on research and development
d base=(product/month) base value for initial conditions
d cumearn=($) cumulative earnings
d dintro=(month) delay in product introductions
d dlife=(month) product lifetime
d dres=(month) delay in product research
d drflow=($/month) disposable revenue flow
d drfpp=(($/month)/product) disposable revenue flow per product
d dpafm=(product) desired products available for market
d dt=(month) solution interval
d factor=(1) ratio of total desired spending to available revenue flow
d length=(month) simulated period
d mpat=(month) marketable products adjustment time
d pafm=(product) products available for marketing
d pam=(product) products at the market
d par=(product/month) product abandonment rate
d pisf=(1) product introduction preference factor. A multiplier to
x   show the preference attached to product introduction
x   when allocating spending.
d pltper=(month) plotting interval
d prat=(month) product research adjustment time
d prcr=(product/month) product research completion rate
d prsr=(product/month) product research start rate
d prtper=(month) printing interval
d pwr=(product/month) product withdrawal rate
d rcpi=(product/month) rate of completing product introductions
d rrspi=($/month) normal rate of spending on product introductions
d rrsr=($/month) normal rate of spending on product research
d rspi=(product/month) rate of starting product introductions
d rsr=(1) ratio of required spending on product introductions to total
x   required spending on research starts and product
x   introductions
d sl=(month) shelf life of a developed product
d tapi=(month) time for averaging product introductions
d tapwr=(month) averaging time for product withdrawal rate
d time=(month) simulated time
d tns=($/month) total required spending
d tots=($/month) total actual spending
d tpisf=(1) table for pisf
d trl=(($/month)/product) table for revenue level

The output is shown in Fig. S.20 for two simulations and in comparative form for each of the three variables. The main point is that, as ever in system dynamics, policy has an enormous effect on system behaviour.

COSMIC supporting documentation and software

INTRODUCTION

This appendix provides information on documentation, software and supporting services available from the COSMIC Holding Company.

TRAINING

Training in system dynamics is available from the author or his associates. It can be a one-week course, as outlined in Chapter 12, or shorter versions for senior managers or analysts who require an appreciation of system dynamics, rather than a full training course. All courses are designed specifically for the customer.

CONSULTANCY

Consultancy in system dynamics is available from the author, supported by his associates where necessary. Fees are charged on a daily basis, plus any significant expenses.

DOCUMENTATION

Several publications are available which support the use of system dynamics.

- An *Introductory Guide* describes the features of COSMIC and COSMOS in more detail.
- A 150-page *User Manual* explains all procedures for using COSMIC and COSMOS.
- *Equations for Systems* provides more than 30 case studies of the use of COSMIC and COSMOS for advanced modelling.

- *System Dynamics Problems* has 45 problems on all aspects of system dynamics, with full solutions.

COSMIC AND COSMOS SOFTWARE PRICES

All prices for COSMIC and COSMOS include a licence for Prospero FOR-TRAN[1], and one copy of the *User Manual* for each licence. COSMIC and COSMOS are supplied on 3.5″ disks as one integrated package. The licence fee covers both packages and the royalty we pay to Prospero Software. Payment may be made in any convertible currency, plus 5% for orders from outside the UK. Order by official purchase order, or cheque, adding postage and packing as specified below.

UK orders for software are subject to value added tax (VAT); the current prices (1995) are shown below. The amount should be added to cheque purchases and a VAT receipt will be sent with the software. Orders from outside the United Kingdom are not subject to VAT. Multiple copy prices apply only to orders placed within 90 days of the first order.

Standard licence

First copy	£800 (£931.25 including VAT[2])
Next four copies	£500 each (£578.75)
Subsequent copies	£350 each (£402.50)

Academic institution licence

First copy	£500 (£578.75)
Next four copies	£300 each (£343.75)
Subsequent copies	£250 each (£285.00)

Software with an academic licence may only be used for teaching and scholarly research and not for commercial consultancy or grant-supported research.

Individual student licences

A charge of £250 (£285.00 including VAT) is made. Orders must be accompanied by a statement from the Head of Department that the software will be used only by a named student for study and research, will not be given or

---

[1] Supplied by Prospero Software Ltd, 190 Castelnau, London SW13 9DH, UK; telephone 0181 741 8531. The licence covers only those parts of Prospero FORTRAN that are necessary for COSMIC and COSMOS. A full FORTRAN package is *not* supplied and there is no FOR-TRAN manual.

[2] Prices allow for the fact that VAT is not charged on documentation.

sold to any other person and will not be used in the student's subsequent employment.

DOCUMENTATION PRICES

The *Introductory Guide* is free. The *User Manual* is £50, which will be deducted from any software purchase made within three months. *Equations for Systems* and *System Dynamics Problems* cost £50 each. VAT is not charged on documentation.

POSTAGE AND PACKING

Postage and Packing for each licence or book are £5 within Europe, £15 in the rest of the world.

GENERAL

Specification and prices may change without notice. We reserve the right to decline orders for software and documentation from competitors. For information or orders please contact:

The COSMIC Holding Company
8 Cleycourt Road
Shrivenham
Swindon
Wilts
SN6 8BN
UK

Telephone +44 (0)1793 782817

Using the disk

INTRODUCTION

The disk at the back of the book contains:

source code for 59 models in COSMIC format;
99 influence diagrams in CorelDRAW Version 3 (CorelDRAW is not
  supplied with the book);
the text for Chapter 10.

The models are all those mentioned in the book and will be invaluable to you in developing your modelling skills through practice. For instructors and advanced students they provide a host of opportunities for model development and revision.

All the models should run with only minor amendments in DYNAMO and DYSMAP2. Users of the other packages referred to in Appendix A will have to rewrite the models but this will be marvellous practice in modelling, requiring close study of the supplied models and thereby developing excellent understanding of the principles of system dynamics, which are the subject of this book.

The influence diagrams will be helpful if you have access to CorelDRAW, a very widely used graphics package, or to any other graphics package which will accept its formats, They will allow you to develop skill in drawing and revising diagrams which is the basis of system dynamics thinking.

The text of Chapter 10 appears as CHAP10.DOC for Microsoft Word 5.5, which may also be readable by other word processing packages, and as CHAP10.TXT, which is in ASCII format and which can be printed by the normal DOS print command. The influence diagrams for Chapter 10 are in the book and the models are on the disk.

READING THE FILES

The model and influence diagram files on the disk are in compressed format. To load them, insert the disk in drive A:and type

A:LOAD

and press enter. This will create a directory called DIAGS as a subdirectory of C:. The compressed file of diagrams and the decompression software will be copied into that directory and the files will then be decompressed after which the compressed file and the decompression software wil be deleted from the subdirectory. Similarly, the models will be decompressed into a subdirectory called MODELS. Finally, the text of Chapter 10 will be copied into a subdirectory called CHAP10 and will exist as CHAP10.DOC and CHAP10.TXT. If you do not wish to use these directory names, you can amend LOAD. BAT on the disk as you wish. If you are not fully conversant with DOS get someone to help you.

Users of COSMIC will have to split the models up into groups of not more than 14 and copy them into directories under the COSMIC main directory.

Index